Sustainable Freight Transport

Sustainable Freight Transport

Special Issue Editors

Lóránt Tavasszy
Maja Piecyk

MDPI • Basel • Beijing • Wuhan • Barcelona • Belgrade

MDPI

Special Issue Editors

Lóránt Tavasszy
Delft University of Technology
The Netherlands

Maja Piecyk
University of Westminster
UK

Editorial Office
MDPI
St. Alban-Anlage 66
Basel, Switzerland

This is a reprint of articles from the Special Issue published online in the open access journal *Sustainability* (ISSN 2071-1050) in 2018 (available at: https://www.mdpi.com/journal/sustainability/special_issues/Sustainable_Freight_Transport)

For citation purposes, cite each article independently as indicated on the article page online and as indicated below:

LastName, A.A.; LastName, B.B.; LastName, C.C. Article Title. *Journal Name* **Year**, *Article Number*, Page Range.

ISBN 978-3-03897-435-2 (Pbk)
ISBN 978-3-03897-436-9 (PDF)

Cover image courtesy of Flickr, Photography: Walmart Inc.

Contents

About the Special Issue Editors

Lóránt Tavasszyis Full Professor in Freight Transportation and Logistics Systems at the Delft University of Technology. He graduated as Transportation Engineer at TU Delft. He has worked with the Dutch national research institute TNO until 2016, holding part-time positions at Radboud University Nijmegen (2004–2009) and TU Delft (2009–2016). His main research topic is freight transportation modelling. Prof. Tavasszy is a fellow of the Netherlands research school TRAIL, member of US Transportation Research Board Freight committees, chair of the World Conference for Transport Research Society's Scientific Committee, and vice chair on Sustainable Transport for the EU Logistics Platform ALICE.

Maja Piecyk is a Reader in Logistics at the University of Westminster. She is a former Deputy Director of the Centre for Sustainable Road Freight, an EPSRC-funded research centre between Heriot-Watt and Cambridge Universities. Her research interests focus on the environmental performance and sustainability of freight transport operations. Much of her current work centres on the optimisation of supply chain networks, GHG auditing of businesses, and forecasting of long-term trends in energy demand and environmental impacts of logistics. Maja is a Chartered Member of the Chartered Institute of Logistics and Transport (UK), and a Fellow of the Higher Education Academy.

Preface to "Sustainable Freight Transport"

This Special Issue of Sustainability reports on recent research focusing on the freight transport sector. This sector faces significant challenges in different domains of sustainability, including the reduction of greenhouse gas emissions and the management of health and safety impacts. In particular, the intention to decarbonise the sector's activities has led to a strong increase in research efforts—this is also the main focus of the Special Issue. We want to thank the authors, the numerous anonymous reviewers, and the publisher, who have contributed to the creation of this issue.

Lóránt Tavasszy, Maja Piecyk
Special Issue Editors

sustainability

MDPI

Editorial

Sustainable Freight Transport

Lóránt Tavasszy [1,*] and Maja Piecyk [2]

[1]   Faculty of Civil Engineering and Geosciences, Transport & Planning Department,
      Delft University of Technology, Stevinweg 1, 2628 CN Delft, The Netherlands
[2]   Faculty of Architecture and the Built Environment, Department of Planning and Transport,
      University of Westminster, 35 Marylebone Road, London NW1 5LS, UK; m.piecyk@westminster.ac.uk
*   Correspondence: l.a.tavasszy@tudelft.nl

Received: 10 October 2018; Accepted: 10 October 2018; Published: 11 October 2018

1. Introduction

This Special Issue of *Sustainability* reports on recent research focusing on the freight transport sector. This sector faces significant challenges in different domains of sustainability, including the reduction of greenhouse gas (GHG) emissions and the management of health and safety impacts. In particular, the intention to decarbonise the sector's activities has led to a strong increase in research efforts, which is also the main focus of the Special Issue.

Sustainable freight transport operations represent a significant challenge with multiple technical, operational, and political aspects; the design, testing, and implementation of interventions require multi-disciplinary, multi-country research. Promising interventions are not limited to introducing new transport technologies, but also include changes in framework conditions for transport, in terms of production and logistics processes [1]. Due to the uncertainty of impacts, the number of stakeholders and the difficulty of optimization across actors, understanding the impacts of these measures is not a trivial problem. Research, therefore, is not just needed on the design and evaluation of individual interventions, but also on the approach of their joint deployment through a concerted, public/private programme. This Special Issue addresses both dimensions, in two distinct groups of papers—the programming of interventions, and the individual sustainability measures themselves.

The first 7 papers, besides offering insights about freight sustainability measures, also address progress in the different, typical stages of programme preparation: (1) defining the objectives and the problem; (2) learning from past experiences; (3) systematic generation of solutions; (4) understanding system behaviour; (5) scenario building; and (6) evaluation of policies. The second group of papers focuses on the evaluation of specific solutions to reduce the carbon content of transport. This concerns a wide range of measures, including improved capacity utilisation, electrification, regulatory measures, alternative fuels and vehicle aerodynamics. We introduce the contributions in more detail below.

Within the first group, the opening paper of *Abiye Tob-Ogu, Niraj Kumar, John Cullen and Erica Ballantyne* reports on a systematic literature review of sustainability intervention mechanisms [2]. Two important findings are: (i) the identification of information and communication technology as an opportunity to drive changes towards sustainable transport; and (ii) the strong geographic compartmentalisation of the literature, confined to continental silos. The authors find that relatively few papers are based on collaborative work across continents.

The contribution of *Hongli Zhao, Ning Zhang and Yu Guan* focuses on the identification of the relative importance of factors determining air cargo safety for dangerous goods [3]. They find that, besides regulation of dangerous goods acceptance, the capacity and quality of equipment and facilities also play a role. A potential implication of these results is that, in the area of safety, benefits of innovations in trade facilitation may be constrained by the available physical infrastructure.

Kinga Kijewska and Mariusz Jedliński introduce the concept of policy durability for sustainable urban freight transport [4]. In urban freight transport, many policies are known to have been abandoned only a few years after their introduction. The authors analyse the causes and provide directions for more robust policy making, focusing on the inclusion of critical stakeholders that need to be involved to make measures succeed. In most roadmaps for decarbonisation in the freight transport sector, a shift of loads is advocated from current trucks to high-capacity vehicles or even other modes of transport.

The recent experiences with these policies in Sweden are evaluated by *Inge Vierth, Samuel Lindgren and Hanna Lindgren* [5]. In their ex post analysis of the impacts of the introduction of longer and heavier vehicles in Sweden, they find that this measure has not had any discernible effect on modal split. The share of different types of emissions of road transport changed, however, leading to a higher share of GHGs.

An important element of discussions about impacts of policy measures concerns the rebound effects of measures. Increased efficiency may reduce emissions per unit moved but may also increase the number of units moved, due to the demand effect, thus partly neutralising the effects of measures. Often, these rebound effects are assessed through the cost and time elasticities of freight transport. In an original contribution, *Franco Ruzzenenti* explains how elasticities as currently used can be misleading [6]. His main assertion is that tabulating flows as is done today neglects the complex interdependence between flows that is present in networks and that is essential for considering rebound effects. Therefore, he develops a new line of thinking using network theory that may prove important for sustainability analyses.

Two further contributions take the perspective of practical solution scenarios at the country and sector level, respectively. For Spain, *Carlos Llano, Santiago Pérez-Balsalobre and Julian Pérez-García* develop scenarios for emission reduction of domestic freight transport [7]. They build up a consistent flow database, develop default emission projections, and study the impacts of a shift of freight from road to railways. Studies that link the analysis of modal shift potential to detailed flow databases are scarce, and may support the development of modal shift policies that take into account the supply chain context of goods flows.

Reporting about Organization for Economic Cooperation and Development (OECD) research aimed at the decarbonisation of the maritime transport services sector, *Ronald Halim, Lucie Kirstein, Olaf Merk and Luis Martinez* develop pathways for emission reduction [8]. In a systematic study, using a global freight transport and emission model, they consider 4 different pathways. Mobilizing all available technologies, these could lead to a reduction of carbon emissions of up to 95% by 2035, well beyond the current commitment of 50% reduction by 2050. The paper describes the approach and assumptions behind this study, which contributed to the formulation of broadly supported decarbonisation targets by the maritime shipping world.

The second group of papers discusses specific interventions that can be implemented to decarbonise the freight transport sector and approaches that can be applied to evaluate their likely effects. *Jessica Wehner* presents the analysis of opportunities to improve the energy efficiency of operations by increasing capacity utilisation in logistics systems [9]. Her research results in the categorisation of factors that cause unutilised capacity within the categories of activities, actors and areas. These factors are then linked to a number of mitigation measures, such as relaxing delivery schedules, training, and off-peak deliveries, among others. The paper also emphasises the need for a standardised approach to the measurement of environmental impacts of logistics to enable meaningful comparisons between companies.

The potential effects of introducing longer and heavier vehicles (LHVs) in the United Kingdom are investigated by *Heikki Liimatainen, Phil Greening, Pratyush Dadhich and Anna Keyes* [10]. The authors estimate that if LHVs were used similarly in Finland in the transport of various commodities, significant savings could be achieved in truck kilometres, transport costs, and CO_2 emissions. Furthermore, lower road freight traffic volumes and reduced emissions are likely to more than offset the possible negative effects of modal shift from rail to road.

Jesko Schulte and Henrik Ny focus on overhead line Electric Road Systems (ERS) as a way to improve the sustainability of transporting goods by road [11]. The research show that although ERS may present some severe violations of the sustainability principles, especially in the raw material extraction, production and use phases, they could still be a valuable element in the transition towards a more sustainable freight transport system.

Based on a case study of a Polish town Gdynia, *Jacek Oskarbski and Daniel Kaszubowski* investigate whether a mesoscopic urban transport model already in use there can be populated with urban freight transport data in order to improve evaluation of potential CO_2 reductions from the designation of dedicated delivery places [12]. They conclude that this approach produces satisfactory results if basic regulatory measures are considered. However, dedicated freight transport models that can take urban supply chain structure into account are more suitable to study more complex policy options.

Tharsis Teoh, Oliver Kunze, Chee-Chong Teo and Yiik Diew Wong demonstrate that opportunity charging offers the potential to significantly reduce the lifecycle costs of using electric vehicles in urban freight transport without increasing related CO_2 emissions [13]. The authors also find that other factors also strongly influencing the lifecycle costs are the use of inductive technology, extension of service lifetime, and reduction of battery price. The use of inductive technology and the carbon intensity of electricity generation are the two other factors with a strong influence on CO_2 emissions from electric vehicles operating in towns and cities.

Ján Ližbetin, Martina Hlatká and Ladislav Bartuška discuss issues related to energy consumption and GHG emissions related to the use of fatty acid methyl esters (FAME) biofuels in road freight transport [14]. They conclude that even though FAME biofuels significantly reduce GHG emissions, their production is highly energy intensive, which translates into steeper fuel prices. Therefore, more research is needed into ways to reduce the energy requirements of FAME biofuels production in order to bring the prices down to an industry-acceptable level.

In the final article, *Erik Johannes, Petter Ekman, Maria Huge-Brodin and Matts Karlsson* focus on aerodynamic improvements for timber trucks in Sweden [15]. While the aerodynamics provide the opportunity to reduce the transport cost of timber in Sweden, the changeover time is found to be the most important parameter to them being economically viable. Hence, in the Swedish timber transport sector aerodynamic kit that does not have to be manually installed is key to the profitability of the investment.

Together the papers in this Special Issue paint a diverse and rich picture of opportunities in the freight transport sector for a transition towards sustainability. They confirm the theoretical availability of a significant and—from the perspective of the global sustainability targets—promising potential for decarbonisation. At the same time, they make us aware of important limitations of policy measures, caveats in our knowledge and weaknesses in our approaches to assess the impacts of policies. All these provide new directions to accelerate R&D, innovation and public policy in the required direction and ultimately create a more sustainable freight transport sector.

Conflicts of Interest: The authors declare no conflict of interest.

References

1. McKinnon, A. *Decarbonizing Logistics: Distributing Goods in a Low Carbon World*; Kogan Page: London, UK, 2018.
2. Tob-Ogu, A.; Kumar, N.; Cullen, J.; Ballantyne, E.E.F. Sustainability Intervention Mechanisms for Managing Road Freight Transport Externalities: A Systematic Literature Review. *Sustainability* **2018**, *10*, 1923. [CrossRef]
3. Zhao, H.; Zhang, N.; Guan, Y. Safety Assessment Model for Dangerous Goods Transport by Air Carrier. *Sustainability* **2018**, *10*, 1306. [CrossRef]
4. Kijewska, K.; Jedliński, M. The Concept of Urban Freight Transport Projects Durability and Its Assessment within the Framework of a Freight Quality Partnership. *Sustainability* **2018**, *10*, 2226. [CrossRef]
5. Vierth, I.; Lindgren, S.; Lindgren, H. Vehicle Weight, Modal Split, and Emissions—An Ex-Post Analysis for Sweden. *Sustainability* **2018**, *10*, 1731. [CrossRef]
6. Ruzzenenti, F. The Prism of Elasticity in Rebound Effect Modelling: An Insight from the Freight Transport Sector. *Sustainability* **2018**, *10*, 2874. [CrossRef]

7. Llano, C.; Pérez-Balsalobre, S.; Pérez-García, J. Greenhouse Gas Emissions from Intra-National Freight Transport: Measurement and Scenarios for Greater Sustainability in Spain. *Sustainability* **2018**, *10*, 2467. [CrossRef]
8. Halim, R.A.; Kirstein, L.; Merk, O.; Martinez, L.M. Decarbonization Pathways for International Maritime Transport: A Model-Based Policy Impact Assessment. *Sustainability* **2018**, *10*, 2243. [CrossRef]
9. Wehner, J. Energy Efficiency in Logistics: An Interactive Approach to Capacity Utilisation. *Sustainability* **2018**, *10*, 1727. [CrossRef]
10. Liimatainen, H.; Greening, P.; Dadhich, P.; Keyes, A. Possible Impact of Long and Heavy Vehicles in the United Kingdom—A Commodity Level Approach. *Sustainability* **2018**, *10*, 2754. [CrossRef]
11. Schulte, J.; Ny, H. Electric Road Systems: Strategic Stepping Stone on the Way towards Sustainable Freight Transport? *Sustainability* **2018**, *10*, 1148. [CrossRef]
12. Oskarbski, J.; Kaszubowski, D. Applying a Mesoscopic Transport Model to Analyse the Effects of Urban Freight Regulatory Measures on Transport Emissions—An Assessment. *Sustainability* **2018**, *10*, 2515. [CrossRef]
13. Teoh, T.; Kunze, O.; Teo, C.-C.; Wong, Y.D. Decarbonisation of Urban Freight Transport Using Electric Vehicles and Opportunity Charging. *Sustainability* **2018**, *10*, 3258. [CrossRef]
14. Ližbetin, J.; Hlatká, M.; Bartuška, L. Issues Concerning Declared Energy Consumption and Greenhouse Gas Emissions of FAME Biofuels. *Sustainability* **2018**, *10*, 3025. [CrossRef]
15. Johannes, E.; Ekman, P.; Huge-Brodin, M.; Karlsson, M. Sustainable Timber Transport—Economic Aspects of Aerodynamic Reconfiguration. *Sustainability* **2018**, *10*, 1965. [CrossRef]

sustainability

MDPI

Review

Sustainability Intervention Mechanisms for Managing Road Freight Transport Externalities: A Systematic Literature Review

Abiye Tob-Ogu [1], Niraj Kumar [2], John Cullen [1] and Erica E. F. Ballantyne [1,*]

[1] Faculty of Social Science, Sheffield University Management School, University of Sheffield, Sheffield S10 2TN, UK; a.tob-ogu@sheffield.ac.uk (A.T.-O.); john.cullen@sheffield.ac.uk (J.C.)

[2] University of Liverpool Management School, Chatham Street, Liverpool L69 7ZH, UK; niraj.kumar@liverpool.ac.uk

* Correspondence: e.e.ballantyne@sheffield.ac.uk

Received: 30 April 2018; Accepted: 4 June 2018; Published: 8 June 2018

Abstract: With road freight transport continuing to dominate global freight transport operations, there is increasing pressure on the freight transport industry and its stakeholders to address concerns over its sustainability. This paper adopts a systematic review to examine the academic literature on road freight transport sustainability between 2001 and 2018. Using content and thematic analysis, the paper identifies and categorises sustainability intervention mechanisms providing useful insights on key research applications areas and continental distribution of sustainable road freight transport (SRFT) research. In addition to the six-overarching sustainability intervention mechanism themes identified: decoupling, Information and Communications Technology (ICT), modality, operations, policy, and other, future research can explore the effectiveness of different interventions mechanisms identified in this study to improve sustainable practices across different continents.

Keywords: road freight; sustainability; intervention mechanisms; systematic review; externalities

1. Introduction

Despite its importance to economic growth and prosperity, there are valid concerns relating to the sustainability of road freight transportation in terms of safety, efficiency, and health implications. These concerns are reflected in the contemporary road freight transport literature [1–6]

Accordingly, there is increasing pressure on stakeholders to address externalities emanating from freight transport operations across a variety of landscapes including urban, inter-urban, and rural landscapes. For example, in Europe, road freight transport sustainability is a priority for the European Commission (EC) with initiatives like MERCURIO, ERTRAC, KOMODA, and FIDEUS highlighting the commitment of the supranational and State level actors to addressing road freight transport sustainability. Academically, authors [6–8] have explored various sustainability initiatives in the road freight sector with insights on policy approaches, multi-stakeholder involvement, and modal integration planning. These initiatives represent some of the different mechanisms employed to intervene and tackle road freight externalities. For example, the literature investigates and discusses the idea of green corridor infrastructure for road freight transportation [9], other studies [1,10,11] have explored applications of information and communication technology (ICT) to aid sustainable road freight operations, whilst other studies discuss policy loopholes and freight energy management strategies [12].

Intervention mechanisms represent efforts, tools, and approaches that are theory or practice informed to address specific challenges. These capture not only the vitality of research inquiries into sustainable road freight transport but also highlight the complexity of the field. A resulting implication of this complexity is a lack of knowledge congruence which can negatively impact the development of research collaboration and efficiency [13]. Further, the literature is yet to address the impact of contextual limitations on the adoption of specific intervention mechanisms and this can have interesting impact for strategic planning amongst freight transport stakeholders. For example, green corridor initiatives can be considered as Pan-European, with conceptual and pragmatic acceptance across the European community. However, limitations relating to infrastructure or regional mobility may affect their adoption outside of Europe, for instance the absence of such regional cooperation in Africa or Southern America limits the pragmatism of such an initiative in these regions and thus highlights potential knowledge gaps concerning relationships between contexts and intervention mechanisms. The purpose of this paper extends to examining the focus of the literature as well as providing some guidance for optimising future research and practice across different regions. In this regard, the objective of this paper is to provide a synthesized account of the literature on sustainable road freight transport (SRFT) interventions offering some insight on the main SRFT research streams, taxonomies, as well as insights on the contextual implications for SRFT intervention mechanisms. Such outcomes can improve future research synergy and collaboration, support strategic planning and offer useful reference for future research.

To achieve our objective, the following research questions were posed:

1. What are the main intervention mechanisms advanced in peer-reviewed publications on sustainable road freight transport?
2. What implications do regional contexts have on the adoption of different intervention mechanisms?

Addressing these questions through a critical review of the literature will advance the significance of sustainable road freight transport as a critical area of research in the logistics and supply chain sustainability literature. Additionally, it will address current knowledge gaps on the relationship between intervention mechanisms and geographical contexts, with implications for future research and practice. The rest of this paper is organised as follows: Section 2 presents a discussion of systematic literature reviews in management research; Sections 3 and 4 describe the methodology and analysis approaches for the study; whilst Section 5 presents the study discussions. Finally, our concluding statements and directions for future research are presented in Section 6.

2. Systematic Literature Reviews

The use of systematic reviews in the social sciences and specifically management research has significantly developed in the last decade with increasing acceptance across ontological and epistemological divides [14]. As knowledge converges and develops towards complementary methods in the social sciences, the pillars of reliability and apposition are increasingly important [15]. It has been advocated that systematic reviews help to map relevant intellectual territories that identify how and where the literature base can benefit from further studies, i.e., the identification of research gaps [14]. Whereas others take a more instructive approach [16], calling for systematic reviews to support the literature's account of contextual factors that need to be integrated into management research.

The importance of these issues is addressed by [17] who underline the use of systematic reviews to enable transparency, inclusivity, heuristics, and explanation in the review process. Accordingly, the importance of systematic reviews of the extant literature on SRFT related studies has been previously emphasized [4] who highlight the benefits to the development of research in this area. However, since Perego's review [4], there has been little done to update the literature in this area and a recent review [18], focuses more on the general urban logistics function rather than road freight transport specifically. SRFT research requires targeted and collaborative synergies to address the ubiquitous challenges faced and a systematic review of the data can give useful funneling for identifying specific trends as well as collaborative scope in SRFT research.

3. Methodology

The importance of a review protocol prior to conducting a systematic review of the literature and cited its usefulness for mitigating biases in the review process has been emphasised [17]. The literature review protocol was implemented in four stages, i.e., design, review, selection and analysis.

3.1. Design of Review Protocol

Accordingly, 3 review team members jointly developed a protocol with inputs from discussions with academic and industry experts in road freight logistics within and outside the UK. The purpose of this was to enhance the rigour and evidence base of the review outcomes. The protocol tied the review objectives to the processes establishing the data sources, plausible databases, inclusion and exclusion criteria, search string techniques and acceptance schedule (Table 1).

Table 1. Review Protocol.

	Inclusion and Exclusion Criteria		Databases	Data Sources
	Inclusion	Exclusion		
Timeframe	Between 2001 and February 2018	Outside 2001–2018	Taylor & Francis Google Scholar Science Direct Web of Science Sage Emerald	Online (Soft) Print (Hard)
Type	Peer reviewed	Non-peer reviewed, books, conference papers		
Topic	Road freight transport, road logistics, sustainable road freight transport	Non-sustainability, Non-road freight transport		
Language	English or English Translate	Non-English		
Reviewer's Initials	Paper no.		Decision (Please tick)	
			Accept/Rationale	Reject/Rationale
Search Technique	Boolean, Verbatim and Word combinations:			

The protocol was not considered a rigid guide and iteration supported modification as the actual review process progressed. Although SRFT publications go back many decades, our focus was on identifying contemporary and updated intervention mechanisms. The cut off timeline for the review was initially set between the years 2001 and 2016 and later extended to 2018 (February), following further reviews and feedback. This period coincided with uptake in technology as well as commencement of the millennium development goals (MDGs), which underlined a global outlook to sustainability across different sectors. Practical constraints relating to time, feasibility, access to materials and review scope also informed the design and modification of the protocol. For example, although we are aware of useful grey publications, we omitted these from the review due to considerations on quality and reliability (peer-review process improves the value of the report) and practicality (impossible to review all publications or gain access to regional publications across different continents).

3.2. Review and Selection

Following the review and affirmation of the agreed review protocol, six databases; ScienceDirect®, Emerald®, Taylor and Francis®, Sage®, Web of Science® and Google Scholar®, were identified as suitable for conducting the literature search. This was informed by learning from similar literature reviews and the need to represent the complexity of SRFT publications. Test searches revealed gaps in scope of individual databases and we observed that the incorporation of more databases offered greater opportunities for capturing the latitude of potential SRFT literature. Simple operator and Boolean search methods were combined to execute the search using different phrases and strings to implement the search.

In the first instance, the review process was designed to follow a funneling procedure, moving from broad references to smaller and restrictive (Boolean) criterion as the review progressed. Search strings and keys works including: "sustainable freight"; "green freight"; "road freight"; "sustain* freight"; "green freight*"; and "road freight*", "sustain* logistics*", were used to search these databases with a combined yield of 2265 hits in 2016 and an extra 88 hits in 2018. After a review of titles and over 300 abstracts from the first searches, a few adjustments were made to the protocol. For example, the phrase "road freight" was removed from the 'list of search strings' due to its extremely large sample when used by itself without 'green' or 'sustainability' included in the search. Boolean logic was applied to combine keywords like "Road freight" and "sustain*", improving the focus of the returned results. In many instances, some of the results from these search strings failed to address any sustainability issues and included other issues besides road freight transport. This led to the rejection of 1158 papers, which were deemed irrelevant based on a 1st screen scanning of the titles and abstracts. An important learning from this process was the critical role that titles, abstracts and keywords play in influencing publication visibility and readership of peer-reviewed material.

Following further searches and 'hit' reviews, a decision was made to exclude Google scholar from the 'search database' because of duplicity and source credibility. For example, a preliminary search conducted using the 'sustainable freight' string returned just over a thousand results with Google scholar accounting for over 90% of the results (Figure 1). Closer scrutiny of the results revealed that over 200 of the results from Google Scholar were repeated on several occasions within the database with varying citations from both peer-review and bogus sources. Furthermore, we established that much of the 'peer-reviewed' references within the Google Scholar batch were already reported by the other databases. Whilst it is plausible to suggest that the exclusion of the Google Scholar database may raise questions about the scope of the evidence incorporated in the review, it was also important that the review was conducted within robust but qualitative parameters. This is particularly important when the "peer review" inclusion criterion is taken into consideration.

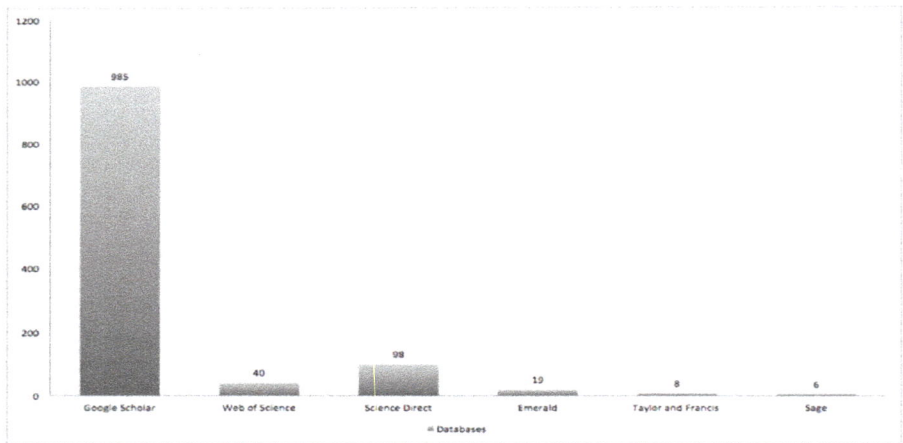

Figure 1. Chart illustrating initial database 'hits' for "Sustainable Freight" string.

Progressively, search terms were replicated across the remaining databases with additional strings used to streamline the searches. As captured by the protocol, the focus was on peer-reviewed material in published sources and a total of 403 hits were returned across 8 re-organized searches. After screening for duplicates and relevance, a total of 168 materials were accepted for further review. A 3rd stage review of the abstracts, introduction and publication type saw an elimination of a further 54 materials which were books, conference proceedings or items that did not materially discuss the related subject of "sustainability in road freight transport". A total of 98 journal articles from 44 different journal titles were finally accepted for inclusion in the review report (Figure 2 and Supplementary Materials: Appendix A1.

Journals

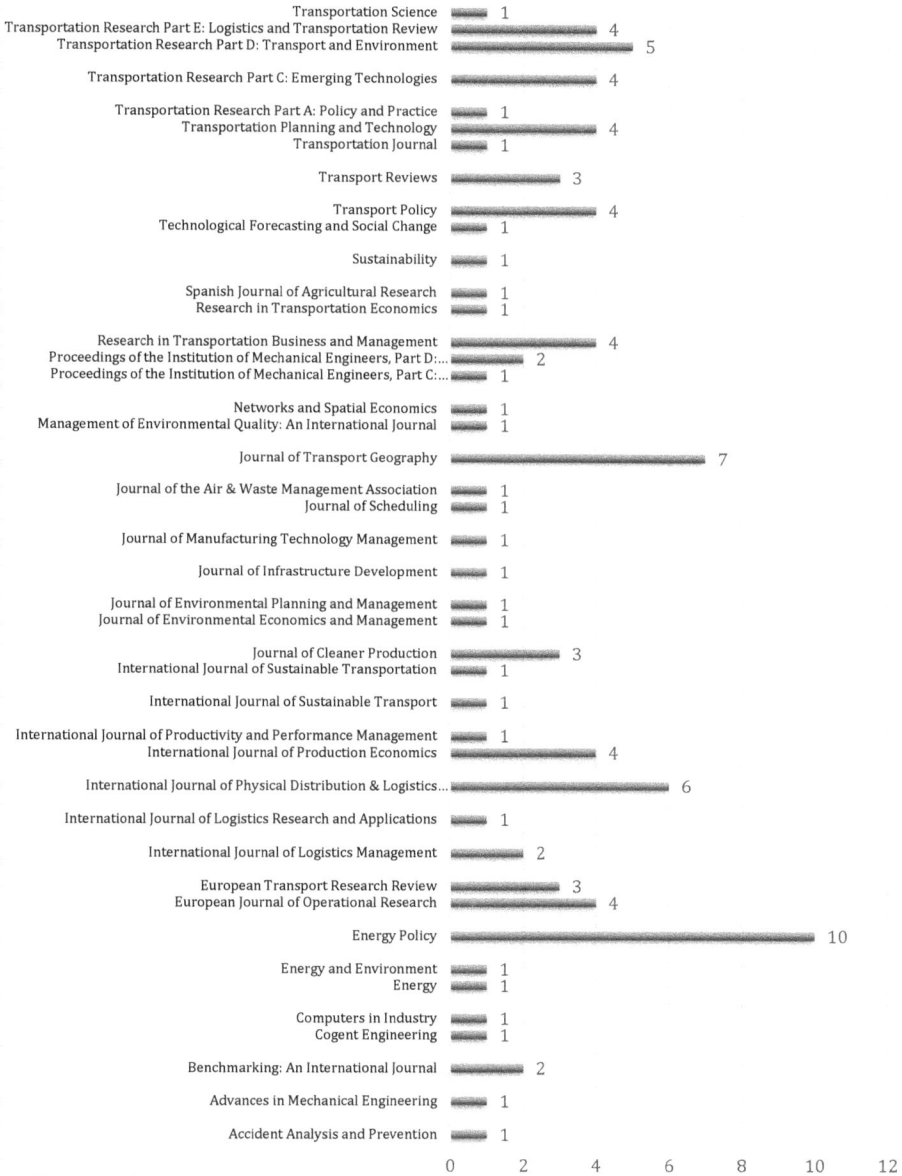

Figure 2. Final Journal Selection.

4. Analysis

Using the context, intervention, mechanisms, and outcome (CIMO) framework [17], each article was carefully evaluated in line with the review objectives and we adopted a combination of content and thematic analyses to review, interrogate, and organise the data for reporting (Supplementary Materials: Appendix A1). Topics covered by the corpus extended across road freight transport performance, design, and policy, highlighting the diverse literature spectrum. NVivo11™ [19] was used to query and review the selected papers, exploring each paper in detail, identifying the principal focus of each paper, key arguments, theoretical underpinnings, methodological design, and key findings. The coding function on NVivo11™, was used to create and further query themes. Where we observed papers as addressing multiple themes, we allocated them to the category where the predominant discourse was aligned based on frequency of keywords used, authors' depiction as content frequency, analogies, and sorting. The use of NVivo11™ and multiple reviewers not only helped reduced perceptive bias, NVivo11™ also supported the speed of the data query process to identify topical links, thematic clusters, and alignments. For example, the frequency and cluster analysis tools in NVivo11™ were used to identify key words and usage contexts, creating an objective output for further analysis. It was also used to support our Jaccard co-efficiency testing to validate the emergent themes from the data.

The process was also influenced by previous knowledge about the literature on road freight transport, for example we are conversant with papers from authors who examined the literature to develop an online benchmarking tool for freight transport operations in the EU, Switzerland, and Norway [20]; papers which investigated the use of ICT in road freight operations, highlighting CO_2 emissions reductions and efficiency gains from the use of ICT in road freight operations [1,3]. This prior knowledge contributed designation of themes although some reported themes were emergent from the coding process.

To support the originality index of the extracted themes, we conducted a Jaccard coefficient similarity test to distinguish the depth of correlation between the different themes [21]. The highest coding similarities involved articles and codes discussing ICT, modality, and operations, with combinations of 0.276 (ICT/modality), 0.143 (ICT/operations), and 0.115 (modality/others), respectively. With the low similarity indexes between the different theme categories, we accepted the interpreted theme categories as distinct themes capturing various intervention mechanisms from the reviewed literature.

In total, six themes were identified in the process: Policy, operations (design and process); modality (uni-modality, co-modality, synchro and inter-modality); decoupling, ICT; and 'others' (land use, UCCs, reporting, and measurement systems). Figure 3 shows the distribution of the articles according to the intervening mechanisms that they addressed. Operations, policy and modality themes commanded higher scholastic attention and accounted the majority of the 98 papers reviewed.

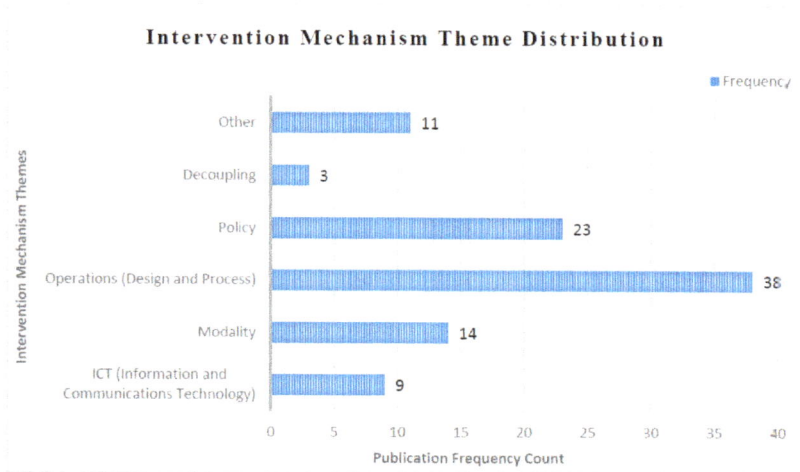

Figure 3. Road Freight Transport Sustainability Intervention Mechanisms.

5. Discussion

There is a growing focus on road freight transport sustainability with a variety of approaches to mitigate its consequences. In Figure 4, the Chart highlights the overall publication trend for papers in this area and although the figures for 2018 suggest a further decline, this is entirely due to the cut off period for the database selection (See Table 1). The analysis of the reviewed literature highlights several interventions, which are classified, summarized (Table 2) and discussed below in themed categories.

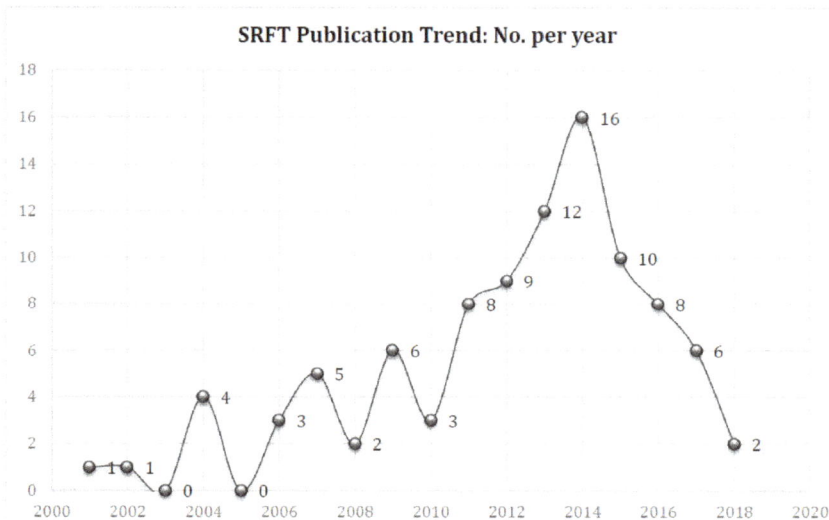

Figure 4. Number of sustainable road freight transport (SRFT) publications per year.

Table 2. Intervention Themes, Key Authors, and Topics Summary.

Theme	Key Authors	Topics	No.
Information and Communications Technology (ICT)	Wang et al., 2015; Sternberg et al., 2014; Marchet et al., 2012; Davies et al., 2007	ICT use for CO_2 reduction, ICT use for intermodal transport, Efficiency benefits of ICT use	9
Decoupling	Alises et al., 2014; Liimatainen and Pollanen, 2013; McKinnon 2007	Decomposition analysis, Impact evidence, Policy roles, and impact	3
Modality	Li et al., 2015; Macharis et al., 2011; Caris et al., 2008; Winebrake et al., 2008	Dynamic modelling for intermodal freight, Decision Support Systems (DSS) for optimising intermodal freight, Energy and emissions trade-offs in road freight, Co-modality	14
Operations	Newnam and Goode, 2015; Li et al., 2015; Midgley et al., 2015; Liimatainen et al., 2014; Schiffer and Walther, 2018, Wang et al., 2014; Palsson and Kovacs, 2014; Ando and Taniguchi, 2006	Socio-technical perspectives of externalities, Alternative fuels, Regenerative braking mechanics, Management strategies, Time travel, reliability, and routing	38
Policy	M'raihi et al., 2015; Stelling, 2014; Ballantyne et al., 2013; Pieyck and McKinnon, 2010; Eom et al., 2009; Dablanc, 2007; Steenhof et al., 2006	Emissions and influencing factors, Stakeholder needs and local council planning, emission ELKS factors and planning horizons, cost measures and practitioner approaches, decomposition analysis, and modal shifts	23
Others	Khorheh et al., 2015; Demir et al., 2014; Islam et al., 2013; Carballo-Panela et al., 2012	Green corridors, congestion planning, land use and urban freight, and performance benchmarking tools	11

5.1. Intervention Themes

5.1.1. Operations

The operations theme represents interventions that focus on optimising SRFT operations through a combination of equipment and process design initiatives. Articles in this category explored intervention mechanisms across strategic, tactical, and operational levels. Topics relating to fleet management strategies [22–25], routing [26–28], vehicular design and load utility [29,30], fuel type trade-offs [31–33], and costs [34], were within this purview. Some of the main contributions in this area include the importance of assigning the 'right' vehicle to the 'right' areas, advocating fleet management models that account for environmental distinctions as a means for addressing CO_2 emissions [22]. Other studies identified significant energy index value (EI_v) gains of 9–17% from modelling hydraulic controls using the greedy optimization technique to investigate driving cycles, highlighting potential benefits heavy goods vehicle (HGV) design as an SRFT intervention [29].

In terms of fuel choices and implications, Li et al.'s study provides useful insight into the potential for alternative fuels in road freight operations [31]. They model consumption and demand using a cost-optimisation strategy to forecast consumption, projecting long-term reliance on diesel and gasoline fuels, which they estimate will still be responsible for over 70% of freight fuel by 2030. Of significant interest and implication for future research in this area, was the identification that resource constraints for other fuel forms remained a principal limitation to bigger decline on gasoline and diesel dependence. Although routing efficiencies remain of key concern to road freight transport scholars, developing contributions in this area include modelling for routing optimisation, Original Equipment Manufacturing (OEM) design insight, energy, and load decisions as well as the development of DSS tools for routing and location planning [26]. The role of decision support tools to aid management decision-making in terms of fleet vehicle selection and optimal combination strategies is still an area with knowledge gaps on applications at different strategic levels.

Overall, we identified that an increasingly salient feature of many articles in this category was the reference to, or combination with information systems technology elements as a fundamental of the operations optimisation models. This was confirmed by the Jaccard coefficient results and underlined elements of interrelationships between different themes [27,35,36].

5.1.2. Policy

Policy captures State driven mechanisms for addressing road freight challenges. Critically, interventions cover local, regional, and national levels of applications and this was interpreted to involve more complexities compared to Decoupling strategies that were specifically national or supranational in scope. Much of the literature in this category focused on the urban freight problem [7,37–40]. Some key topical issues under this theme explored robustness of policy mechanisms, for example one study advocate a 'new stakeholder' approach for addressing the urban freight problem at local council levels [7]. This was to cater for conflicting objectives that often pitch businesses and councils at opposing divides. Like Klumpp's application of the Jevon paradox (rebound) theory to examine SRFT failure reasons from an operations perspective [41], a previous study had reviewed the USA environment and highlighted some crucial policy misconstructions in terms of the efficiency metrics for road freight management advising;

> "*Policymakers should be careful when using existing freight elasticity estimates in the literature to estimate the HGV rebound effect. Aside from general caveats associated with these indirect measures of the rebound effect, freight elasticity estimates are influenced by a number of "factors of variability" categorized by the specific nature of the shipping activity, the macroeconomic influences involved, and the measurement tools used to assess elasticities. Ignoring these factors may lead to biased results when applying the literature to a specific policy analysis case*"—[12], (pp. 258)

The rebound effect refers to increased resource consumption because of relative efficiencies in performance, i.e., the difference between projected and actual energy savings as a direct correlate of increased efficiencies [12,42]. The arguments put forward suggests that policy makers need to go beyond energy efficiency saving metrics to actual energy demand reduction measures. It is advocated that measurement adjustments be made to policy projections for energy efficiencies in the road freight sector, where rebound effects can be as high as 24%.

Similarly, another study models the same problem in Tunisia and explores policy strategies for addressing the road freight emissions challenge [43]. They proffer a combination of incentivising arrangements and fiscal strategies as useful for addressing these challenges. Furthermore, they compare policy options in terms of decoupling as a mechanism for intervention as opposed to other incentivising and fiscal arrangements, with a conclusion that the peculiar economic and political realities in the context would significantly affect the viability of such a strategy. This point is particularly instructive in the evaluation of strategic options for different countries, with developing economies less likely to effectively pursue policy strategies that de-emphasize their main revenue and growth processes. Another research saw modelled a policy quadrant to advance some policy directions for road freight transport planning using empirics from Sweden [37]. Of keynote is the requirement for a combination of legal, economic, societal, and knowledge instruments at national/local levels, which will support direction, income, infrastructure, and behavioural adaptations, respectively required to meet future targets. Some key contributions in this area include best practice collations, strategic planning tools, and incentivising approaches for SRFT and cooperation amongst stakeholders.

5.1.3. Modality

The modality theme addresses the means and mode of transport employed to effect freight mobility. Under this theme, the main topics focus on the combination with or substitution of road freight transport with other modes of transport. In terms of substitution, the literature acknowledges the critical qualities of flexibility, speed, and time from road freight transport, with implications for last mile dependency on road freight for the foreseeable future [44–48]. However, the literature presents a variety of modal combinations for addressing congestion, emissions, and cost concerns unimodal road freight transport [46,49,50]. The terms '*co-modality*', '*multimodality*' and '*synchro or intermodality*' are used to represent modal options within the literature. Co-modality is defined as the efficient use of different modes [51], whilst Ruiz-Garcia et al. differentiate between intermodality and multimodality, stating that multimodality implies using different transport modes and administration, whilst intermodality refers to the integration of administrative and transfer process of freight shipment across different transport modes [52]. Some operations aspects addressed in this category include packaging designs, intermodal component requirements, and modal integration [47,50].

Intermodality is a central theme of the literature in this category; all the thirteen papers reviewed under this theme discussed intermodality in some degree. This is perhaps driven by the rationale that intermodality provides the most reasonable compromise for managing the emissions and congestion challenge of road freight [53]. Rail and water modal combinations are considered as best complements or alternatives for road freight transport although the literature acknowledges that for many shipments from international sources, water freight transport is already an inalienable part of the freight transport chain since it is the common export and import option for shipments between countries [48]. However, the literature also notes intermodality as a complex model and highlights some common constraints to its operationalisation: infrastructure [37,43,54]; decision support systems [50,54]; interoperability and planning [10,52,55]; and transitioning implementation [48]. Contributions in this area include costs modelling for different modal combinations, environmental benefits, and integration efficiencies stemming from modal combinations.

Overall, the availability or investment in infrastructure like railways, jetties, hubs, and freight corridors are prerequisites for modality-based interventions. The absence of these can undermine or restrict the usefulness of modal interventions. Future directions in this area may focus on decision support tools to aid transition, interoperability, and planning with significant elements of policy drivers in this regard.

5.1.4. Decoupling

Unlike policy initiatives, decoupling as a policy strategy can only be pursued as a national or supranational strategy mechanism and therefore excludes independent interventions at local council levels [56]. Decoupling strategies established as national or supranational policy approaches, aimed at separating economic growth from freight as a measure of curbing externalities from freight [6].

Traditionally, decoupling measures have focused on freight intensity (tonne-km), using modal split, vehicle utilisation and emissions as metric units for GDP comparisons, economic planning and forecasting [6,57,58]. Although results have been positive in countries like the UK and Spain, this is still an emerging area within the literature and key concerns extend to its *'emissions-shifting'* and *measurement metrics ambiguities*. Additionally, its applicability as a viable mechanism in developing countries has been rejected and recent political upheavals in Europe and the USA could further exert limitations on strategies that de-emphasize manufacturing as a means of curbing freight externalities [43,57].

Perhaps a major contribution of studies in this area is the development of decomposition analysis frameworks for investigating road freight and GDP correlations, contributing to progressive insight and alignment between specific industry and freight intensity [6,56,57]. Freight policy strategists at regional and national levels can benefit from these studies, with prospects of integrating context specific economic structures for carbon reporting, haulage distances and modal choice splits into existing GDP aggregate measures.

5.1.5. Information and Communications Technology

ICT accounted for ten (10%) of the articles reviewed although many of the other articles were cross-themed with ICT. We constructed ICT to encompass both information systems (IS) and information technology (IT), referring to combinations of hard and soft connectivity tools that support communication exchanges, remote monitoring and performance management within freight transport operations [1,3,59]. Some papers adapt taxonomies for identifying and classifying road freight transport ICT systems, although Wang et al.'s taxonomy provides the most comprehensive overview for deconstructing ICT mechanisms for road freight transport operations [1,59].

Accordingly, ICT is conceptualised as consisting of three main components, the software components applications, including operating systems; the hardware components; and the information component [1,51,59]. Addressing issues around connectivity, network relationship management, enterprise processes, and asset management [3,60,61], ICT is commonly presented as positively impacting road freight transport through operational efficiencies in road freight [1,62], providing social benefits [3,10], cost reduction and effectiveness [55], driver working time, and administration time reductions [3]. Critically, only Button et al. [10] and Sternberg et al.'s [3] papers explicitly addresses the social aspects of ICT's potential in terms of road freight transport sustainability. There is perhaps more need for targeted research enquiries on the application of ICT to address social issues in road freight transport.

The bulk of the literature focuses on environmental and economic aspects ICT use for road freight transport sustainability [10,55,61–63]. The topical issues of ICT adoption drivers and barriers is explored within the literature, with size, management capabilities, topologies, interoperability, and industry structure emerging as some critical areas of concurrent research inquiries [51,59,63]. Some key contributions under this theme include safety, emissions modelling, and operations integrations among freight stakeholders.

Despite the increased research uptake in this area, significant opportunities exist for contributions around ICT mechanisms for achieving social sustainability measures as well as the development of decision support system (DSS) tools for road freight efficiency and emission planning. We identified that although there is a growing interest in the area of "big data" and "automated or driverless freight transport", none of the results in our search discussed these as key topics. This perhaps points to gaps in the literature or limitations of publication abstracts and it is hoped that future research will address these gaps.

5.1.6. Others

This generic category encompasses studies that focus on, performance and reporting tools [64,65], land use and infrastructure [38,66], and freight transport reviews [44,67,68]. For example, the concept of 'green corridors' as a Pan-European intervention mechanism for road freight transport sustainability is addressed [66]. Green corridors require dedicated infrastructure for freight mobility, each of which would incorporate inland waterways, road, rail, and shipping. As a strategy, 'green corridors' encompasses all of policy, ICT, intermodal and operations mechanisms that create dedicated freight infrastructure frameworks that are ecologically and environmentally friendly. Additionally, one of the papers, considers the infrastructure challenge from a more social perspective, exploring the illegal use of parking bays and the implications for policy makers and managers, where illegal demand is fuelling unauthorised parking with disruptive outcomes [38].

Finally, in terms of reviews, Khorheh et al., introduce an interesting perspective to the externality problem, highlighting some direct and indirect impacts of road freight transport. They also highlight taxation and incentive planning as some socio-economic mechanisms, in addition to information technology and cultural instrumentations [44]. Their paper provides an extensive review of emissions, discussing concurrent operational framework tools for managing emissions in road freight transport and is comparable to a previous work [69]. Studies under this theme have contributed to research guidance, strategic conceptualisations, and urban consolidation centre strategies.

5.2. Regional Context Implications

In terms of contexts and implications a coordinate analysis of the papers focused on identifying the empirics of the papers or stated geographical locality of the papers reviewed (Table 3 and Figure 5).

Table 3. Geographic distribution according to continental regions.

Distribution of SFT Research Focus According to Continents	Count of Distribution of SFT Research Focus According to Continents
Africa	4
Australia	4
Europe	63
Generic	13
North America	8
South America	1
South East Asia	5
Total	98

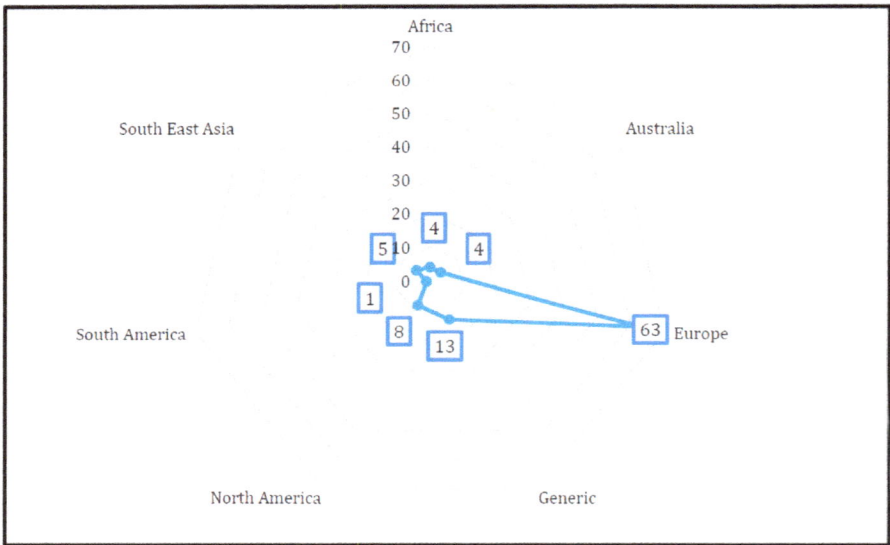

Figure 5. Radar Chart showing coordinates of publications.

As depicted by Figure 5, there is significant disproportionality in the regional coordinates of the articles reviewed. Perhaps influenced by database locations and web analytics settings, Europe unsurprisingly accounted for the majority (63) of the papers reviewed, however, there were interesting patterns observed across the different categories reviewed. All the different intervention mechanism themes had been examined within European contexts, with decoupling being the exclusive preserve of Europe. All three papers that examined decoupling as a subject matter were based on European empirics [6,57,70]. It may be useful for future research to explore how regional frameworks like the European Commission or European Union are influencing policy at State levels in comparison to other regional blocs outside of Europe. Further, the geographical differences observed provide a useful justification for future research around benchmarking and best practice sharing from one country or region to another. This area is yet to be explored in the literature and there may be opportunities for impact in terms of transferable exchanges between firms in this area.

With regards to publication trend analysis, besides the geographical distribution trend discussed above, no other specific predictive trend was apparent in the papers reviewed. For example, although the earliest paper focused on ICT [10], there was a 7-year gap between that and the next ICT paper, however subsequent ICT papers did not follow any timeline specification. For the other themes, we observed a closer distribution of publications across the different years. From 2002, operations focused research seemed the most stable stream of research with an average of two papers per year, although we observed that publications in this area have averaged four papers per year since 2014. The decoupling stream seems to have lost traction, as there has been no publication since 2014. Perhaps the notable pattern seems to be the steady decline in publications from a peak of 16 papers in 2014 to six in 2017 (Figure 4). This underlines the need for more focused SRFT research.

Our categorisation also highlighted some heterogeneity between modality and operations. For operations, routing and scheduling, facility planning, fleet design, and energy consumption were the most common topics within the European literature [30,36,71]. In contrast, the only paper from Africa that was reviewed under these categories explored a myriad of bottlenecks such as corruption, insecurity, and infrastructure limitations to road freight operations in Nigeria [24]. There is scope for future studies to explore and model optimal operations and modality frameworks for countries in Africa and South America.

As depicted by Figure 3 and Table 4, policy and operations are the most common intervention mechanisms for road freight sustainability. Although the literature suggests that European and American contexts are more likely to produce SRFT research initiatives compared to Asian and African nations [43,72], there are still knowledge gaps in relation to establishing factors that drive sustainable road freight policies by way of comparative studies across different continental regions. Additionally, whilst some policy papers highlight stakeholder engagement decision challenges [7], none of the papers we reviewed were focused on addressing multi-criteria decision making problems at policy level. This is an important yet emerging area of interest and future studies in this area may hold useful learning for policy makers and researchers in terms of decision making optimization, knowledge transfer, and cross-national collaboration.

Table 4. Pivot matrix of intervention themes per continental regions.

Intervention Mechanisms	Continental Regions							
	AF	AS	AU	EU	GN	NA	SA	Total
Decoupling	0	0	0	3	0	0	0	3
ICT (Information & Communications Technology)	0	0	0	8	0	1	0	9
Modality (Inter and Co-modality)	1	0	0	8	3	2	0	13
Operations (Design and Process)	1	3	1	26	4	2	0	38
Other	0	1	2	5	2	1	0	10
Policy	2	1	1	13	4	2	1	23
Grand Total	4	5	4	63	13	8	1	98

ICT is increasingly gaining preference amongst management and researchers, who identify its potential to support sustainable road freight across social, environmental, and economic frameworks [1,3]. Whilst papers that focus specifically on ICT as a freight intervention mechanism are limited, much of operations and modality themed papers recognize the propensity of ICT to support initiatives in these areas [61,73]. For example, Harris et al. linked the success of 33 EU intermodal framework projects to ICT technology [51]. This position is consistent with the findings from previous studies [48,52], where the successes of intermodal interventions were project as dependent on ICT breakthroughs. ICT offers a robust scope for exploring the multifaceted challenges associated with road freight transport in terms of both existing technology and the range of problems addressed. Whilst requiring significant cost investment to implement, ICT offers benefits in terms of performance control and monitoring [59]. Control in the sense that management have the complete command over its deployment and usage within their operations, with extended benefit for society. Despite the costs, it provides a more attractive option for addressing performance and sustainability issues within road freight operations.

As a developing pathway within the interventions approaches, the literature in this area is still relatively sparse, particularly regarding social outcomes from ICT deployment for road freight transport sustainability. Case studies and related in-depth methodologies may be adopted to help promote understanding on adoption drivers, barriers and derived benefits from ICT use for road freight transport operations. Our findings highlight potential areas for future research contributions by way of extending current models and approaches to South American, African, and Asian contexts. Future research may adopt exploratory approaches to understand drivers and barriers to interventions or their effectiveness in relation to contexts.

6. Conclusions

The aim of this paper has been to provide a concise overview of the extant literature on road freight transport sustainability, identifying and categorising intervention mechanisms as well as reporting on intervention alignments with continental regions. In addressing the two research questions, this study has identified six theme categories (decoupling, ICT, modality, operations, policy, and others) within which the extant literature on road freight transport sustainability can be characterized. Combining content and thematic analyses approaches, extracted themes were subjected to Jaccard's coefficient similarity test in order to validate the originality of each identified theme. In this regard, we believe our study has contributed to the future research design agenda in this area with clear pathways for future studies to explore and make contributions in this contemporary, yet complex area of academic interest.

Most notably, the results of the systematic literature review revealed that over a third (thirty-eight) of the papers reviewed featured research around operations (design and process). Often the main contributions of articles falling under the operations theme were around matching the vehicle to the specific area in which it operates, and utilising fleet management models for addressing emissions. Similarly, policy driven mechanisms featured highly in the articles reviewed (twenty-three), although with urban freight dominating the academic literature in this area, there is clearly opportunity for future research to expand beyond the urban context.

The geographical distribution of the articles reviewed (2001–2018) was also particularly revealing, highlighting that sixty-five percent of papers reviewed identified with Europe as their geographic region. Furthermore, papers that identify with decoupling as a policy strategy intervention are exclusively associated with Europe, although there has been no recent publication under this theme. Perhaps there is opportunity for enquiries in this area, exploring the potential for implementing decoupling strategies in North America and Asia. Also, future studies may investigate the effectiveness of decoupling strategies across Europe as the UK prepares to depart from the European Union.

Furthermore, as per regional contexts and mechanisms, we noted correlations between mechanisms and continental coordinates. Our continental analysis suggests that SRFT research has relatively low international collaborative applications, a common problem with many sustainability practices that are occurring in continental silos. However, we recognise that externality impacts are not always local and perhaps more needs to be done to improve sustainable practices across different continents to drive collective and effective impact that will improve our understanding of different interventions across different contexts. We are confident that our findings make significant contributions in a complex field of study by categorising the extant literature in some simple yet objective modus that will support the development of the field as well as support future research classifications. These findings will act as further stimulus for research in this area of SRFT.

Supplementary Materials: The following are available online at http://www.mdpi.com/2071-1050/10/6/1923/s1. Appendix A (1 and 2) contain details (schedule of reviewed papers and search log records) that are supplemental to the main text and have been referenced in the discussion. The information within the appendix can be crucial to the understanding of the themes discussed in the paper.

Acknowledgments: This research was partially supported by the project "Promoting Sustainable Freight Transport in Urban Contexts: Policy and Decision-Making Approaches (ProSFeT)", "funded by the H2020-MSCA-RISE-2016 programme (Grant Number: 734909)" and the EU-India Research & Innovation Partnership for efficient and sustainable freight transportation (REINVEST) project "funded by the European Union 'EU-India Research and Innovation Partnership' (Grant Number: R/142842)".

Conflicts of Interest: The authors declare no conflict of interest.

References

1. Wang, Y.; Sanchez-Rodrigues, V.; Leighton, E. The use of ICT in road freight transport for CO2 reduction— An exploratory study of the UK's grocery retail industry. *Int. J. Logist. Manag.* **2015**, *26*, 2–29. [CrossRef]
2. Palsson, H.; Kovacs, G. Reducing transport emissions: A reaction to stakeholder pressure or a strategy to increase competitive advantage. *Int. J. Phys. Distrib. Logist. Manag.* **2014**, *44*, 284–304. [CrossRef]
3. Sternberg, H.; Prockl, G.; Holmström, J. The efficiency potential of ICT in haulier operations. *Comput. Ind.* **2014**, *65*, 1161–1168. [CrossRef]
4. Perego, A.; Perotti, S.; Mangiaracina, R. ICT for logistics and freight transportation: A literature review and research agenda. *Int. J. Phys. Distrib. Logist. Manag.* **2011**, *41*, 457–483. [CrossRef]
5. McKinnon, A.C.; Piecyk, M.I. Measurement of CO2 emissions from road freight transport: A review of UK experience. *Energy Policy* **2009**, *37*, 3733–3742. [CrossRef]
6. McKinnon, A.C. Decoupling of Road Freight Transport and Economic Growth Trends in the UK: An Exploratory Analysis. *Transp. Rev.* **2007**, *27*, 37–64. [CrossRef]
7. Ballantyne, E.E.F.; Lindholm, M.; Whiteing, A.E. A comparative study of urban freight transport planning: Addressing stakeholder needs. *J. Transp. Geogr.* **2013**, *32*, 93–101. [CrossRef]
8. Richardson, B.C. Sustainable transport: Analysis frameworks. *J. Transp. Geogr.* **2005**, *13*, 29–39. [CrossRef]
9. Clausen, I.U.; Geiger, C.; Behmer, C. Green Corridors by Means of ICT Applications. *Procedia Soc. Behav. Sci.* **2012**, *48*, 1877–1886. [CrossRef]
10. Button, K.; Doyle, E.; Stough, R. Intelligent transport systems in commercial fleet management: A study of short term economic benefits. *Transp. Plan. Technol.* **2001**, *24*, 155–170. [CrossRef]
11. Tob-Ogu, A.; Kumar, N.; Cullen, J. ICT adoption in road freight transport in Nigeria—A case study of the petroleum downstream sector. *Technol. Forecast. Soc. Chang.* **2018**, *131*, 240–252. [CrossRef]
12. Winebrake, J.J.; Green, E.H.; Comer, B.; Corbett, J.J.; Froman, S. Estimating the direct rebound effect for on-road freight transportation. *Energy Policy* **2012**, *48*, 252–259. [CrossRef]
13. Grant, R.M.; Baden-Fuller, C. A Knowledge-Based Theory of Inter-Firm Collaboration. *Acad. Manag. Proc.* **1995**, 17–21. [CrossRef]
14. Tranfield, D.; Denyer, D.; Smart, P. Towards a methodology for developing evidence-informed management knowledge by means of systematic review. *Br. J. Manag.* **2003**, *14*, 207–222. [CrossRef]
15. Van Wee, B.; Banister, D. How to Write a Literature Review Paper? *Transp. Rev.* **2016**, *36*, 278–288.
16. Morrell, K. The narrative of "evidence based" management: A polemic. *J. Manag. Stud.* **2008**, *45*, 613–635. [CrossRef]
17. Denyer, D.; Tranfield, D. Producing a systematic literature review. In *The SAGE Handbook of Organisational Research Methods*; Buchanan, D., Bryman, A., Eds.; Sage Publications: London, UK, 2009; pp. 671–689.
18. Lagorio, A.; Pinto, R.; Golini, R. Research in urban logistics: A systematic literature review. *Int. J. Phys. Distrib. Logist. Manag.* **2016**, *46*, 908–931. [CrossRef]
19. *NVivo Qualitative Data Analysis Software*; QSR International Pty Ltd.: Doncaster, VIC, Australia, 2016.
20. Islam, D.M.Z.; Zunder, T.H.; Jorna, R. Performance evaluation of an online benchmarking tool for European freight transport chains. *Benchmark. Int. J.* **2013**, *20*, 233–250.
21. Bouchard, M.; Jousselme, A.L.; Doré, P.E. A proof for the positive definiteness of the Jaccard index matrix. *Int. J. Approx. Reason.* **2013**, *54*, 615–626. [CrossRef]
22. Velazquez-Martinez, J.C.; Fransoo, J.C.; Blanco, E.E.; Valenzuela-Ocana, K.B. A new statistical method of assigning vehicles to delivery areas for CO2 emissions reduction. *Transp. Res. Part D Transp. Environ.* **2016**, *43*, 133–144. [CrossRef]
23. Allen, J.; Browne, M.; Cherrett, T. Investigating relationships between road freight transport, facility location, logistics management and urban form. *J. Transp. Geogr.* **2012**, *24*, 45–57. [CrossRef]

24. Ubogu, A.E.; Ariyo, J.A.; Mamman, M. Port-hinterland trucking constraints in Nigeria. *J. Transp. Geogr.* **2011**, *19*, 106–114. [CrossRef]

25. McKinnon, A.C.; Ge, Y. The potential for reducing empty running by trucks: A retrospective analysis. *Int. J. Phys. Distrib. Logist. Manag.* **2006**, *36*, 391–410. [CrossRef]

26. Ehmke, J.F.; Campbell, A.M.; Thomas, B.W. Vehicle Routing to Minimize Time-Dependent Emissions in Urban Areas. *Eur. J. Oper. Res.* **2015**, *251*, 478–494. [CrossRef]

27. Fleischmann, B.; Gnutzmann, S.; Sandvoß, E. Dynamic Vehicle Routing Based on Online Traffic Information. *Transp. Sci.* **2004**, *38*, 420–433. [CrossRef]

28. Haughton, M.A. Route reoptimization's impact on delivery efficiency. *Transp. Res. Part E Logist. Transp. Rev* **2002**, *38*, 53–63. [CrossRef]

29. Midgley, W.J.; Cebon, D. Control of a hydraulic regenerative braking system for a heavy goods vehicle. *Proc. Inst. Mech. Eng. Part D J. Automobile Eng.* **2015**, *230*, 1338–1350. [CrossRef]

30. Olsson, J.; Woxenius, J. Localisation of freight consolidation centres serving small road hauliers in a wider urban area: Barriers for more efficient freight deliveries in Gothenburg. *J. Transp. Geogr.* **2014**, *34*, 25–33. [CrossRef]

31. Li, W.; Dai, Y.; Ma, L.; Hao, H.; Lu, H.; Albinson, R.; Li, Z. Oil-saving pathways until 2030 for road freight transportation in China based on a cost-optimization model. *Energy* **2015**, *86*, 369–384. [CrossRef]

32. Gilpin, G.; Hanssen, O.J.; Czerwinski, J. Biodiesel's and advanced exhaust aftertreatment's combined effect on global warming and air pollution in EU road-freight transport. *J. Clean. Prod.* **2014**, *78*, 84–93. [CrossRef]

33. Demir, E.; Bektas, T.; Laporte, G. A comparative analysis of several vehicle emission models for road freight transportation. *Transp. Res. Part D Transp. Environ.* **2011**, *16*, 347–357. [CrossRef]

34. Lammgard, C.; Andersson, D. Environmental considerations and trade-offs in purchasing of transportation services. *Res. Transp. Bus. Manag.* **2014**, *10*, 45–52. [CrossRef]

35. Sternberg, H.; Germann, T.; Klaas-Wissing, T. Who controls the fleet? Initial insights into road freight transport planning and control from an industrial network perspective. *Int. J. Logist. Res. Appl.* **2013**, *16*, 493–505. [CrossRef]

36. Pérez-Martínez, P.J. The vehicle approach for freight road transport energy and environmental analysis in Spain. *Eur. Transp. Res. Rev.* **2009**, *1*, 75–85. [CrossRef]

37. Stelling, P. Policy instruments for reducing CO_2-emissions from the Swedish freight transport sector. *Res. Transp. Bus. Manag.* **2014**, *12*, 47–54. [CrossRef]

38. Alho, A.R.; de Abreu e Silva, J. Analyzing the relation between land-use/urban freight operations and the need for dedicated infrastructure/enforcement—Application to the city of Lisbon. *Res. Transp. Bus. Manag.* **2014**, *11*, 85–97. [CrossRef]

39. Liimatainen, H.; Stenholm, P.; Tapio, P.; McKinnon, A.C. Energy efficiency practices among road freight hauliers. *Energy Policy* **2012**, *50*, 833–842. [CrossRef]

40. Dablanc, L. Goods transport in large European cities: Difficult to organize, difficult to modernize. *Transp. Res. Part A Policy Pract.* **2007**, *41*, 280–285. [CrossRef]

41. Klumpp, M. To Green or Not to Green: A Political, Economic and Social Analysis for the Past Failure of Green Logistics. *Sustainability* **2016**, *8*, 441. [CrossRef]

42. Matos, F.J.F.; Silva, F.J.F. The rebound effect on road freight transport: Empirical evidence from Portugal. *Energy Policy* **2011**, *39*, 2833–2841. [CrossRef]

43. M'raihi, R.; Mraihi, T.; Harizi, R.; Taoufik Bouzidi, M. Carbon emissions growth and road freight: Analysis of the influencing factors in Tunisia. *Transp. Policy* **2015**, *42*, 121–129. [CrossRef]

44. Khorheh, M.A.; Moisiadis, F.; Davarzani, H. Socio-environmental performance of transport systems. *Manag. Environ. Qual. Int. J.* **2015**, *26*, 810–825.

45. Winebrake, J.J.; Corbett, J.J.; Falzarano, A.; Hawker, S.J.; Korfmacher, K.; Ketha, S.; Zilora, S. Assessing Energy, Environmental, and Economic Tradeoffs in Intermodal Freight Transportation. *J. Air Waste Manag. Assoc.* **2008**, *58*, 37–41. [CrossRef]

46. Kim, N.S.; Van Wee, B. Assessment of CO_2 emissions for truck-only and rail-based intermodal freight systems in Europe. *Transp. Plan. Technol.* **2009**, *32*, 313–333. [CrossRef]

47. Arnold, P.; Peeters, D.; Thomas, I. Modelling a rail/road intermodal transportation system. *Transp. Res. Part E Logist. Transp. Rev.* **2004**, *40*, 255–270. [CrossRef]

48. Bontekoning, Y.M.; Priemus, H. Breakthrough innovations in intermodal freight transport. *Transp. Plan. Technol.* **2004**, *27*, 335–345. [CrossRef]

49. Li, L.; Negenborn, R.R.; De Schutter, B. Intermodal freight transport planning—A receding horizon control approach. *Transp. Res. Part C Emerg. Technol.* **2015**, *60*, 77–95. [CrossRef]

50. Macharis, C.; Caris, A.; Jourquin, B.; Pekin, E. A decision support framework for intermodal transport policy. *Eur. Transp. Res. Rev.* **2011**, *3*, 167–178. [CrossRef]

51. Harris, I.; Wang, Y.; Wang, H. ICT in multimodal transport and technological trends: Unleashing potential for the future. *Int. J. Prod. Econ.* **2015**, *159*, 88–103. [CrossRef]

52. Ruiz-Garcia, L.; Barreiro, P.; Rodriguez-Bermejo, J.; Robla, J.I. Review. Monitoring the intermodal, refrigerated transport of fruit using sensor networks. *Span. J. Agric. Res.* **2007**, *5*, 142–156. [CrossRef]

53. Sanchez-Rodrigues, V.; Cowburn, J.; Potter, A.; Naim, M.; Whiteing, A. Developing "Extra Distance" as a measure for the evaluation of road freight transport performance. *Int. J. Product. Perform. Manag.* **2014**, *63*, 822–840. [CrossRef]

54. Caris, A.; Macharis, C.; Janssens, G.K. Planning Problems in Intermodal Freight Transport: Accomplishments and Prospects. *Transp. Plan. Technol.* **2008**, *31*, 277–302. [CrossRef]

55. Marchet, G.; Perotti, S.; Mangiaracina, R. Modelling the impacts of ICT adoption for inter-modal transportation. *Int. J. Phys. Distrib. Logist. Manag.* **2012**, *42*, 110–127. [CrossRef]

56. Liimatainen, H.; Pollanen, M. The impact of sectoral economic development on the energy efficiency and CO_2 emissions of road freight transport. *Transp. Policy* **2013**, *27*, 150–157. [CrossRef]

57. Alises, A.; Vassallo, J.M.; Guzman, A.F. Road freight transport decoupling: A comparative analysis between the United Kingdom and Spain. *Transp. Policy* **2014**, *32*, 186–193. [CrossRef]

58. Steenhof, P.; Woudsma, C.; Sparling, E. Greenhouse gas emissions and the surface transport of freight in Canada. *Transp. Res. Part D Transp. Environ.* **2006**, *11*, 369–376. [CrossRef]

59. Marchet, G.; Perego, A.; Perotti, S. An exploratory study of ICT adoption in the Italian freight transportation industry. *Int. J. Phys. Distrib. Logist. Manag.* **2009**, *39*, 785–812. [CrossRef]

60. Crainic, T.G.; Ricciardi, N.; Storchi, G. Models for Evaluating and Planning City Logistics Systems. *Transp. Sci.* **2009**, *43*, 432–454. [CrossRef]

61. Ando, N.; Taniguchi, E. Travel time reliability in vehicle routing and scheduling with time windows. *Netw. Spat. Econ.* **2006**, *6*, 293–311. [CrossRef]

62. Davies, I.; Mason, R.; Lalwani, C. Assessing the impact of ICT on UK general haulage companies. *Int. J. Prod. Econ.* **2007**, *106*, 12–27. [CrossRef]

63. Walker, G.H.; Manson, A. Telematics, urban freight logistics and low carbon road networks. *J. Transp. Geogr.* **2014**, *37*, 74–81. [CrossRef]

64. Kinnear, S.; Rose, A.; Rolfe, J. Emissions Reporting in the Australian Road Freight Transport Sector: Is There a Better Method than the Default Option? *Int. J. Sustain. Transp.* **2014**, *9*, 93–102. [CrossRef]

65. Islam, D.M.Z.; Fabian Meier, J.; Aditjandra, P.T.; Zunder, T.H.; Pace, G. Logistics and supply chain management. *Res. Transp. Econ.* **2013**, *41*, 3–16. [CrossRef]

66. Carballo-Penela, A.; Mateo-Mantecon, I.; Domenech, J.L.; Coto-Millán, P. From the motorways of the sea to the green corridors' carbon footprint: The case of a port in Spain. *J. Environ. Plan. Manag.* **2012**, *55*, 765–782. [CrossRef]

67. Marchet, G.; Melacini, M.; Perotti, S. Environmental sustainability in logistics and freight transportation: A literature review and research agenda. *J. Manuf. Technol. Manag.* **2014**, *25*, 775–811. [CrossRef]

68. Demir, E.; Bektas, T.; Laporte, G. A review of recent research on green road freight transportation. *Eur. J. Oper. Res.* **2014**, *237*, 775–793. [CrossRef]

69. Demir, E.; Huang, Y.; Scholts, S.; Van Woensel, T. A selected review on the negative externalities of the freight transportation: Modeling and pricing. *Transp. Res. Part E Logist. Transp. Rev.* **2015**, *77*, 95–114. [CrossRef]

70. Liimatainen, H.; Arvidsson, N.; Hovi, I.B.; Jensen, T.C.; Nykänen, L. Road freight energy efficiency and CO_2 emissions in the Nordic countries. *Res. Transp. Bus. Manag.* **2014**, *12*, 11–19. [CrossRef]

71. Morrison, G.; Roebuck, R.L.; Cebon, D. Effects of longer heavy vehicles on traffic congestion. *Proc. Inst. Mech. Eng. Part C J. Mech. Eng. Sci.* **2013**, *228*, 970–988. [CrossRef]

72. Agbo, A.A.; Zhang, Y. Sustainable freight transport optimisation through synchromodal networks. *Cogent Eng.* **2017**, *4*, 1421005. [CrossRef]
73. Crainic, T.G.; Ricciardi, N.; Storchi, G. Advanced freight transportation systems for congested urban areas. *Transp. Res. Part C Emerg. Technol.* **2004**, *12*, 119–137. [CrossRef]

sustainability

MDPI

Article

Safety Assessment Model for Dangerous Goods Transport by Air Carrier

Hongli Zhao [1,2,*], Ning Zhang [1,3] and Yu Guan [4]

1   School of Economy and Management, Beihang University, Beijing 100191, China; nzhang@buaa.edu.cn
2   Civil Aviation Management Institute of China, Beijing 100102, China
3   Collaborative Innovation Center for Aviation Economy Development, Zhengzhou University of aeronautics, Zhengzhou 450015, China
4   China International Engineering Consulting Corporation, Beijing 100048, China; gy_buaa@163.com
*   Correspondence: zhaohongli@buaa.edu.cn; Tel.: +86-10-5825-0589

Received: 15 March 2018; Accepted: 20 April 2018; Published: 24 April 2018

Abstract: The safety of dangerous goods transport by air is directly related to human health and environmental pollution. This paper investigates a model to evaluate the safety performance of the transport of dangerous goods by air carriers. Based on a literature review, international regulations related to dangerous goods air transportation, and expert opinions, this paper identifies an assessment factor system with five drivers: organization/regulations, equipment/facilities, operations, emergency, and training. A hybrid evaluation method of a joint analytical hierarchy process and entropy weight is used to determine the importance of each factor and driver. The results suggest that the regulation of dangerous goods acceptance, sufficient equipment/facilities, and the condition of the equipment/facilities are the most important factors affecting the safety performance of dangerous goods transportation by air. An empirical study reveals that the proposed model is stable and reliable; thus, the model can guide resource allocation for air carriers to improve safety management of dangerous goods transportation.

Keywords: air transportation; assessment model; analytic hierarchy process (AHP); entropy weight; dangerous goods

---

1. Introduction

The safe transport of dangerous goods is of paramount importance to the government and enterprises in any country. The type and quantity of dangerous goods transported via air continue to increase due to new technologies and the use of new types of hazardous materials [1]. Dangerous goods include explosives, flammables, oxidizing substances, toxins, radioactive materials, and corrosive materials. If these hazardous substances are not properly handled, risks such as leakage, fire, or explosions may lead to air accidents or incidents, threatening the safety of air transport. These consequences may cause personal injury, property damage, and especially, environmental pollution [2]. For example, on 28 July 2011, a Boeing 747-48EF cargo aircraft owned by Asiana Airlines traveling from Seoul to Shanghai caught fire and crashed into the sea 107 km west of the Jeju Island [3]. An investigation of the accident indicated that the cargo aircraft was carrying a total of 58 tons of newly developed electronic products, including mobile phones and lithium batteries, which are classified as dangerous goods [3]. This accident caused two personal deaths, expensive losses of cargo and an aircraft, and sea pollution. The extent of consequences of such incidents depends on the type and quantity of the dangerous goods and the circumstances of the release. The pollution will be more serious if toxic substances, corrosive materials, or radioactive materials are being carried and then released. Although serious accidents resulting in heavy pollution during the air transportation of dangerous goods have not occurred in recent years, each company that handles dangerous goods, including air carriers, is at risk of accidents or of other unsafe events that may cause great damage to the economy and peoples' lives as well as to the property and the environment. Therefore, from the perspective of the government and air carriers, ensuring the safety and minimizing the risk and potential losses caused by such incidents is highly important.

Research referring to dangerous goods transportation has addressed different aspects of these problems. Routing choice or road selection have long been areas of interest in the road transport of dangerous goods, aiming to reduce the potential negative environmental and public health impacts [4–6]. The safety analysis and a risk assessment approach comprise the other focuses of the research on road tunnels, railways, and sea transport of dangerous goods [7–11]. The methodology used in risk-related research can be classified as qualitative, quantitative, or a combination thereof. Qualitative approaches mainly summarize risk hazard identification from historical data of accidents, incidents, and unsafe events to identify control measures for reducing accident rates [8,9], relying on sharp insight and experience. Some studies have proposed specific mathematical formulas to calculate the accident rate, damage rate, release rate, and concentration level of released dangerous goods in railway transportation [10,12], but no empirical application currently exists. In terms of the combined qualitative and quantitative approach, a popular tool is the risk matrix, which couples hazard severity levels with likelihood levels to determine a cumulative risk level based on an expert's score on each risk factor [13,14]. Another representative decision-making method combining qualitative and quantitative techniques is the analytic hierarchy process (AHP). Different types of risk factors are identified hierarchically in the first step via qualitative analysis and then prioritized in order of importance as calculated using a quantitative method [11,14,15]. The advantage of AHP lies in the use of expert experience to quantify the relative importance of factors at different levels. A disadvantage is that the factor weights are easily affected by expert subjectivity.

Prior studies have highlighted the importance of safety analysis or risk assessment when transporting dangerous goods by road, railway, and sea; however, few scholars have discussed these topics with respect to air transport. No research appears to have focused specifically on the safety evaluation or risk assessment of dangerous goods transported by air. Hsu et al. [14] established 14 indicators and established a risk matrix to evaluate the operational safety of dangerous goods transported via air using fuzzy AHP in the Taiwan region. Chang et al. [15] identified 17 risk factors using expert interviews and prioritized the order of management problems associated with air transport of dangerous goods using AHP in the Taiwan region. However, both studies were limited to Taiwan. Furthermore, the risk factors were identified from an industry development perspective, including policies and regulations, safety audits and supervision, cargo agents, air police stations, and customs airline personnel; only three indices were geared toward air carriers, which is hardly sufficient to guide the management of dangerous goods in such settings. Research in this field began in China in 2000, and over 10 papers on risk analysis have been published up to this point, but few have dealt with air carriers based on the evaluation methodology. Du [16] established 10 indices based on personnel, equipment, environment, and management to evaluate the safety of dangerous goods transport activities among air carriers, but these factors lacked the necessary detail to guide air carriers in improving their management of dangerous goods transport. A vulnerability assessment of a ground emergency system pertaining to the air transport of dangerous goods was studied, including emergency system construction and system implementation vulnerability [17]; unfortunately, the research did not extend to other activities. Therefore, studies of risk assessment in dangerous goods transport focusing on air carriers are needed. The present study seeks to fill this gap.

An increasing volume of hazardous materials has been transported via air in China in recent years. By 2016, 26 out of 59 domestic airlines (44%) held permits for the air transport of dangerous goods as cargo [18]. Dangerous goods transported by air can be found in air freights as well as items carried by passengers or in checked baggage. Statistics show that the risk of unsafe incidents caused by luggage is larger than that caused by cargo [19]. The safety of dangerous goods air transportation is an unavoidable and pressing issue for airlines. Safety always comes first. If an air carrier encounters an unacceptable risk, no passenger will be willing to board its flight and the sustainable development of the airline will likely be questionable. If the air transportation industry is exposed to many enduring potential risks of dangerous goods, then the corresponding negative impacts on the economy, society, and environment cannot be ignored.

To ensure and improve the safety level, a safety management system (SMS) was introduced and has become increasingly popular among governments and air enterprises [20]. Beginning in 2015, the International Civil Aviation Organization (ICAO) required dangerous goods safety management to include air carrier SMS [21]. The core of the SMS is risk management, and the most crucial component of risk management is the identification of safety factors to monitor potential risks prior to accidents and incidents, which is the foundation of control measure development.

In the previous literature, accident or fatality data were often investigated and used to measure risk and/or safety of dangerous goods transport by road, railway, and sea [6–9]. However, from a sustainability perspective, safety refers to preventing historical accidents and incidents from occurring again while ensuring a timely response to such events. It is more important to take corrective action to prevent future errors by emphasizing proactive safety measures, including adequate funds, resources, and manpower [22]. This study aims to assess the proactive safety performance of dangerous goods air carriers to prevent accidents before they occur. The overall goals of this study were to establish a model to assess the safety performance of air transport companies. The proposed model includes two key issues: factors affecting the safety sustainability of air transport enterprises and how to assign weights to these factors. The main objectives of this study are as follows:

— To identify and categorize the main contributing factors in dangerous goods transport that affect the safety and sustainability of air enterprises.
— To assign weights to these factors using a reasonable method.
— To test the model stability through an empirical study.

2. Research Method

The risk management process is used as a reference in this paper to establish the safety assessment model for air transport of dangerous goods. The first step of the risk management process is the identification of all potential risks. The next step is the assessment of identified risks to select suitable and effective safety control measures leading to risk reduction. Thus, risk factor identification and assessment are the most vital components of the entire risk management process.

The framework of this research process is shown in Figure 1. Several methods were used to achieve the research objectives. To compile a comprehensive list of risk factors, interviews with dangerous goods air transportation experts, using the Delphi method, were conducted to validate safety factors identified based on a literature review and to explore additional factors. In the assessment model, weight assignment is an important part of the evaluation result. At present, subjective and objective assignment methods constitute the major approaches. In the second step, to measure the weight of each identified safety factor, a mixed analysis method combining the AHP method and the entropy method was applied. The former involves expert-based weight attribution and the latter can compensate for the deficiencies of subjective opinion to some extent. The assessment model was established by calculation and analysis. Finally, an empirical study was used to apply a fuzzy synthetic evaluation (FSE) method to the model, and the combined method of the weight assignment proved to be stable.

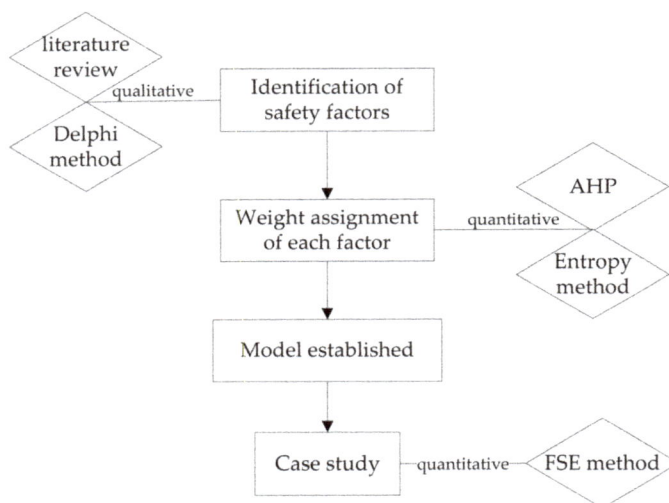

Figure 1. The research method framework.

3. Model Established

3.1. Identification of Safety Factors

Establishing factors for air carrier safety assessment is a critical feature of this study and provides a foundation for subsequent research. A literature review and expert interviews were conducted for this paper. The literature review surveyed previous research on dangerous goods transported by road, railway, sea, and air. A universal definition of safety research on a metro railway was used for reference, which outlined six preliminary categories composed of human factors, facilities, and management actions [23]. Though that research was aimed at the safety management of metro enterprises rather than dangerous goods transport, some attributes are transferable to dangerous goods transportation via air; certain factors, such as investment and infrastructure, are essential to any business seeking to ensure safety.

Dangerous goods air transport must be carried out according to regulations, including specific operation and training requirements. These include Annex 18 to the Convention on International Civil Aviation (Chicago Convention), The Safe Transport of Dangerous Goods by Air [24], which outlines the responsibilities of dangerous goods operators, including operations, information, and training; and Technical Instructions for the Safe Transport of Dangerous Goods by Air (Doc 9284) [25], issued by the ICAO, which stipulates specific operational responsibilities for airline operators, including stowing, segregation, and documentation. These regulations are legal requirements for member states, and China is no exception.

Based on the extant literature and international regulations, 19 factors influencing the safety performance of air transport of dangerous goods were collected. The list of factors was then examined by 15 experts on dangerous goods air transportation. The professionals came from enterprises, government, and research institutes, and all are knowledgeable in the field of air safety with at least 10 years of experience. The detail information of all the 15 experts is listed in Table 1.

Table 1. The detail information of the 15 experts.

Items	Description	Number
Gender	Male	7
	Female	8
Age	36–45 years old	7
	46–55 years old	8
Education	Bachelor degree	9
	Graduate and above	6
Years of experience	10–15 years	4
	15–20 years	8
	over 20 years	3
Service institution	enterprises	8
	government	2
	research institutes	5

After reviewing the list, some experts suggested adding the factor of "quality control of outsourcing" to the index set because some air carriers outsource dangerous goods business to ground handling agents; hence, a good quality control program is needed to ensure the outsourcer party complies with all safety requirements of the carrier. Then, the list of factors was updated and re-distributed to the 15 experts, who agreed that all factors derived from the literature, regulations, and expert opinion were reasonable and important. The final 20 factors are presented in Table 2, grouped into five dimensions based on their properties and attributes: organization/regulations, equipment/facilities, operations, emergency, and training. Table 2 also provides an explanation of each factor and corresponding references.

Table 2. The safety factors of dangerous goods air carriers.

Drivers	Factors	Explanation
Organization and regulations (E1)	Organizational structure (E_{11})	A good organizational and managerial structure delineates clear responsibilities and a reasonable division of labor [16]
	Quality control of outsourcing (E_{12})	The quality control system is effective if dangerous goods business is outsourced [Expert opinion].
	Communication and coordination (E_{13})	Smooth and effective communication and coordination between company departments are essential for daily work [8,16].
	Safety investment (E_{14})	Safety investment is essential funding that ensures safe operation of dangerous goods, such as by introducing new technology, training, safety incentives, or other activities [23].
	Rules and regulations (E_{15})	Rules and regulations delineate clear responsibilities for staff, thereby improving safety overall [15,16].
	Self-supervision (E_{16})	A clear dangerous goods self-supervision and inspection system with well-defined responsibilities is necessary for proper implementation [16,24].
Equipment and facilities (E2)	Sufficient equipment/facilities (E_{21})	Accidents are likely to occur if equipment and facilities are inadequate [8,14,15].
	Equipment/facilities conditions (E_{22})	The condition of equipment and facilities depends on service times and maintenance [16,23].
	Equipment/facilities performance (E_{23})	Equipment and facilities should be reliable, and advanced technology should be adopted to meet increasing freight volume [16,23].

Table 2. *Cont.*

Drivers	Factors	Explanation
	Luggage safety operations (E_{31})	Responsibilities such as sharing information with passengers and pre-checking and receiving luggage must be in place and performed properly [24,25].
	Ordinary cargo safety operations (E_{32})	Responsibilities such as sharing information with shippers and pre-checking and receiving cargo must be in place and performed properly [24,25].
Operations (E3)	Dangerous goods acceptance (E_{33})	Checking and receipt of dangerous goods must be consistent, and all transport documents, packaging, and so forth must comply with regulations [24,25].
	Dangerous goods storage (E_{34})	The storage and stacking of dangerous goods must conform to regulatory requirements [24,25].
	Dangerous goods loading (E_{35})	Dangerous goods allocation, aircraft commander notice, apron loading, and other ground supports must conform to regulatory requirements [14,24,25].
	Emergency management plan (E_{41})	The emergency management plan is an action guide to minimize potential event damage [17].
Emergency (E4)	Emergency-handling measures (E_{42})	Emergency-handling personnel and equipment must be adequate; efficient and timely actions contribute to safety [17].
	Emergency drilling plan (E_{43})	The emergency drilling plan should be complete and conducted regularly. Summarizing problems after drilling will help to improve safety [17].
	Training organization (E_{51})	A specific department should be responsible for organizing staff training to improve operational capabilities [24,25].
Training (E5)	Training program (E_{52})	The dangerous goods training program should be up-to-date and compliant with ICAO requirements [15,24,25].
	Training quality control (E_{53})	The training quality depends on the instructor, training method, training environment and location, and so forth [16,24,25].

3.2. Weight Assignment

The safety factors discussed in the previous section may not equally affect the safety of dangerous goods air carriers. A method of weight assignment must, therefore, be introduced to reflect respective contributions to each safety factor and driver. A hybrid evaluation method based on AHP and entropy weight is proposed in this study.

AHP is a structured technique for organizing and analyzing complex decisions. It provides a comprehensive and rational framework for group decision making and is widely used around the world [26]. However, the disadvantage of AHP is that it is influenced easily by expert knowledge and experience or the preferences of decision makers. The entropy method is mainly based on the correlation among the indicators, using a certain mathematical model, to calculate the index weights. The advantage is that it fully taps into the information implied in the raw data and the evaluation results are backed by a strong mathematical theory [27]. However, it ignores the knowledge and experience of decision makers, and sometimes, the weight obtained from them may not match the actual importance.

Given the advantages and disadvantages of these two methods, this study attempts to combine AHP and entropy by adopting the latter to complement the functions of the former. The two methods can thus overcome their shortcomings and make the results more accurate. Notably, this research is not the first to combine the entropy weights with AHP to determine index weight. The entropy method first appeared in thermodynamics and was incorporated into information theory by American mathematician Shannon [28]. The earliest application of the entropy weighting method in conjunction with AHP was a study of ship investment decision making [29]. Nowadays, the AHP-entropy method has been known and used in the assessment of various industries, including the safety assessment of food-waste feed [27], the safety evaluation of smart grids [30], the risk assessment in banks [31], and in community sustainability assessments [32]. These studies demonstrated that this integrated method is scientific and effective.

3.2.1. Steps of AHP

The basic process to obtain the weights is detailed below (adapted from Reference [33]):

Step 1: Construct a set of relative weight matrices (RWMs).

This paper uses a 1–9-point scale to score the relative importance of each driver and factor individually. For instance, if driver E1 and E2 are measured, and E1 is 5 times more important than E2 to the goal of safely transporting dangerous goods by air carriers, then the relative weight of E1 to E2 is denoted as 5; if E2 is 5 times more important than E1, the relative weight of E1 to E2 is denoted as 1/5. As illustrated above, each driver is assigned a global priority. This process is also used to weight the priorities on the upper level (its driver) for each factor.

Step 2: Hierarchical ordering.

After establishing the RWMs, the maximum eigenvalues and corresponding eigenvectors of each RWM can be calculated. Various hierarchy factors are ordered by their importance relative to other factors from the previous hierarchy (that is, hierarchical ordering).

$$E \times h = \lambda_{\max} \times h, \ \sum_{i=1}^{n} h_i = 1 \tag{1}$$

where n is the size of the matrix, $E = (e_{ij})_{n \times n}$ is the RWM, $\lambda_{\max}$ is the maximum eigenvalue of matrix E, and h denotes the eigenvectors of E.

Step 3: Examine the consistency of the hierarchy.

The consistency index (CI) is used to determine the consistency of the hierarchy. It is calculated as follows:

$$CI = \frac{(\lambda_{\max} - n)}{(n - 1)} \tag{2}$$

Then, the random consistency ratio (CR) is obtained from

$$CR = \frac{CI}{RI} \tag{3}$$

The RI is the average random consistency index. The value of RI for different matrix orders appears in Table 3.

Table 3. The average random consistency index of the 1–10 matrices.

n	1	2	3	4	5	6	7	8	9	10
RI	0	0	0.58	0.90	1.12	1.24	1.32	1.41	1.45	1.49

When the CR is less than 0.1, the weight coefficient distribution is reasonable and the matrix is considered consistent; otherwise, the RWM must be revised and the weight coefficient should be re-distributed.

3.2.2. Steps of Entropy Weight

The entropy method can measure the degree of disorder in a system. When the indicator provides more useful information, the difference in values among the evaluated objects on the same indicator is high and the entropy is small; thus, the weight of the selected indicator should be set correspondingly high. On the contrary, if the difference is small and the entropy is high, then the relative weight should be smaller [34]. The entropy weight method can reduce the impact of the subjective arbitrariness in the empowerment, making the evaluation result more objective. The steps of the entropy weight method are as follows:

Step 1: Normalize the elements of $E = (e_{ij})_{n \times n}$ RWM and obtain the standard matrix $F = (f_{ij})_{n \times n}$

$$f_{ij} = \frac{e_{ij}}{\sum_{j=1}^{n} e_{ij}} \tag{4}$$

Step 2: Calculate the entropy G_j, variation coefficient K_j, and weight L_j of each index:

$$G_j = -\frac{1}{\ln n} \sum_{i=1}^{n} f_{ij} \ln f_{ij} \tag{5}$$

$$K_j = 1 - G_j \tag{6}$$

$$L_j = \frac{K_j}{\sum_{j=1}^{n} K_j} \tag{7}$$

Step 3: Use the entropy weight L_j of the jth index to revise the weight vector h_j obtained via AHP to derive the comprehensive weight of the jth evaluation index W_j:

$$W_j = \frac{L_j h_j}{\sum_{j=1}^{n} L_j h_j} \tag{8}$$

When using the AHP-entropy method to evaluate the safety vulnerability of dangerous goods air carriers, the weights are calculated twice: AHP gives subjective results and the entropy weight provides an objective evaluation. This integrated method ensures the scientific reliability of the weight assigned to each factor that is combined with expert experience and the original objective data.

3.2.3. Calculation

The 15 experts were invited to determine the contribution of each safety factor to each driver using AHP. These experts included scholars and government administrators, the latter of whom possessed greater authority in dangerous goods air transportation but did not have sufficient understanding of AHP. To comprehensively assess this scoring method, we distributed e-mail questionnaires to obtain the factor weights and the respondents then provided an oral explanation of the scoring method by telephone.

Five drivers of organization/regulations (E1), equipment/facilities (E2), operations (E3), emergency (E4), and training (E5) were placed at the criteria level. The sub-criteria level was composed of 20 safety factors. Each of the 15 experts provided six RWMs containing five matrices at the sub-criteria level and one matrix at the criteria level; a total of 90 RWMs were collected. Then, the arithmetical average values of the 15 experts' RWMs were calculated to obtain six final RWMs to proceed to step 2 of the AHP method. Taking the 15 RWMs of the criteria level (five drivers) as an example, the average results are shown in Table 4.

Table 4. The relative weight matrix of the five drivers.

Drivers	E1	E2	E3	E4	E5
E1	1	2.391111	2.055238	2.976296	2.344444
E2	0.418216	1	2.082222	2.896296	2.211111
E3	0.486562	0.480256	1	2.874074	2.34
E4	0.335988	0.345269	0.347938	1	1.516296
E5	0.42654	0.452261	0.42735	0.659502	1

Next, we calculated the maximum eigenvalue $\lambda_{max} = 5.2269$ and obtained the eigenvector. $h = (0.3597, 0.2501, 0.1921, 0.0981)^T$. The *CI* equaled

$$CI = \frac{\lambda_{max} - n}{n - 1} = \frac{5.2269 - 5}{5 - 1} = 0.0567$$

As shown in Table 3, when $n = 5$, $RI = 1.12$, from which we determined that $CR = \frac{CI}{RI} < 0.1$. Therefore, the consistency of this RWM is satisfactory, indicating that the distribution of weights is reasonable. Next, we calculated the entropy, variation coefficient, entropy weight, and comprehensive weight according to Formulas (4)–(8), constructing Table 5:

Table 5. The weight results of the five drivers.

Drivers	AHP Weight	Entropy	Variation Coefficient	Entropy Weight	Comprehensive Weight
E1	0.3597	0.994834	0.005166	0.09836	0.188832
E2	0.2501	0.988141	0.011859	0.225782	0.301383
E3	0.1921	0.983953	0.016047	0.305511	0.313234
E4	0.1	0.986302	0.013698	0.260785	0.139187
E5	0.0981	0.994245	0.005755	0.109562	0.057365

Similarly, the weight at the sub-criteria level was obtained by the AHP method and revised using the entropy method. All the *CR* values shown in Table 6 are less than 0.1, suggesting that all RWMs were sufficiently consistent.

Table 6. The consistency check of the five RWMs of the 20 factors.

Factor Matrix	λ_{max}	CI	RI	CR
$E_{11} \sim E_{16}$	6.2282	0.0456	1.24	0.0368
$E_{21} \sim E_{23}$	3.0689	0.0344	0.58	0.0594
$E_{31} \sim E_{35}$	5.2511	0.0628	1.12	0.0561
$E_{41} \sim E_{43}$	3.0642	0.0321	0.58	0.0553
$E_{51} \sim E_{53}$	3.034	0.017	0.58	0.0293

The AHP weight, entropy weight, and comprehensive weight results for each factor are listed in Table 7.

Table 7. The weight results of the 20 factors.

Factors	AHP Weight		Entropy Weight		AHP-Entropy Weight	
	Local Weight	Global Weight	Local Weight	Global Weight	Local Weight	Global Weight
E_{11}	0.388	7.3%	0.113502	1.1%	0.282154	5.3%
E_{12}	0.1752	3.3%	0.21852	2.1%	0.245289	4.6%
E_{13}	0.1333	2.5%	0.217898	2.1%	0.186095	3.5%
E_{14}	0.0902	1.7%	0.185966	1.8%	0.107471	2.0%
E_{15}	0.1027	1.9%	0.161273	1.6%	0.106117	2.0%
E_{16}	0.1106	2.1%	0.102842	1.0%	0.072875	1.4%
E_{21}	0.5056	15.2%	0.369197	8.3%	0.542323	16.3%
E_{22}	0.2513	7.6%	0.510081	11.5%	0.372413	11.2%
E_{23}	0.2431	7.3%	0.120723	2.7%	0.085264	2.6%
E_{31}	0.185	5.8%	0.012719	0.4%	0.010947	0.3%
E_{32}	0.2577	8.1%	0.077747	2.4%	0.09321	2.9%
E_{33}	0.3008	9.4%	0.438010	13.4%	0.612948	19.2%
E_{34}	0.1343	4.2%	0.263426	8.0%	0.164588	5.2%
E_{35}	0.1222	3.8%	0.2081	6.4%	0.118306	3.7%
E_{41}	0.5236	7.3%	0.210608	5.5%	0.352677	4.9%
E_{42}	0.333	4.6%	0.470493	12.3%	0.50107	7.0%
E_{43}	0.1434	2.0%	0.318899	8.3%	0.146253	2.0%
E_{51}	0.4824	2.8%	0.464489	5.1%	0.640977	3.7%
E_{52}	0.2182	1.3%	0.429028	4.7%	0.267794	1.5%
E_{53}	0.2995	1.7%	0.106482	1.2%	0.091229	0.5%

3.3. Model

The safety assessment model for dangerous goods transport by air was established after identifying the safety factors and assigning a weight to each. The safety assessment model as described in Table 8 was refined from the results of Tables 2, 5 and 7. The model can be used to assess the safety level of dangerous goods transport by air for airlines.

Table 8. The safety assessment model for dangerous goods transport by air carriers.

5 Drivers	Weights (W1)	20 Factors	Weights (W2)
Organization and regulations (E1)	0.188832	Organizational structure (E_{11})	0.282154
		Quality control of outsourcing (E_{12})	0.245289
		Communication and coordination (E_{13})	0.186095
		Safety investment (E_{14})	0.107471
		Rules and regulations (E_{15})	0.106117
		Self-supervision (E_{16})	0.072875
Equipment and facilities (E2)	0.301383	Sufficient equipment/facilities(E_{21})	0.542323
		Equipment/facilities conditions (E_{22})	0.372413
		Equipment/facilities performance (E_{23})	0.085264
Operations (E3)	0.313234	Luggage safety operations (E_{31})	0.010947
		Ordinary cargo safety operations (E_{32})	0.09321
		Dangerous goods acceptance (E_{33})	0.612948
		Dangerous goods storage (E_{34})	0.164588
		Dangerous goods loading (E_{35})	0.118306
Emergency (E4)	0.139187	Emergency management plan (E_{41})	0.352677
		Emergency-handling measures (E_{42})	0.50107
		Emergency drilling plan (E_{43})	0.146253
Training (E5)	0.057365	Training organization (E_{51})	0.640977
		Training program (E_{52})	0.267794
		Training quality control (E_{53})	0.091229

4. Case Study

After establishing the safety assessment model for dangerous goods transport by air, we used an empirical study to examine the model stability. Many evaluation methodologies are available, such as the grey incidence analysis, artificial neural networks, and others [35]. The fuzzy set theory is suitable for risk assessment and has been adopted in many risk management studies [36]. The FSE method is a particularly useful tool to manage uncertainty and multiple attributes in group decision-making theories. FSE is defined by different fuzzy operators, which may produce different results even when using the same assessment model [37]. We used four fuzzy operators to test the stabilization of the model proposed in this paper.

We selected one mid-scale airline that has operated a dangerous goods transport business for over five years. Ten experienced experts (2/3 of the 15 experts mentioned above) offered individual evaluations of the safety performance (actual state) of this airline according to the factors listed in Table 2. The evaluation was divided into two stages: in the first, experts reviewed all the relevant documents in the office; in the second, they observed the actual process/situation in the field. According to the factors listed in Table 2, 12 factors related to organization and regulation (E1), emergency (E4), and training (E5) were examined in the first stage. Taking E_{11} (organizational structure) as an example, we provided the experts with the organizational chart containing the department and divisions responsible for dangerous goods safety in this airline to facilitate the scoring process. Eight total factors spanning equipment and facilities (E2) and operations (E3) were evaluated in the second stage. Taking E_{21} (sufficient equipment/facilities) for instance, the experts went to the warehouse and the ramp to determine whether the equipment and facilities of dangerous goods were adequate.

Sustainability **2018**, *10*, 1306

The evaluation set consisted of V = {excellent, good, ordinary, poor, bad}. Taking E_{11} for example, after reviewing the organizational chart for dangerous goods safety of this airline, two of the 10 experts assigned a rating of "excellent"; four said "good"; three said "ordinary"; and one said "poor". Therefore, the evaluation results of E_{11} were {0.2, 0.4, 0.3, 0.1, 0}. The 20 factors were assigned individually, and results appear in Table 9.

Table 9. The evaluation values of the 20 factors.

Assessment Level	Excellent	Good	Ordinary	Poor	Bad
E_{11}	0.2	0.4	0.3	0.1	0
E_{12}	0.3	0.3	0.2	0.2	0
E_{13}	0.1	0.1	0.3	0.3	0.2
E_{14}	0.1	0.2	0.3	0.2	0.2
E_{15}	0.5	0.2	0.3	0	0
E_{16}	0.2	0.3	0.3	0.2	0
E_{21}	0.1	0.1	0.5	0.3	0
E_{22}	0.1	0.4	0.3	0.1	0.1
E_{23}	0.2	0.3	0.2	0.2	0.1
E_{31}	0.1	0.4	0.4	0.1	0
E_{32}	0.2	0.5	0.2	0.1	0
E_{33}	0.2	0.2	0.4	0.2	0
E_{34}	0.2	0.4	0.3	0.1	0
E_{35}	0	0.4	0.4	0.1	0.1
E_{41}	0.1	0.4	0.3	0.2	0
E_{42}	0.2	0.3	0.4	0.1	0
E_{43}	0.1	0.2	0.5	0.1	0.1
E_{51}	0.1	0.3	0.3	0.2	0.1
E_{52}	0.3	0.3	0.2	0.2	0
E_{53}	0	0.2	0.6	0.2	0

The weights of the 5 drivers in Table 8 is denoted as vector W1; the weights of the 20 factors is denoted as vector W2; the assessment set in Table 8 is denoted as matrix R. Then the evaluation results Q can be calculated by

$$Q = W1^T \times R \times W2 \tag{9}$$

In fuzzy evaluation, the commonly used operators include the minimum and maximum operator $(Z(\wedge, \vee))$, the multiplication and maximum operator $(Z(\bullet, \vee))$, the minimum and bounded operator $(Z(\wedge, \oplus))$, and the multiplication and bounded operator $(Z(\bullet, \oplus))$ [38,39]. To compare the discrepancy of the evaluation results based on the AHP weights and comprehensive weights, eight evaluation results (Table 10) were calculated using four different fuzzy operators, respectively according to Formula (9). Under the AHP weights, different operators produced different results: the evaluation results of the two operators were "good" and those of the other two operators were "ordinary". Under the weights revised by the entropy method, different operators had the same results ("ordinary" for all four operators). As such, the comprehensive weights demonstrated better weight stability than the AHP weights, and the model developed in this paper seems to be robust and reliable because the evaluation results did not vary by the operator.

Table 10. The evaluation results of the different fuzzy operators.

	Operator	Excellent	Good	Ordinary	Poor	Bad	Results
Fuzzy evaluation results of AHP weights	$Z(\wedge, \vee)$	0.1513	0.2987	0.2856	0.1707	0.0936	Good
	$Z(\bullet, \vee)$	0.1370	0.2960	0.3395	0.1762	0.0513	Ordinary
	$Z(\wedge, \oplus)$	0.1778	0.2999	0.2858	0.1811	0.0555	Good
	$Z(\bullet, \oplus)$	0.1670	0.2919	0.3325	0.1716	0.0370	Ordinary
Fuzzy evaluation results of AHP-entropy weights	$Z(\wedge, \vee)$	0.1488	0.2403	0.3282	0.1878	0.0949	Ordinary
	$Z(\bullet, \vee)$	0.1503	0.2354	0.3680	0.1985	0.0477	Ordinary
	$Z(\wedge, \oplus)$	0.1747	0.2529	0.3174	0.1929	0.0622	Ordinary
	$Z(\bullet, \oplus)$	0.1596	0.2710	0.3569	0.1782	0.0343	Ordinary

The empirical results show that the efficiency and stability of the AHP-entropy method are better than that of AHP alone and the evaluation results are more scientific and reliable according to the model and algorithm established in this paper.

5. Results and Discussion

The model proposed in this paper aims to provide support for analyzing the safety factors of dangerous goods transport by air carriers. The 20 safety factors listed in Table 2, collected from a literature review and field experts' opinions, have three features. First, the factor system is comprehensive, incorporating safety assurance into human resources (organization, E1), finances (investment, E1), and infrastructure (equipment and facilities, E2) along with the safety promotion of professional operations (E3), emergency management (E4), and training (E5). This system is thorough and provides enhanced guidance to air carriers to improve managerial oversight related to dangerous goods. Second, although some factors such as E_{12} (quality control of outsourcing) and E_{14} (safety investment) were proposed and used in dangerous goods air transportation initially, the list of factors was examined twice by industry experts with different occupational backgrounds, all of whom pointed out that the factors are essential for the safety management of dangerous goods transport by air.

The weights reflect the importance of each driver and factor. Judging from the weight results in Table 5, the comprehensive priority of the five safety drivers are E3 > E2 > E1 > E4 > E5. The comprehensive priorities of the five safety drivers are E3 > E2 > E1 > E4 > E5. In all cases, the importance (that is, weight) of E1, E2, and E3 were higher than E4 and E5, and operations (E3), with a weight of 0.313234, was identified as the most important driver affecting the safety of dangerous goods transport by air. As such, dangerous goods operations should be prioritized first to guarantee safety, followed by equipment and facilities. The operation of dangerous goods air transportation not only involves accepting, storing, and loading declared dangerous goods according to the ICAO requirements, it also requires the identifying of undeclared dangerous goods from ordinary cargo and luggage to prevent potential risks, which may lead to more serious accidents and incidents [1]. In fact, it is difficult for air carriers to distinguish hidden dangerous goods from ordinary cargo and luggage without using security inspection machines, which has been a complicated proposition in China for quite some time.

The AHP weight, entropy weight, and comprehensive weight results for each factor are listed in Table 7. These three weights are subdivided by local weight (that is, the priority of each factor in its own driver) and global weight (that is, its relative importance among all 20 factors). By comparing the changes in global weights before and after revision using the entropy method, it is found that: (i) according to the global weight results calculated by the AHP method, the importance of each factor was nearly equally matched. The contribution of only one factor, E_{21} (sufficient equipment/facilities), exceeded 10%; the weight distributions of the other indices were balanced. The AHP method alone cannot determine the key activities on which air carriers should focus to ensure safety, especially when resources are limited. (ii) After revision by the entropy method, three factors had global contributions above 10%: dangerous goods acceptance (E_{33}), sufficient equipment/facilities

(E_{21}), and equipment/facility conditions (E_{22}). The global weight of each was 19.2%, 16.3%, and 11.2%, respectively, accounting for 46.8% of the total. In other words, along with sufficient equipment/facilities, air carriers should also focus on regulating dangerous goods acceptance and equipment/facility conditions. These three aspects collectively determine the safety and sustainability of the transport of dangerous goods by air. Therefore, after revising the AHP method via the entropy method, the obtained weight set is more scientific and has practical value for industry work.

Dangerous goods acceptance (E_{33}) was found to be the most important factor affecting the safety of dangerous goods transport by air as revealed in Table 7. Du [40] indicated that acceptance is an essential component of the safe transport of dangerous goods. In our research, we found dangerous goods acceptance to be the most important factor among the 20 factors. The main task of the dangerous goods acceptance for air carriers is verifying the regulatory compliance, including the classification, packaging, marking, labeling, and all associated documents, a task that is completed by the shipper or cargo agent. Dangerous goods acceptance transfers risk from the shipper or cargo agent to the carrier. In the event of an incomplete investigation during the dangerous goods acceptance procedure, the carrier is held accountable even if either the shipper or cargo agent is at fault [25]. Therefore, a specialized team of air carriers is often responsible for dangerous goods acceptance in actual operations.

Equipment and facilities (E2) were found to have high priority as indicated in Tables 5 and 7. The equipment and facilities for dangerous goods transport by air include, but are not limited to specialized warehouses, storage racks, unit load devices, forklift trucks, safety defense equipment, inspection equipment, and so on. They constitute the essential hardware to ensure the proper handling of dangerous goods. Compared to the performance (E_{23}) of equipment/facilities, sufficiency (E_{21}; global weight = 16.3%) and conditions (E_{22}; global weight = 11.2%) take precedence. Air carriers are encouraged to maintain and upgrade equipment and facilities in a timely manner to minimize the potential risks associated with damage and degradation.

The weights of the drivers and factors in Tables 5 and 7 were calculated with AHP and the entropy method based on a pair-wise comparison of the relative importance of each driver and factor, judged by 15 Chinese experts using a 9-point scale. Therefore, the findings of the key drivers and factors detailed herein are highly relevant to the actual conditions in China. Although the data in Table 5 show that dangerous goods operations (E3) and equipment/facilities (E2) are key drivers behind the safety performance of air carriers in China, organization and regulations (E1), emergency (E4), and training (E5) cannot be ignored. Rather, the management, organization, and training surrounding dangerous goods constitute strong and indispensable support for the infrastructure and operations business in China. Without organization and training, any infrastructure, operations, and emergency handling of dangerous goods are impossible.

We kept international applicability in mind during this study, including refining safety factors from the literature when choosing an empirical case. The 20 factors summarized in Table 2 were adopted from previous studies conducted around the world or from ICAO international regulations, indicating that these factors are suitable for dangerous goods transport by air in China as well as in other countries. Additionally, while the model investigated in this paper depended partly on the judgment of Chinese experts and revealed some key drivers and factors useful for the development of risk control measures in China, the 20 safety factors identified can also be used to assess the safety situations regarding dangerous goods air transport of air carriers around the world. The assessment model was verified using a case analysis combined with the FSE method. When selecting a mid-scale airline as an empirical case, the representativeness and typicality were emphasized. The case study indicates that the model obtains reliable assessment results: the findings show that the evaluation results acquired through the AHP-entropy method are more stable than those calculated by the AHP method. The proposed method is also more efficient and reasonable in identifying air carrier safety levels. Therefore, the safety assessment model proposed in this paper is reliable and has good feasibility and practicality for dangerous goods transport by air carriers.

Sustainability **2018**, *10*, 1306

6. Conclusions

In contrast to road, railway, and marine transport methods, air transport is more international and the goods it carries are of higher universal value. The impact scope of the occurrence of accidents and incidents involving dangerous goods is wider and the consequences are even worse than with other modes of transportation. The transport safety of dangerous goods is an important aspect of aviation safety. China plays an important role in the chain of dangerous goods air transportation. Shipping dangerous goods is one of the most complex airline tasks, requiring careful safety measures and transportation technologies. Therefore, studies concerning the safety management of dangerous goods air transportation are necessary.

The main contributions of the paper are summarized below:

(i)   Based on a literature review and interviews with industry experts, a novel index system was established to assess the safety of dangerous goods transport activities by air carriers, including 20 factors related to organization and regulations, equipment and facilities, operations, emergency, and training. Compared with other studies on dangerous goods air transportation [14–17,40], the factors proposed in this paper focus on the risks air transport enterprises can control to achieve safer, greener sustainable development, reflecting the comprehensive safety status of air carriers.

(ii) AHP and entropy methods were used jointly to determine factor weights. By comparing the changes in factor weights before and after the entropy revision, the proposed method appears to reconcile the influence of subjective preferences from AHP method experts and objective data deviation in the entropy method. The weights were also more scientific in reflecting the important safety factors related to dangerous goods air transportation and hence can guide air carrier management.

(iii) A case study was used to apply an FSE method to the model. The combined method for weight assignment proved to be stable. To our knowledge, this study is the first to apply a combined qualitative and quantitative approach to study the safety assessment of dangerous goods transport by air carriers. Its findings provide ways to differentiate risk factors in dangerous goods transport and enrich the application of safety evaluation techniques.

(iv) The findings reveal that for operations and infrastructure, especially in terms of dangerous goods acceptance, the sufficiency and condition of infrastructure are the most important factors affecting the safety performance of dangerous goods air transportation in China. The results provide a suggested scheme for air carrier resource allocation to achieve better safety performance and sustainable development.

A number of future research directions could be pursued from this study. Some factors identified in this paper had been previously incorporated into safety studies on dangerous goods air transport while others had not. The newly introduced factors were drawn from two sources: literature related to other transport modes (that is, metro railway) and expert opinion. Although all factors were further reviewed by 15 experts who were experienced and had worked in dangerous goods air transportation for over 10 years, some factors affecting the safety of air carrier transport may have been overlooked. As such, additional research is warranted to examine the factors affecting the safety of dangerous goods air transportation.

Author Contributions: The author Hongli Zhao drafted the manuscript. Ning Zhang and Yu Guan contributed to the research methodology and polished the language.

Acknowledgments: This research was supported by the Security Capacity Construction Foundation of Civil Aviation Authority of China (Project Number: 14000900100016J013). The authors would like to acknowledge the experts who participated in the study and those who provided suggestions.

Conflicts of Interest: The authors declare no conflict of interest.

References

1. Ellis, J. Undeclared dangerous goods—Risk implications for maritime transport. *WMU J. Marit. Aff.* **2010**, *9*, 5–27. [CrossRef]
2. Forigua, J.; Lyons, L. Safety analysis of transportation chain for dangerous goods: A case study in Colombia. *Transp. Res. Procedia* **2016**, *12*, 842–850. [CrossRef]
3. 1001 Crash. Asiana Cargo-Boeing 747-48EF off Jeju, South Korea, 28 July 2011. 2015. Available online: http://www.1001crash.com/index-page-description-accident-Asiana_B747-lg-2-crash-299.html (accessed on 24 August 2016).
4. Fabiano, B.; Currò, F.; Palazzi, E.; Pastorinov, R.A. framework for risk assessment and decision-making strategies in dangerous good transportation. *J. Hazard. Mater.* **2002**, *93*, 1–15. [CrossRef]
5. Torretta, V.; Rada, E.C.; Schiavon, M.; Viotti, P. Decision support systems for assessing risks involved in transporting hazardous materials: A review. *Saf. Sci.* **2017**, *92*, 1–9. [CrossRef]
6. Conca, A.; Ridella, C.; Sapori, E. A risk assessment for road transportation of dangerous goods: A routing solution. *Transp. Res. Procedia* **2016**, *14*, 2890–2899. [CrossRef]
7. Gheorghe, A.V.; Birchmeier, J.; Vamanu, D.; Papazoglou, I.; Kroger, W. Comprehensive risk assessment for rail transportation of dangerous goods: A validated platform for decision support. *Reliab. Eng. Syst. Saf.* **2005**, *88*, 247–272. [CrossRef]
8. Batarliene, N.; Jarasuniene, A. Analysis of the accidents and incidents occurring during the transportation of dangerous goods by railway transport. *Transport* **2014**, *29*, 395–400. [CrossRef]
9. Ellis, J. Analysis of accidents and incidents occurring during transport of packaged dangerous goods by sea. *Saf. Sci.* **2011**, *49*, 1231–1237. [CrossRef]
10. Verma, M.; Verter, V. Railroad transportation of dangerous goods: Population exposure to airborne toxins. *Comp. Oper. Res.* **2007**, *34*, 1287–1303. [CrossRef]
11. Molero, G.M.; Santarremigia, F.E.; Aragonés-Beltrán, P.; Pastor-Ferrando, J.P. Total Safety by design: Increased safety and operability of supply chain of inland terminals for containers with dangerous goods. *Saf. Sci.* **2017**, *100*, 168–182. [CrossRef]
12. Chen, J.; Wen, C. Risk assessment model approach for dangerous goods transported by railway. *J. Transp. Secur.* **2011**, *4*, 351–359. [CrossRef]
13. Benekos, I.; Diamantidis, D. On risk assessment and risk acceptance of dangerous goods transportation through road tunnels in Greece. *Saf. Sci.* **2017**, *91*, 1–10. [CrossRef]
14. Hsu, W.K.K.; Huang, S.H.S.; Tseng, W.J. Evaluating the risk of operational safety for dangerous goods in airfreights—A revised risk matrix based on fuzzy AHP. *Transp. Res. Part D* **2016**, *48*, 235–247. [CrossRef]
15. Chang, Y.H.; Yeh, C.H.; Liu, Y.L. Prioritizing Management issues of moving dangerous goods by air transport. *J. Air Transp. Manag.* **2006**, *12*, 191–196. [CrossRef]
16. Du, W.B. Risk Analysis and Control Method of Dangerous Goods Air Transportation of Some Company. Master's Dissertation, Fudan University, Shanghai, China, 2010. (In Chinese)
17. Yang, W. Research on Vulnerability Assessment of Ground Emergency System in Air Transport of Dangerous Goods. Master's Dissertation, Civil Aviation University of China, Tianjin, China, 2015. (In Chinese)
18. Civil Aviation Authority of China. The Annual Report of Dangerous Goods Air Transport in China of the Year 2016. Available online: http://www.caacdgc.org/tzgg/zhxw/201705/t20170527_6390.html (accessed on 12 December 2017).
19. Xu, J. Analysis and control measures for unsafe events in air transportation of dangerous goods. *China Civ. Aviat.* **2016**, *4*, 31–32. (In Chinese)
20. Chen, W.; Li, J. Safety performance monitoring and measurement of civil aviation unit. *J. Air Transp. Manag.* **2016**, *57*, 228–233. [CrossRef]
21. International Civil Aviation Organization. *Annex 19 to the Convention on International Civil Aviation: Safety Management*; International Civil Aviation Organization: Montreal, QC, Canada, 2016.
22. Shyur, H.J. A quantitative model for aviation safety risk assessment. *Comput. Ind. Eng.* **2008**, *54*, 34–44. [CrossRef]
23. Kyriakidis, M.; Hirsch, R.; Majumdar, A. Metro Railway safety: An analysis of accident precursors. *Saf. Sci.* **2012**, *50*, 1535–1548. [CrossRef]

24. International Civil Aviation Organization. *Annex 18 to the Convention on International Civil Aviation: The Safe Transport of Dangerous Goods by Air*; International Civil Aviation Organization: Montreal, QC, Canada, 2016.

25. International Civil Aviation Organization. *Technical Instructions for the Safe Transport of Dangerous Goods by Air (Doc 9284)*, version 2017–2018; International Civil Aviation Organization: Montreal, QC, Canada, 2017.

26. Song, L.L.; Li, Q.M.; George, F.L.; Deng, Y.L.; Lu, P. Using an AHP-ISM Based Method to study the vulnerability factors of urban rail transit system. *Sustainability* **2017**, *9*, 1065. [CrossRef]

27. Chen, T.; Jin, Y.Y.; Qiu, X.P.; Chen, X. A hybrid fuzzy evaluation method for safety assessment of food-waste feed based on entropy and the analytic hierarchy process methods. *Expert Syst. Appl.* **2014**, *41*, 7328–7337. [CrossRef]

28. Shannon, C.E. A mathematical theory of communication. *Bell Syst. Tech. J.* **1948**, *27*, 379–423. [CrossRef]

29. Huang, W.Z. AHP method on the weight of entropy and its application in ship investment decision making. *J. Shanghai Marit. Univ.* **2000**, *21*, 97–101. (In Chinese)

30. Xie, C.S.; Dong, D.P.; Hua, S.P.; Xu, X.; Chen, Y.J. Safety evaluation of smart grid based on AHP-entropy method. *Syst. Eng. Proceida* **2012**, *4*, 203–209.

31. Wang, T. The Research of electronic banking risk evaluation based on comprehensive assessment AHP-entropy. *Int. J. Sci. Technol.* **2014**, *7*, 413–422. [CrossRef]

32. Wu, G.D.; Duan, K.F.; Zuo, J.; Zhao, X.B.; Tang, D.Z. Integrated sustainability assessment of Public Rental Housing Community Based on a Hybrid Method of AHP-Entropy Weight and Cloud Model. *Sustainability* **2017**, *9*, 603. [CrossRef]

33. Saaty, T.L. An exposition of the AHP in reply to the paper 'Remarks on the analytic hierarchy process'. *Manag. Sci.* **1990**, *36*, 259–268. [CrossRef]

34. Zou, Z.H.; Yun, Y.; Sun, J.N. Entropy method for determination of weight of evaluating indicators in fuzzy synthetic evaluation for water quality assessment. *J. Environ. Sci.* **2006**, *18*, 1020–1023. [CrossRef]

35. Peng, Z.L.; Zhang, Q.; Yang, S.L. Overview of comprehensive evaluation theory and methodology. *Chin. J. Manag. Sci.* **2015**, *11*, 245–256. (In Chinese)

36. Zhao, X.B.; Hwang, B.G.; Gao, Y. A fuzzy synthetic evaluation approach for risk assessment: A case of Singapore's green projects. *J. Clean. Prod.* **2016**, *115*, 203–213. [CrossRef]

37. Suer, G.A.; Arikan, F.; Babayigit, C. Effects of different fuzzy operators on fuzzy bi-objective cell loading problem in labor-intensive manufacturing cells. *Comput. Ind. Eng.* **2009**, *56*, 476–488. [CrossRef]

38. Liu, Y.T.; Hu, J.B. A mathematical model and its parameters estimation for fuzzy evaluation. *J. Beijing Polytech. Univ.* **2001**, *3*, 112–115. (In Chinese)

39. Shen, J.H.; Fu, X.Y.; Zhao, Y.X. Improvement of the Fuzzy Comprehensive Evaluation Model. *Fuzzy Syst. Math.* **2011**, *6*, 127–132. (In Chinese)

40. Du, J. Research on the Risk Evaluation Index System for Collection in Air Transport of Dangerous Goods. *Saf. Environ. Eng.* **2012**, *5*, 77–79. (In Chinese)

![sustainability logo] *sustainability*

MDPI

Article

The Concept of Urban Freight Transport Projects Durability and Its Assessment within the Framework of a Freight Quality Partnership

Kinga Kijewska * and Mariusz Jedliński

Faculty of Economics and Engineering of Transport, Maritime University of Szczecin, 11 Pobożnego Str., 70-507 Szczecin, Poland; m.jedlinski@am.szczecin.pl
* Correspondence: k.kijewska@am.szczecin.pl

Received: 6 May 2018; Accepted: 26 June 2018; Published: 28 June 2018

Abstract: This article focuses on the role of Urban Freight Transport (UFT) projects in improving the life quality of city inhabitants. The main focus of the deliberations is the aspect of UFT projects' durability. The authors take an original approach to the definition of UFT project durability and also provide the results of a research study carried out in 2018. This made it possible to furnish an answer to the research questions that boiled down to the analysis of the current status of the relevant academic literature, to attempt to define the total durability of a UFT project, and to indicate the critical gaps in perception among the key stakeholders of the projects. In this study, particular attention is paid to the terminological synthesis and the conclusion resulting from adopting induction and deduction as the methods of solving research study problems. A novelty is the approach adopted in the project evaluation emphasising the mentioned durability aspect as one of the major success factors. This is particularly important for implementation of a Freight Quality Partnership as a solution enabling development of sustainable systems of urban logistics. The solution was treated as a specific implementation project for which the issue of key importance is the identification of success factors in the context of satisfying the needs of diverse groups of UFT stakeholders. It should be stressed that durability of projects in the area of UFT is critically important, even though there is a significant conceptual gap in that regard. The research study involved the originally developed concept of the Pyramid of Stakeholders Survey. By means of this concept, FQP durability was analysed on the example of the experience gained in the course of the solution functioning in Szczecin.

Keywords: logistics; urban freight transport; Freight Quality Partnership; project durability

---

1. Introduction

An Urban Freight Transport (UFT) system may be characterised as a sociotechnical system consisting of a compilation of infrastructural (technical, social) systems and interdependence networks of stakeholders. It includes elements such as e.g., technologies—vehicles, ICT solutions, logistics infrastructure, legal regulations or market factors—supply and demand for the distributed goods [1–5]. In a holistic perspective, in accordance with the idea of sustainable development, it should also account for social and environmental aspects recommended by the European Commission (the concept of zero-emission urban logistics by 2030).

In the academic literature, urban logistics (in the social logistics dimension—this concept was introduced among others in [6]) is often perceived as conflicting with the activities connected with goods deliveries, passenger transport and the life quality of city inhabitants [7–11]. Logistics as an area of knowledge is perceived in three aspects: business, military, and social [6]. In view of the challenges of sustainable development, it is the social aspect that becomes the key challenge for contemporary logistics systems. The goals set for contemporary logistics systems, and in particular for transport subsystems, even force a pro-social approach. This results from the need to reduce negative environmental impacts on the one hand and the need to ensure logistics systems users a high level of satisfaction with received service on the other hand. Therefore, the logistic approach should be treated as a specific kind of art of managing conflicting connections. This gains particular importance in the context of urban logistics systems. This is because the often raised issues, being the consequence of a given freight transport system functioning in a city, include phenomena such as congestion, road accidents, noise or environmental pollution [5,12–16]. Delivery vehicles are responsible for ca. 50% pollutant emissions, even though their share in the city road traffic is only from 20% to 30% [17]. However, due to delays in planned trips or deliveries as a result of congestion or lack of parking spaces, delivery vehicles often stop in traffic lanes, blocking them and consequently decreasing the effectiveness of logistic operations performed within the city. Therefore, as already mentioned, recent years have seen an increased interest in comprehensive streamlining of goods deliveries in city areas [18]. Still, increasing the logistic efficiency of a city, while mitigating the negative environmental impacts of the logistics, is a challenging task [19,20]. Additionally, the growing number of UFT stakeholders who are mostly characterised by diversity in terms of the structure of their needs, objectives and expectations, leads to a critical lack of "common operational picture" or even lack of "shared situational awareness" [4,21].

This is because the implemented initiatives often solve the problems of one group of stakeholders, while significantly infringing on the interests of others. This discrepancy results mainly from planning the solutions implementation without taking into account e.g., the city characteristics or opinions of all the stakeholders of urban freight transport [20]. What is more, observations made over recent years prove that many solutions, though positive for the society, cannot function on the fully commercial principles, and most often after the pilot phase, they must be subsidised by local authorities (which can be exemplified by e.g., urban consolidation centres) [22].

In European countries, the concept of involving possibly all UFT stakeholders in the decision-making process has been in place since the 1990s. This is done via associations taking the form of Freight Quality Partnerships (FQP) [23]. The main task of FQP is to involve the interested parties—in a conscious and equal manner—in the process of managing cargo flows in a city. In 2010, there were 38 Freight Quality Partnerships in Great Britain, and their activities contributed to increasing the interactions between private stakeholders of freight transport [24,25]. It turns out, however, that the necessary condition to avoid implementations that are not economically viable, have a low level of acceptability or are unsuitable due to the specific nature of the city, is their systematic evaluation that takes into account the complexity of the city and also expectations of senders, recipients, shippers, local and regional authorities, and citizens [26].

The research studies carried out in research units [27,28] prove that the process of establishing and successful functioning of FQP is a complex and long-lasting process consisting of

- appointing a team to lead the FQP activities;
- indicating groups of UFT stakeholders in a selected area;
- convincing the stakeholders of the need to establish a FQP and to actively participate in meetings;
- developing some solutions to support UFT streamlining, which are tailored to the specificity of the area where they are to function;
- implementing the solutions;
- monitoring of the implemented solutions.

Unfortunately, while it is relatively easy to specify the expected usefulness of a given project, to identify its structure and the sources of potential conflicts, the area of the greatest research potential is monitoring the effects of implemented solutions, especially in the aspect of long-term durability of their effects.

In view of the above, it is legitimate to formulate a definition of the very term "FQP project durability", to search for its praxeological sources, to distinguish durability phases and to conduct empirical studies involving the key stakeholders of UFT projects, which specify subjective and generalised perception of project durability. The article attempts at filling the conceptual gap within the scope specified above. The proposed methodology may constitute a significant tool for FQP evaluation as a solution aimed at combining the interests of various stakeholder groups. The presented research study is a summary of an experiment regarding FQP functioning in the city of Szczecin.

2. Urban Freight Transport Project Durability

The concept of Sustainable Urban Freight Transport has become a permanent element of the broader idea of sustainable development. According to [29], it can be defined as a set of logistics and freight transport activities of the city area that are economically viable and contribute to the improvement of environment, quality of life and social issues, conform to the logic of the "four As" and have a vision of continuous improvement, take into account the interactions between the different stakeholders concerned and proposed solutions that are appropriate to the different stakeholders, and in which sustainability, in terms of earning relative to a certain benchmark, must be quantifiable and qualifiable. More and more often cities notice that it is necessary to take more intensive measures in that regard, so as to find the fine balance between the residents' life quality (in terms of, *inter alia*, the need to ensure deliveries to shops, hotels, restaurants and service outlets) and the simultaneously incurred external costs resulting from e.g., pollution, noise, congestion, which are negative environmental impacts specifically caused by organised transport. However, based on more than 20 years of experiences in the development of Sustainable Urban Logistics projects, it should be stated that many achievements and results are abandoned after the project period (some examples can be found in [5,10,27,29]). Accordingly, the research perspective adopted by the Authors focuses on the proposal to strive to maintain and protect, for as long as possible, each of the achieved individual effects as well as their bundles via project initiatives taken up by cities in order to optimise the solutions regarding freight transport in the city. This is because, on the one hand, there is a discerrible problem of inertia affecting every solution, but on the other hand, it is possible to notice a too instrumental approach to maintaining the effects, which is a result of a need to meet the institutional requirements connected with financing the project rather than actual strategic thinking focused on permanent improvement of the city inhabitants' holistic well-being. Following the above, the difference between the sustainability and the durability of urban logistics activities should be emphasized.

The sustainability of urban logistics is related to the realization of the freight flows inner the city area. It is directly and mostly connected with the functioning of the supply chains, especially taking to the account the interdependencies between them and the city attractiveness [30]. In the result, the idea of Sustainable Supply Chain Management (SuSCM) has been established. This concept is based on the environmental and societal influences of supply chains [31]. The durability of urban logistics activities is related to their specificity as projects. It is the crucial challenge in terms of the functioning of urban logistics measures during and (most importantly) after the project period. This issues is related to the all spheres of sustainability (environment, society, economics). However, it is critically important from the economic point of view [29].

As found in the literature, the term "project" is most often defined as a unique undertaking limited in terms of time, scope, costs, and customer satisfaction [32]. An important aspect that should be noted here: so far, the key issue in project evaluation was first and foremost its performance [33]. However, according to T. Kotarbiński [34], performance of each action in the universal sense is expressed by its effectiveness, profitability, and cost efficiency. These are the three explicit practical merits of this universal measure of performance of each action, which can be expressed in a synthetic and quantified form.

Therefore, each action is the most "effective" if it makes it possible to achieve (in whole or to some extent) the intended purpose or at least enables its implementation in a specific future. The second aspect, i.e., "profitability", is always a feature that describes an action assessed positively due to the prevalence of received results (E) compared to the incurred outlays (N), where (E-N > 0). However, it is mostly the third merit, i.e., the "cost efficiency" aspect, that verifies the achieved effect of each project (E) in relation to the outlays incurred for that purpose (N), and the ratio should always be greater than one (N/E > 1).

Yet, in the academic literature, according to the traditional approach, the measures of a project success are only limited to the scope of performed tasks, their costs, or implementation time [35–38]. Thus, so far, not enough heed was paid to the durability of effects of a given project after its completion. Currently, evaluation of so-called "project success" also involves the durability of its outcomes, which is mainly decided by the stakeholders, i.e., the contractors, sponsors, and users [39]. Nevertheless, the meaning of project durability is more often than not understood too narrowly: as ensuring the functioning of the outcomes after the project completion, thus meeting its business goals [36], which in a sense is understandable due to the specific relevance of this aspect in implementing any ideas (projects) co-financed with the EU funds. For that reason, a contractor is obliged (usually in the guidelines regarding the project implementation) to ensure the functioning/making use of the results after the project completion and for a specified period of time. The duty to maintain project durability is stipulated in Art. 71 of the Regulation (EU) No. 1303/2013 of the European Parliament and of the Council dated 17 December 2013.

To sum up, in the context of studies of the relevant literature, it is currently possible to specify three major factors that, in the Authors' opinion, have an impact on project durability, and which are relevant from the point of view of the project specificity in the scope of deliveries performed in an urban area [35–43]. Thus, from a perspective of "practical values" of effectiveness of any actions, these mainly include: "usefulness" (understood as the extent to which the stakeholders' expectations were met and the extent to which the problems they had voiced were solved), "effectiveness" (which boils down to the extent to which the assumed objective was met as a result of the project implementation), and "efficiency" (as an indicator of utilizing all the possessed resources in the project, within a specific normative time), in view of the solutions being the result of the actions taken so far to deliver goods in the city area.

Unfortunately, it seems that the concept of "UFT project durability" as such, analysed in economic terms, still poses considerable problems of academic nature. From the point of view of praxeology, and therefore in the dogma of systematics of the three aforementioned dimensions of performance, it turns out that, first, each project should be required, understood and accepted by possibly maximally extensive circle of stakeholders, which will ensure its fullest "usefulness" function. Second, if the assumed effects have been achieved in full and, additionally, are maintained over the assumed period of time after the project completion, the UFT project may be considered fully "effective". Third, the normative time required for specifying the "efficiency" is always connected with two aspects, i.e., the project itself (its duration and the required period of maintaining the project), and the time of protecting the structure and minimal amounts of its effects after the project has been formally completed and accounted for.

It turns out, therefore, that with respect to a project in the area of urban freight transport (UFT), when discussing its durability it is reasonable to mention the universal and synthetic measures of the action, relevant from the point of view of praxeology, due to their mutual terminological convergence (Figure 1). Thus, the specified usefulness of UFT project effects from the point of view of stakeholders is a manifestation of profitability of such an action, and efficiency is based on the effectiveness of the action taken for the project itself and its maintaining.

It should be noted that the project phase duration which covers implementation of planned effects of a UFT project and maintaining them over the project accounting period is only an element of a broader understanding of such project durability. The authors are of the opinion that in order to grasp the overall meaning of the term "UFT project durability" it is necessary to take into account the post-project phase and predominantly the aspect of "protecting", for as long as possible, the effects already achieved (and sustained) in the course of the project duration.

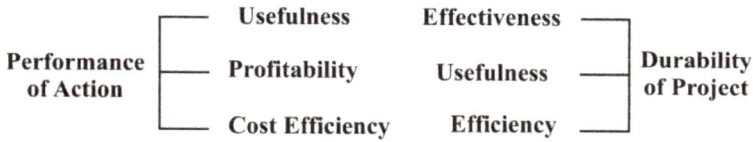

Figure 1. The convergence of terminology.

Therefore, the specific (the main and the auxiliary) goals of a UFT project are achieved with a specific effectiveness within the framework adopted for its implementing, maintaining and protecting the project effects, with a possibly maximal use of (mainly material, personnel, and information) resources assigned to it in the particular phases (Figure 2).

Phase		
Project		Post-Project
Implementing the Project	Maintaining the Project	Consolidation Effects of the Project
Building the Effects	Maintainig the Effects	Protecting the Project Effects

Figure 2. The phases of project durability.

Therefore, in order to better grasp the concept of "UFT project durability", it is necessary to interpret it as "preserving, over the longest time possible, the obtained effect, in terms of both its total size and the structure of effects obtained as the final outcome, and also maintaining the minimal representation of each of the single elements of the final effect". Only such a synergistic approach makes it possible to notice the frequent phenomenon of project effects "fading out" over a long term (after the project has been formally completed and accounted for), as due to its nature it is not capable of self-regulating, especially when financial leverage does not apply any more (which usually requires continuing the financing with funds other than those related to the project). Additionally, this highlights the need to preserve "the system of the original structure of effects" (in particular with regard to the individual elements of the system) and "the minimum values" adopted for each of such specialised elements of the original structure of effects. This is because the shrinking size of the effects is followed by the "transposition" phenomenon or—to put it in simple terms—the rule of "substitutability versus complementarity "of the project effects. The higher the durability of effects, the greater their complementarity, and the lower the durability, the more important the substitutability of the effects. Therefore, the problem lies with the relations between the expected substitutability and complementarity of a UFT project, which may, in an extreme situation, take the form of "the cannibalism of effects" (Figure 3).

As the graph shows, an increase in Complementarity is accompanied by a decrease in Substitutability, and vice versa. Consequently, in the situation when the Substitutability grows, the Instability of UFT project effects grows, too. An increase in the level of Complementarity, in turn, is accompanied by an increase in UFT project Durability. Therefore, the most desirable situation is the one where an increase in Complementarity of UFT project effects leads to enhancing the synergy effect which translates directly into an increase in project effects Durability, via their protection in the post-project phase.

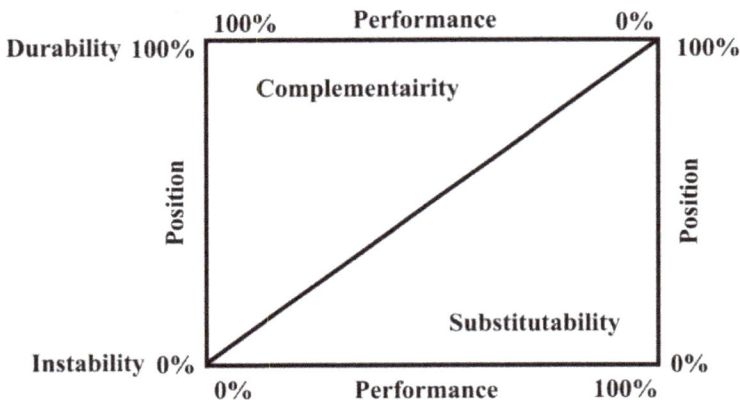

Figure 3. Complementarity vs. Substitutability principle.

To conclude, from this perspective, "durability of a UFT project effects" should be considered as "the strategy for improving the life quality of city inhabitants within the limits determined by the permitted volumes of the major partial effects that make up the total effect of a given project", because only this approach represents the synergistic perspective. Additionally, a UFT project evaluation under an FQP in terms of durability should be based on assessing the aforementioned factors by the stakeholders, i.e., the city authorities, inhabitants, and representatives of the businesses operating in the area of interest of the FQP.

3. The "Pyramid of Stakeholders" Survey

One of the key issues in FQP functioning is appropriate selection of stakeholders who will be cooperating to initiate effective measures in the area of sustainable development of urban freight transport. The stakeholders are grouped in accordance with various criteria. The classical breakdown suggested by E. Taniguchi includes shippers, freight carriers, residents, and city administrators [44]. Under the CityMove and CityLog projects, an additional category was proposed—truck and vehicle manufacturers [45]. However, in order to evaluate FQP functioning, the matter of key importance is focusing on the goals which the individual stakeholder groups want to achieve [1,46,47]. Taking to the account the expectations and objectives of the UFT stakeholders groups, following [30], two major area of interest should be emphasized: public sphere and private sphere. Both are directly connected but their major aims and point of views are different. In this context, as proposed in [3], it is necessary to emphasise three points of reference: private interest, public interest, and inhabitants' expectations. Accordingly, in view of the proposed methodology, it is reasonable to break down the stakeholders into groups that represents the three points of reference: "Inhabitants and Community Councils" to represent the residents' interests, "Business" to represent the private interest, and "City Authorities" to represent the public interest. It should be stressed that the stakeholder group that represents Business covers any and all entities engaged in urban freight transport functioning, i.e., carriers, commercial entities, production plants, HoReCa sector, etc.

In contemporary cities, there are three key groups of UFT stakeholders making up the so-called "Pyramid of Stakeholders" (Figure 4). The first group includes Inhabitants (I) who represent specific consumer needs with regard to goods and services—the demand side—but also represent a specific level of maturity in terms of accepting the adopted principles of sustainable development. The second group is Business (B) which is the market response to the inhabitants' demand by offering the supply of specific goods and services. Finally, the third group is City Authorities (CA) that implement the specific municipal strategy by, *inter alia*, the logistic policy (e.g., shaping the urban logistic infrastructure) or implementing initiatives such as "Smart City" (e.g., introducing modern logistic solutions), while seeing to a stable and economically reasonable balance between the reported needs and the corresponding supply.

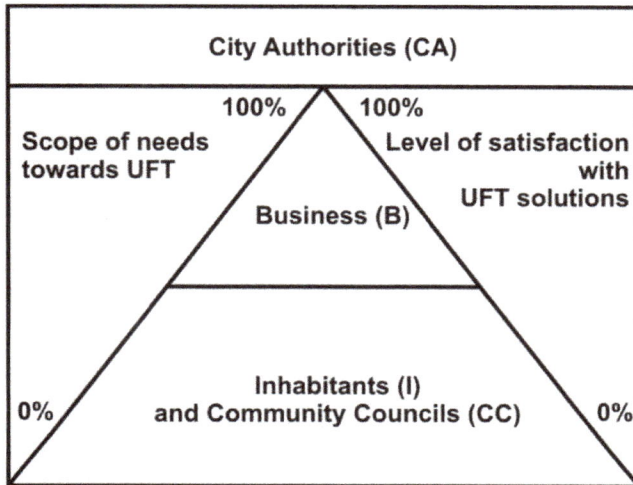

Figure 4. The pyramid of stakeholders.

It turns out, therefore, that the scope of needs in relation to UFT projects is the resultant of the needs of the Inhabitants (I) and Business (B). The problem, however, is to identify any "discrepancies (gaps)" and to assign them appropriate weights.

In order to find out the FQP stakeholders' opinions, in the period from January to March 2018 the Authors carried out a survey based on a questionnaire made up of 28 questions. Three questionnaires were applied, which included mainly closed-ended and semi-open questions and one additional open question, for each of the three studied groups of stakeholders. The research process was based on the direct interview method. The questions were addressed to all the Community Councils existing within the central part of the city of Szczecin, and to three departments of the Municipal Office of Szczecin, which were involved in the FQP functioning in Szczecin. In the case of Inhabitants and Business groups, the survey was carried out on a random sample. The survey involved 215 representatives (including 105 women and 110 men) representing Inhabitants (I) and 10 organisational units of Community Councils (CC) who represent the first group of Stakeholders, 150 entities from the second group being Business (B), and three organisational units representing the third group, i.e., City Authorities (CA). The structure of the aforementioned groups of respondents is presented in Figure 5.

It should be stressed that in view of the specific nature of the research, it was not reasonable to single out any individual groups of stakeholders from within the Business category. The analysis presented further on shows the aggregated values of the results obtained for the subgroups.

The results of the survey, presented and discussed further on in this article, were focused on three areas, i.e., evaluation of the quality of living in the city, managing the information on handled projects, and evaluation of durability (described in terms of fading out of effects) of completed projects (Figure 6).

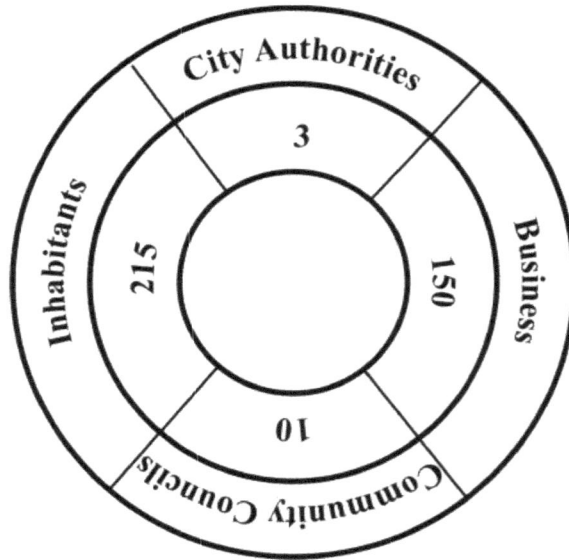

Figure 5. The structure of respondents.

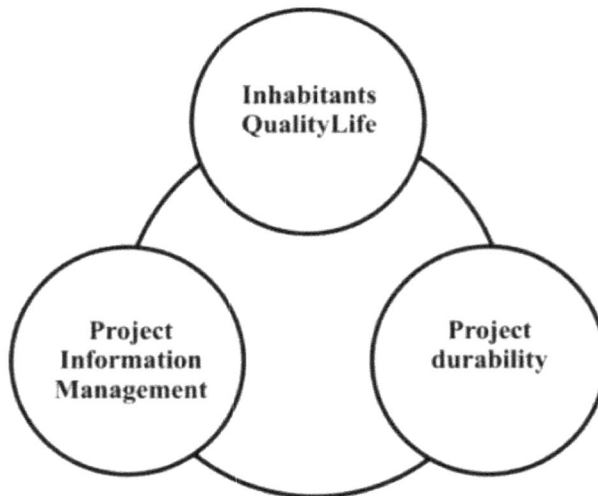

Figure 6. Three focuses of the analysis and survey results assessment.

Taking up the analysis of the survey results it should be noted that the respondent group from the first group of stakeholders was dominated by middle-aged people (54% of them fell within the range from 41 to 50 years of age, whereas 42% of them were from 31 to 40 years old). The full age structure is presented in Figure 7.

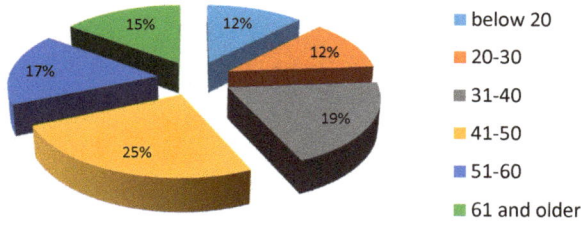

Figure 7. The age structure of the respondents from the "Inhabitants" group.

Analysing the responses concerning the first area, i.e., the answers to the question regarding the Inhabitants' (I) perception of quality of life in the city over the past five years (Figure 8), explicit scepticism was shown by the inhabitants aged 31–40 years (15 per cent), 41–50 years (8 per cent), and the seniors, i.e., people aged over 61 (8 per cent), who pointed out that the situation was worse.

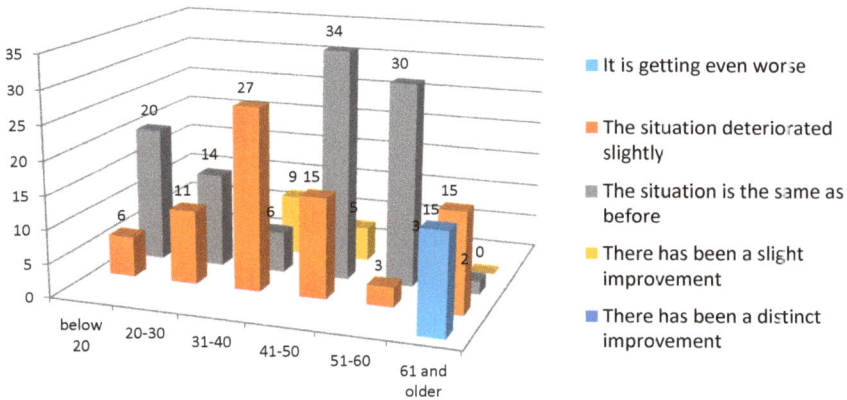

Figure 8. Quality of life in the city perceived by the Inhabitants (I), broken down into age groups.

Summing up the city inhabitants' perception of quality of life in the city (Figure 9), it turned out that seven per cent of them thought it was worse, 36 per cent thought it deteriorated slightly, and 49 per cent did not notice any changes in the five-year perspective. Only eight per cent of the respondents were of the opinion that there was a slight improvement.

Much more sceptical opinions were expressed by the Community Councils: as many as 62 per cent asserted that the situation deteriorated slightly, whereas 38 per cent did not perceive any positive changes and thought it was the same as before (Figure 10).

The Business (B) group representatives, in turn, stressed that any positive changes were imperceptible (60 per cent of the respondents), while the others thought that the situation deteriorated slightly (10 per cent) or it was getting even worse (13 per cent). An indiscernible improvement was observed by merely 17 per cent of the respondents (Figure 11).

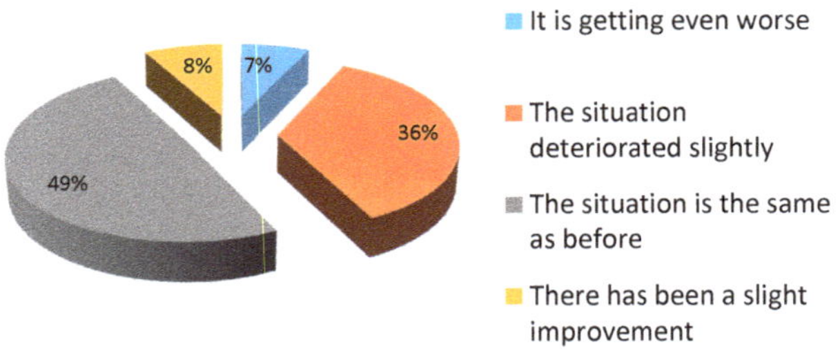

Figure 9. Quality of life in the city perceived by the Inhabitants (I)—the synthesis.

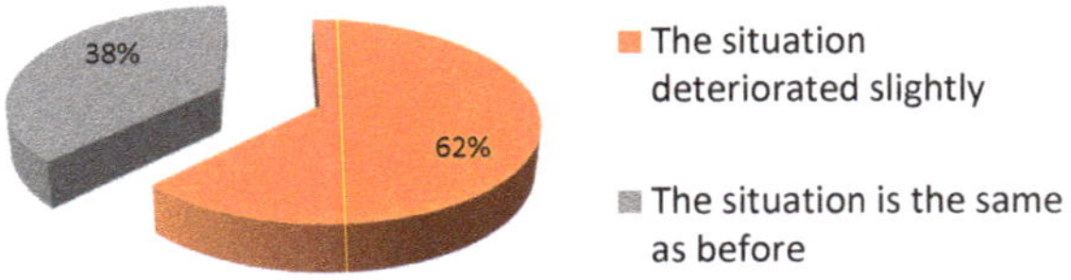

Figure 10. Quality of life in the city perceived by the Community Councils (CC).

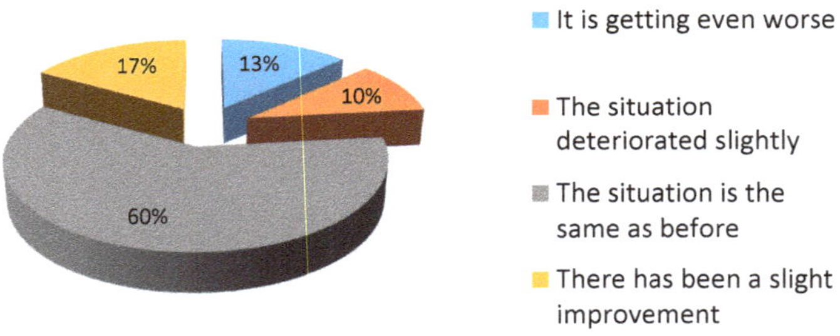

Figure 11. Quality of life in the city perceived by the Business (B).

Summing up, there is a discernible gap between the perception of the situation by the City Authorities (CA). Namely, 100 per cent of CA representatives asserted that the situation improved slightly ($\nearrow$), while among the representatives of the Inhabitants (I) and the Business (B), no-one felt an improvement, and only 49 and 60 per cent of them, respectively, thought that the situation was unchanged ($\rightarrow$), whereas 36 and 10 per cent thought it deteriorated slightly ($\searrow$), while 7 and 13 per cent respectively expressed an opinion that it was getting even worse ($\downarrow$). The aggregated opinions are presented in Table 1.

Analysing the second area, i.e., the answers to the question regarding the perception of the information policy by the City Authorities (CA), i.e., the communications on commencing, continuing or completing projects taken up in the area of deliveries of goods in the city of Szczecin, which were adopted in order to contribute to increasing the perception of increased quality of life by the Inhabitants (I) of Szczecin over the past five years (Figure 12), it turned out that nearly one quarter of the respondents (24 per cent) thought that there was no information at all, 12 per cent assessed that the information was limited, and nearly one half (43 per cent) reckoned that communications were available only to interested parties.

Table 1. Quality of life in the city perceived by the stakeholders—the synthesis [%].

	City Authorities (CA)			
100		↗		100
	Inhabitants (I)	Business (B)		
	49	→	←	60
92	36	↘	↙	10	83
	7	↓	↓	13

Legend: → The situation was unchanged; ↘ The situation has deteriorated slightly; ↓ The situation was getting even worse.

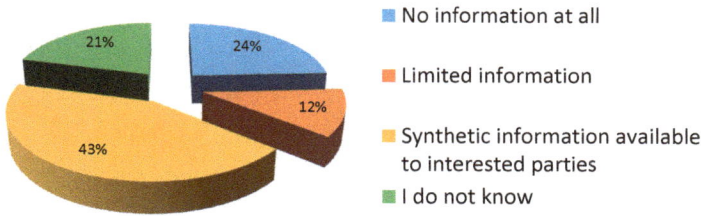

Figure 12. The evaluation of the City Authorities' (CA) communications on commencing, continuing or completing projects taken up in the area of deliveries of goods in the city of Szczecin—the opinions of the Inhabitants (I).

Similar opinions were voiced by the respondents from the Community Councils (CC), as 34 per cent agreed that there was a complete lack of information, and 22 per cent thought the adopted system of providing information was limited. Similarly, they also indicated that communications or their synthetic forms were available only to interested parties (33 and 11 per cent, respectively). The structure of the opinions is presented in Figure 13.

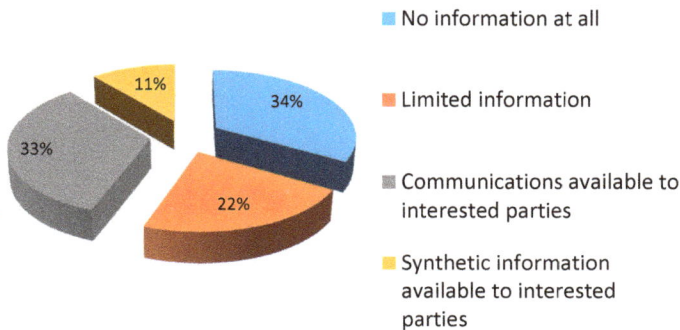

Figure 13. The evaluation of the City Authorities' (CA) communications on commencing, continuing or completing projects taken up in the area of deliveries of goods in the city of Szczecin—the opinion of the Community Councils (CC).

As for the opinions of the Business (B), pursuant to Figure 14 they are even more critical when it comes to evaluation of the information policy run by the City Authorities (CA). As many as 66 per cent of the respondents believed that such a policy did not exist, and only 14 per cent of them indicated that it was there, though they found it very limited. Thirtreen and 7 per cent of the surveyed entrepreneurs, respectively, were sure that communications and/or synthetic information were available.

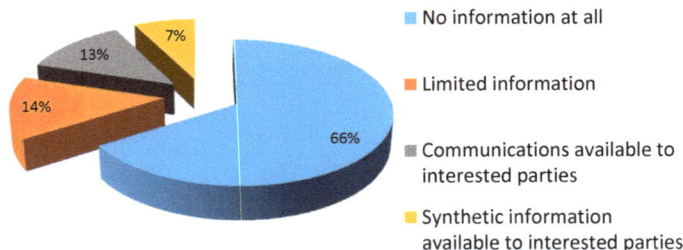

Figure 14. The evaluation of the City Authorities' (CA) communications on commencing, continuing or completing projects taken up in the area of deliveries of goods in the city of Szczecin—the opinion of the Business (B).

Regrettably, this perception of the Inhabitants (I), Community Councils (CC) and Business (B) is confirmed by the responses obtained from the City Authorities (CA) admitting that they supplied information to a limited extent (33 per cent) or none at all (33 per cent) to the Inhabitants (I), Community Councils (CC) and Business (B) with regard to commencing, continuing or completing projects taken up in the area of deliveries of goods in the city of Szczecin, and only synthetic information was available to interested parties. Therefore, summing up, 33 per cent of the City Authorities' (CA) representatives asserted that they did not inform (↓) the Inhabitants (I) at all, and other 33 per cent did it only to a limited extent (↘). Lack of information (↓) was found by as many as 60 per cent of the Business (B) representatives and 30 per cent of the Inhabitants (I), and limited information—by 14 per cent of the Business (B) and 12 per cent of the Inhabitants (I), respectively. A much bigger percentage of the Inhabitants (I)—as many as 30 per cent—found that the communication was limited (→), whereas in the case of the Business (B) this view was shared by as few as 13 per cent, but the latter saw the possibility of accessing the communications (↗) (though for seven per cent of the respondents—B). The synthesised opinions are presented in Table 2, and the data shown in it confirm the need to prioritise the information policy with regard to UFT projects.

Table 2. Synthesised evaluation of the City Authorities' information policy by the stakeholders [%].

		City Authorities (CA)			
	33			33	
		↓	↘		
		Inhabitants (I)	Business (B)		
	0	↗	↖	7	
72	30	→	←	13	100
	12	↘	↙	14	
	30	↓	↓	66	

Legend: ↗ Possibility of accessing the information; → Communication was limited; ↘ Limited scope of information; ↓ No communication.

Finally, analysing the third aspect, the representatives of the City Authorities (CA) estimated that effects of each project are preserved for a period of up to one year, as after a period of one to two years, nobody will remember any initiatives taken up in the city in order to improve the organisation of goods deliveries. However, according to the responses of the Inhabitants (I), as many as 44 per cent of them thought that the effects would fade away as soon as the project is completed, whereas 14 per cent of them believed that would happen within half a year at the latest. So, the optimistic view of the City Authorities (CA) was shared by merely two per cent of the surveyed Inhabitants (I). Regrettably, almost one third of them (30 per cent) were unable to provide any time estimates for the project durability. Nevertheless, there is some potential for trust among the Inhabitants (I), as seven per cent of them believed the time horizon from two to five years should be regarded as the basis for the decision-makers' purposefulness and far-sightedness in relation to any projects to be implemented, and three per cent of them thought the time horizon should exceed six years (Figure 15).

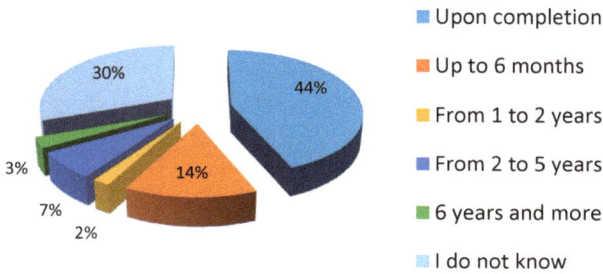

Figure 15. The estimated time of fading away of effects of initiatives implemented in the city in order to improve the organisation of goods deliveries—(I).

However, the Community Councils (CC) represented definitely more pessimistic views, as 62 per cent of them asserted that the project "dies" as soon as it is completed, while 38 per cent estimated the survivability of its effects for the period from seven months to one year (Figure 16).

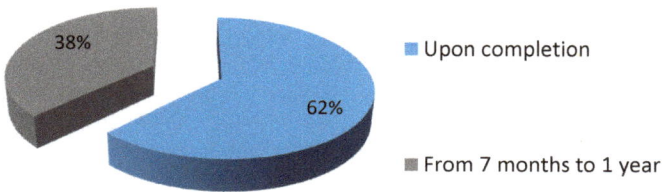

Figure 16. The estimated time of fading away of effects of initiatives implemented in the city in order to improve the organisation of goods deliveries—(CC).

Similarly, 43 per cent of the Business (B) representatives estimated the project longevity to be from seven months to one year, but only five per cent of this group thought the effects would disappear right after the project completion. According to 15 per cent of them, the effects would be maintained for up to six months, but unfortunately 37 per cent of respondents from this group were unable to specify even approximate time horizons (Figure 17).

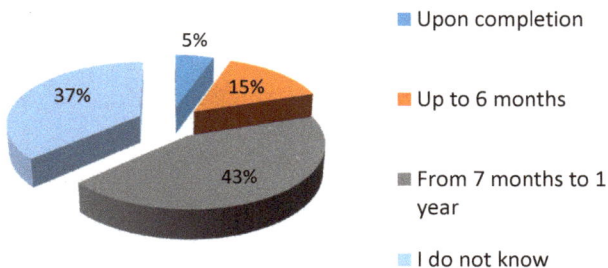

Figure 17. The estimated time of fading away of effects of initiatives implemented in the city in order to improve the organisation of goods deliveries—(B).

To synthesise the results presented in Table 3: the Inhabitants (I) were more reluctant to estimate the time of fading away UFT project effects compared to the representatives of the Community Councils (CC) and the Business (B), whereas the most optimistic views were shared by the City Authorities (CA).

Table 3. Synthesised estimated time of fading away of effects of initiatives implemented in the city in order to improve the organisation of goods deliveries [%].

	Upon Completion	Up to 6 Months	Up to 1 Year	Up to 2 Years	Over 6 Years
Inhabitants (I)	44	14	-	2	3
Community Councils (CC)	62	-	38	-	-
Business (B)	5	15	43	-	-
City Authorities (CA)	-	-	-	66	-
UFT project effects durability	●	●	●	●	→

It is easy to notice the "ladder of optimism" which indicate that Inhabitants evaluate the project durability as maximum one year, whereas Community Councils, and first and foremost Business, tend to extend the period up to two years. Definitely more optimistic attitude was shown by City Authorities that assumed the effects of each UFT project would be sustained for more than two years.

4. Conclusions and Recommendations

The analysis of the relevant academic literature as well as the empirical studies have proved that, first, the issue of key importance is to correctly define the dimensions of "UFT project durability", and second, for the purposes of a comprehensive analysis it is necessary to focus not only on phases I and II but also on phase III of the analysis, i.e., "protection of effects" of UFT projects (Figure 18). This is because it turned out that the traditional approach was based on "building the effects" in the course of the project implementation (phase I) and then "maintaining" them upon the project completion (phase II). A success of a project is when the whole structure of effects has been built and then the whole of it (100%) has been preserved over a specified period of time. But the "durability of project effects" and their profitability to all the stakeholders can only be attained when the structure and the minimum values of the effects are protected for the longest time possible.

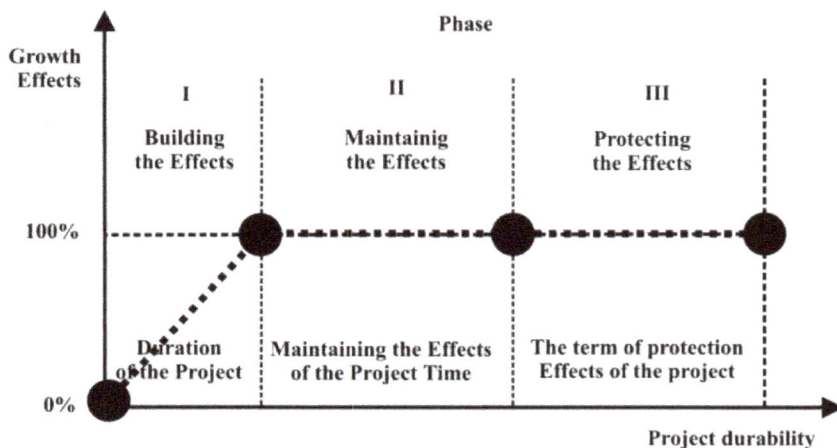

Figure 18. The critical points of the three project phases.

Based on this approach, it becomes particularly important to exceed the "enchanted" boundary of one year after the project completion, which is visible in the results of the survey. Whereas the Inhabitants (I) were mostly (83.2 per cent) convinced that UFT projects are forgotten after 1 year, (while 20.13 per cent believed it was as early as after six months), the Business (B) was slightly more optimistic: even though 68 per cent of its representatives asserted that it was one year at the most, merely 0.22 per cent of them pointed out to the period of six months, which could be considered a slightly optimistic view.

Additionally, there is a discernible asymmetry between the perspective of the decision-makers (the City Authorities) and the beneficiaries (the Inhabitants, Community Councils and Business) with regard to the criteria of UFT projects durability. It is possible to notice certain optimism ($\nearrow$) on the side of the decision-makers, and moderate pessimism ($\searrow$) on the side of the stakeholders, which is synthesised in Table 4.

Table 4. Asymmetry between the perspective of the decision-makers (the City Authorities) and the beneficiaries (the Inhabitants, Community Councils and Business).

Durability Criterion	City Authorities	Inhabitants	Community Councils	Business
Usefulness	$\nearrow$	$\rightarrow / \searrow$	$\rightarrow$	$\rightarrow / \searrow$
Effectiveness	$\searrow$	$\searrow$	$\searrow$	$\searrow$
Efficiency	$\nearrow$	$\rightarrow / \searrow$	$\searrow$	$\rightarrow / \searrow$

Summing up,

- There is a distinct "gap" in the perception of the quality of life in the city as well as the durability of effects of UFT projects taken up so far in order to optimise goods deliveries in the city between the City Authorities (CA) and the Inhabitants (I) and their representatives, i.e., the Community Councils (CC), and the Business (B);
- To level it off, it is necessary to take coordinated and integrated measures, mainly in the area of information and communication, so that the knowledge on any completed and pending projects—and its effects—is easily accessible to all the Stakeholders;
- There is a considerable though so far neglected "trust potential" especially among the Inhabitants (I) regarding the purposefulness of initiatives aimed at improving the goods deliveries in the city, taken up by the City Authorities (CA);
- The opinions formulated by the Inhabitants (I) and Business (B) on improving the quality of life in the city undoubtedly constitute an objective verification of the practical usefulness of the initiatives (projects) taken up by the City Authorities (CA) with regard to urban freight transport (UFT) optimisation.

Author Contributions: K.K. contributed to the data collection process as well as the final edition of the paper. She is the author of Sections 1, 2 and 4. M.J. was responsible for assumptions for the analysis as well as the final edition of the paper He is the author of Sections 2–4.

Funding: This paper has been founded under grant No 1/S/IZT/2018 "Analysis of logistic determinants of sustainable urban development" financed with a subsidy from the Ministry of Science and Higher Education for statutory activities.

Acknowledgments: This research outcome has been achieved under grant No 1/S/IZT/2018 "Analysis of logistic determinants of sustainable urban development" financed with a subsidy from the Ministry of Science and Higher Education for statutory activities.

Conflicts of Interest: The authors declare no conflict of interest.

Abbreviations

UFT   Urban Freight Transport
FQP   Freight Quality Partnership

References

1. Ballantyne, E.; Lindholm, M.; Whiteing, T. *A Comparative Study of Urban Freight Transport Planning: Addressing the Issue of Stakeholders*; Unpublished Work; University of Leeds: Leeds, UK; Chalmers University of Technology: Gothenburg, Sweden, 2012.
2. Instituto Superiore per la Protezionee la Ricerca Ambientale (ISPRA). *Qualita Dell Ambiente Urbano—V Rapporto*; Instituto Superiore per la Protezionee la Ricerca Ambientale (ISPRA): Rome, Italy, 2008.
3. Nesterova, N.; Quak, H.A. City Logistics Living Lab: A Methodological Approach. *Transp. Res. Procedia* **2016**, *16*, 403–417. [CrossRef]
4. Quak, H.; Lindholm, M.; Tavasszy, L.; Michael, B. From freight partnerships to city logistics living labs—Giving meaning to the elusive concept of living labs. *Transp. Res. Procedia* **2004**, *12*, 461–473. [CrossRef]
5. Quak, H.; van Duin, R.; Visser, J. City logistics over the years . . . Lessons learned, research directions and interests. In *Innovations in City Logistics*; Nova Science: New York, NY, USA, 2008; pp. 37–54.
6. Szołtysek, J.; Sadowski, A.; Kalisiak-Mędelska, M. *Social Logistics. Theory and Application*; Wydawnictwo Uniwersytetu Łódzkiego: Łódź, Poland, 2017.
7. Carlsson, C.-M.; Janné, M. Sustainable urban transport in the Øresund Region. In *Rethinking Transport in the Øresund Region*; Carlsson, C.-M., Emtairah, T., Gammelgaard, B., Vestergaard Jensen, A., Thidell, Å., Eds.; Lund University: Lund, Sweden, 2012.
8. Dablanc, L. Urban goods movement and air quality policy and regulations issues in European cities. *J. Environ. Law* **2008**, *20*, 245–267. [CrossRef]
9. Dablanc, L. *Freight Transport for Development Toolkit: Urban Freight*; The International Bank for Reconstruction and Development/The World Bank: Washington, DC, USA, 2009; Available online: http://www.ppiaf.org/freighttoolkit/sites/default/files/pdfs/urban.pdf (accessed on 20 April 2018).
10. Iwan, S. *Implementation of Good Practices in the Area of Urban Delivery Transport*; Wydawnictwo Naukowe Akademii Morskiej w Szczecinie: Szczecin, Poland, 2013.
11. Witkowski, J.; Kiba-Janiak, M. Correlation between city logistics and quality of life as an assumption for referential model. *Procedia-Soc. Behav. Sci.* **2012**, *39*, 568–581. [CrossRef]
12. Dablanc, L. City distribution, a key element of the urban economy: Guidelines for practitioners. In *City Distribution and Urban Freight Transport*; Macharis, C., Melo, S., Eds.; Edwards Elgar Publishing Ltd.: Cheltenham, UK, 2011.
13. Lindholm, M.; Browne, M. Local authority cooperation with urban freight stakeholders: A comparison of partnership approaches. *Eur. J. Transp. Infrastruct. Res.* **2013**, *13*, 20–38.
14. Taniguchi, E.; Thompson, R.G. (Eds.) *City Logistics: Mapping the Future*; CRC Press: Boca Raton, FL, USA, 2014.
15. Vaghi, C.; Percoco, M. City logistics in Italy: Success factors and environmental performance. In *City Distribution and Urban Freight Transport: Multiple Perspectives*; Emerald: Northampton, UK, 2011; pp. 151–175. Available online: www.bestufs.net (accessed on 10 February 2016).
16. Yannis, G.; Golias, J.; Antoniou, C. Effects of urban delivery restrictions on traffic movements. *Transp. Plan. Technol.* **2006**, *29*, 295–311. [CrossRef]
17. AR Dablanc, L. Goods transport in large European cities: Difficult to organize, difficult to modernize. *Transp. Res. Part A Policy Pract.* **2007**, *41*, 280–286. [CrossRef]
18. Bozzo, R.; Conca, A.; Marangon, F. Decision support system for city logistics: Literature review, and guidelines for an ex-ante model. *Transp. Res. Procedia* **2014**, *3*, 518–527. [CrossRef]
19. Iwan, S.; Kijewska, K. The Integrated Approach to Adaptation of Good Practices in Urban Logistics Based on the Szczecin Example. *Procedia-Soc. Behav. Sci.* **2014**, *125*, 212–225. [CrossRef]
20. Macharis, C.; Melo, S. (Eds.) Introduction—City distribution: Challenges for cities and researchers. In *City Distribution and Urban Freight Transport*; Edward Elgar: Cheltenham, UK, 2011; pp. 1–9.
21. Małecki, K.; Iwan, S.; Kijewska, K. Influence of Intelligent Transportation Systems on reduction of the environmental negative impact of urban freight transport based on Szczecin example. *Procedia-Soc. Behav. Sci.* **2014**, *151*, 215–229. [CrossRef]

22. UN Habitat. *Planning and Design for Sustainable Urban Mobility: Global Report on Human Settlements*; UN Habitat: New York, NY, USA, 2013.
23. Browne, M.; Nemoto, T.; Visser, J.; Whiteing, T. Urban freight movement and public private partnerships. Presented at the City Logistics Conference, Madeira, Portugal, 25–27 June 2003.
24. Allen, J.; Browne, M.; Piotrowska, M.; Woodburn, A. Freight Quality Partnerships in the UK—An analysis of their work and achievements. In *Green Logistics Project*; Transport Studies Group, University of Westminster: London, UK, 2010.
25. Allen, J.; Browne, M.; Cherrett, T. Investigating relationships between road freight transport, facility location, Logistics Management and Urban Form. *J. Transp. Geogr.* **2012**, *24*, 45–57. [CrossRef]
26. Behrends, S. *Urban Freight Transport Sustainability—The Interaction of Urban Freight and Intermodal Transport*; Department of Technology Management and Economics, Chalmers University of Technology: Gothenburg, Sweden, 2011.
27. Lindholm, M. Successes and Failings of an Urban Freight Quality Partnership—The Story of the Gothenburg Local Freight Network. *Procedia-Soc. Behav. Sci.* **2014**, *125*, 125–135. [CrossRef]
28. Lindholm, M.E.; Browne, M. *Freight Quality Partnerships around the World: 1st Report on a Survey*; VREF Centre of Excellence for Sustainable Urban Freight Systems: Troy, NY, USA, 2014.
29. Gonzalez-Feliu, J. *Sustainable Urban Logistics: Planning and Evaluation*; John Wiley & Sons: Hoboken, NJ, USA, 2018.
30. Boudoin, D.; Morel, C.; Gardat, M. Supply chains and urban logistics platforms. In *Sustainable Urban Logistics: Concepts, Methods and Information Systems*; Springer: Berlin/Heidelberg, Germany, 2014; pp. 1–20.
31. Morana, J. Sustainable supply chain management in urban logistics. In *Sustainable Urban Logistics: Concepts, Methods and Information Systems*; Springer: Berlin/Heidelberg, Germany, 2014; pp. 21–35.
32. Atkinson, R. Project management: Cost, time and quality, two best guesses and a phenomenon, its time to accept other success criteria. *Int. J. Proj. Manag.* **1999**, *17*, 337–342. [CrossRef]
33. Project Management Institute. *A Guide to the Project Management Body of Knowledge (PMBOK®Guide)*, 3rd ed.; Project Management Institute: Newtown Square, PA, USA, 2008.
34. Kotarbiński, T. *Wybór Pism. Myśli o Działaniu. T.1.*; PWN: Warszawa, Poland, 1974.
35. Iwan, S. Adaptative approach to implementing good practices to support environmentally friendly urban freight transport management. *Procedia-Soc. Behav. Sci.* **2014**, *151*, 70–86. [CrossRef]
36. Serradora, J.; Turner, R. The Relationship between Project Success and Project Efficiency *Procedia-Soc. Behav. Sci.* **2014**, *119*, 75–84. [CrossRef]
37. Sundqvist, E.; Backlund, F.; Chronéer, D. What is Project Efficiency and Effectiveness? *Procedia-Soc. Behav. Sci.* **2014**, *119*, 278–287. [CrossRef]
38. Turner, R.; Zolin, R. Forecasting Success on Large Projects: Developing Reliable Scales to Predict Multiple Perspectives by Multiple Stakeholders over Multiple Time Frames. *Proj. Manag. J.* **2012**, *43*, 87–99. [CrossRef]
39. Dvir, D.; Lipovetsky, S.; Shenhar, A.J.; Tishler, A. What is really important for project success? A redefined, multivariate, comprehensive analysis. *Int. J. Manag. Decis. Mak.* **2003**, *4*, 382–404.
40. Gonzalez-Feliu, J.; Morana, J. Are City Logistics Solutions Sustainable? The Cityporto case. *Trimest. Lab. Territ. Mob. Ambient.* **2010**, *3*, 55–64.
41. Iwan, S. Building a Consensus between the Needs of Urban Freight Transport Stakeholders. In Proceedings of the Carpathian Logistics Congress, Kraków, Poland, 9–11 December 2013.
42. Shenhar, A.; Dvir, D. *Reinventing Project Management: The Diamond Approach to Successful Growth and Innovation*; Harvard Business Press: Brighton, MA, USA, 2007.
43. Wysocki, R.K. *Effective Project Management: Traditional, Agile, Extreme*, 6th ed.; Wiley: Hoboken, NJ, USA, 2011.
44. Taniguchi, E.; Thompson, R.G.; Yamada, T.; van Duin, R. *City Logistics. Network Modelling and Intelligent Transport Systems*; Pergamon: Oxford, UK, 2001.
45. Lepori, C.; Banzi, M.; Konstantinopoulou, L. *Stakeholders' Needs*; CITYLOG Deliverable D1.2; Centro Ricerche Fiat S.c.p.A.: Turin, Italy, 2010.

46. Hofenk, D. Making a Better World—Carrier, Retailer, and Consumer Support for Sustainable Initiatives in the Context of Urban Distribution and Retailing. Ph.D. Thesis, Open University of the Netherlands, Heerlen, The Netherlands, 2012.

47. Holguín-Veras, J.; Sánchez-Díaz, I.; Browne, M. Sustainable Urban Freight Systems and Freight Demand Management. *Transp. Res. Procedia* **2016**, *12*, 40–52. [CrossRef]

sustainability

MDPI

Article

Vehicle Weight, Modal Split, and Emissions—An Ex-Post Analysis for Sweden

Inge Vierth *, Samuel Lindgren and Hanna Lindgren

Swedish National Road and Transport Research Institute, P.O. Box 55685, 102 15 Stockholm, Sweden; samuel.lindgren@vti.se (S.L.); hanna.lindgren@vti.se (H.L.)
* Correspondence: inge.vierth@vti.se

Received: 27 April 2018; Accepted: 23 May 2018; Published: 25 May 2018

Abstract: This study combines official statistics on freight transportation and emissions to present the long-run development of the use of longer and heavier road vehicles (LHVs), modal split, road freight efficiency, and GHG emissions and air pollution following the increase in the maximum permissible vehicle weight in Sweden in 1990 and 1993. We find that LHVs were quickly incorporated in the vehicle fleet and that road freight efficiency of the largest vehicles increased after the reforms. There was no discernable break in modal split trends as the modal share for road continued its long-run development. We show that road transportation contributes by far the most to emission costs. The composition of the emissions from road freight changed after the weight reforms, with an increasing share of GHG-emissions.

Keywords: longer heavier vehicles; road freight transport; GHG emissions; environmental impact; modal shift

---

1. Introduction

The past decade has seen an active debate among policy-makers and researchers about the cost and benefits associated with allowing larger-than-conventional road freight vehicles [1,2]. This topic has gained much attention, not least considering the freight sector's contribution to the emissions of greenhouse gases (GHG) and air pollution. Allowing higher road vehicle dimensions is hypothesized to influence many important outcomes, including infrastructure, traffic safety, modal split, vehicle operating cost, as well as emissions of GHG and air pollution. The impacts of LHVs on traffic safety, road infrastructure and investments are covered in Steer et al. [3], Ortega et al. [1], and Ericsson et al. [4] but are outside the scope of this study. However, despite the wide-spread interest in these impacts there is limited empirical evidence on the consequences of longer and heavier vehicles (LHVs).

The purpose of this study is to provide an ex-post analysis of the use of LHVs, modal split, road freight efficiency, and emissions of GHG and air pollution following the increased maximum truck weights in Sweden in 1990 and 1993. Sweden offers an ideal setting to study this topic since it is one of few countries in the world where full-scale implementation of LHVs has been in place for several decades. Table 1 shows the development of maximum vehicle weight and length in Sweden. Restrictions on the dimensions were first set in 1968 (37 tonnes, 24 m) and subsequently increased in 1974, 1990, 1993, 1997, and 2015. Our analysis focus on the reforms increasing the dimensions from 51.4 tonnes (and 24 m) to 60 tonnes (24 m) which we will refer to as the introduction of longer and heavier vehicles (LHVs). We refrain from analyzing the length reform in 1996 which increased the maximum length of vehicle combinations from 24 to 25.25 m.

The analysis is based on official statistics of annual domestic freight transportation by road, rail, and water covering the period 1985 to 2013, as well as calculations of GHG and air pollution spanning from 1990 to 2013. The regulation framework and our wide-spanning data coverage offer a unique setting for studying the large-scale implementation of LHVs over time. We find that, after the weight reforms, there was a substantial increase in the share of road transport performance by trucks with a load capacity above 40 tonnes. The development mainly came at the expense of the vehicles with the lowest capacity. This shows the high degree of incorporation of LHVs in the Swedish vehicle fleet, which may be explained by the relatively large vehicle dimensions in Sweden that existed prior to the first reform in 1990 (see Table 1).

Table 1. Timeline of maximum vehicle dimensions in Sweden.

Year	Max. Length (m)	Max. Weight (t)
1968	24	37
1974	24	51.4
1990	24	56
1993	24	60
1996	25.25	60
2015	25.25	64
2018 [1]	25.25	74

[1] To be implemented.

Our analysis of the modal split shows that the road share in Sweden increased steadily before, during, and immediately after the weight reforms, whereas the rail and water shares were decreasing during this period. We document increases in the levels of tonne-km by both road and rail during this time interval, which implies that the falling rail share was driven by relatively higher tonne-km growth for road than for rail transportation. After a decade, rail had regained its pre-form share of the market. There was no discernable break in modal split trends at the time of the weight reforms. On the contrary, the road share continued on its long-run development.

We find modest evidence that the efficiency of road freight improved. The transport performance per vehicle-km rose in the years during and after the weight reform, particularly among vehicles of the highest maximum load capacity.

Finally, we show that the composition of emissions from road freight changed in the years following the weight reforms. While greenhouse gases increased, emissions of particulate matter (PM), nitrogen dioxides (NO_x), sulfur dioxide (SO_2), and non-methane volatile organic compounds (NMVOC) had dropped significantly at the end of the 1990s. The consistent increase in road transport performance during this period led to falling emission factors (emissions per tonne-km) for each pollutant. We synthesize these findings by showing that, in the 1990s, the cost of road freight emissions remained constant while the cost of emissions from all modes decreased. Overall, our results suggest that allowing higher weight dimensions on a large scale would increase the use of those vehicles fairly quickly but not lead to considerably adverse environmental impacts, at least not in terms of the outcomes and time period considered in this study.

The rest of this study is outlined as follows. The next section provides a literature review on the consequences of LHVs, while Section 3 describes the methodology and data used in the analysis. Section 4 presents the main results, and Section 5 concludes.

2. Related Studies

A large and growing body of literature examines the costs and benefits of allowing LHVs (see overviews in Ortega et al. [1] and Sanchez et al. [2]). There is widespread agreement in these studies that LHVs bring about reductions in the operating cost of road freight, fuel consumption, and emissions per tonne-km [3]. Increased truck capacity enables companies to consolidate loads and reduce the number of vehicle journeys needed to move a given amount of freight [5]. This gives rise to fuel savings which in turn translates into reduced cost for operating the vehicles. Several studies point out that although LHVs increase emission per vehicle-kilometre driven, they will bring about a reduction in emissions per tonne-kilometre [6–8]. Gutberlet et al. [9] show that the fuel and emission savings can be significant for individual companies, while the overall impact is limited if LHVs are restricted to certain areas or road segments.

A main concern of LHVs is that the cost reduction will attract freight from less environmentally-damaging modes and induce additional demand for freight transportation. In other words, there is a risk that increasing road freight transportation off-sets the social benefits stemming from the reduced environmental impact per tonne-km. The extent of the modal shift is subject to wide debate and research [10]. Many analyses of the modal shift effects of LHVs use modelling approaches which require assumptions about input parameters, including price elasticities, reductions in road haulage cost, and load factors. The results from the modelling studies vary considerably. Some predict substantial modal shifts [6,11,12] whereas others forecast only minor changes [13]. Steer et al. [3] summarize this strand of literature and conclude that the variation in predicted modal shift from adopting LHVs indeed tends to stem from different assumptions regarding own- and/or cross-price elasticities of the modes.

Empirical evidence of the effects of LHVs is relatively scarce and covers either full-scale implementation or temporary trials. In Australia, high capacity vehicles with a maximum length of 26 m and weight of 68.5 tonnes were introduced in 1984 and were permitted extensive network access, including main roads in urban areas [14]. These vehicles were gradually incorporated in the Australian vehicle fleet during the 1990s and 2000s. Their share of road freight went from practically zero in 1991 to a third in 2007 [15]. During this period, both road and rail freight transportation experienced an increase in transport performance at the expense of coastal shipping. Rail increased or maintained its share in the market for long-distance transport and bulk products but lost shares on shorter distances for non-bulk goods [16]. The growth in road freight was also promoted by increased demand for reliable and timely delivery as well as improvements in road infrastructure and vehicle technology [15]. The introduction of high capacity vehicles was estimated to have reduced fuel consumption in the articulated fleet by at least 11% [17].

In Finland, permission was granted to operate trucks of 76 tonnes and 30 m on designated parts on the road network. An evaluation of the trial showed that as of 2017, some 40 heavy trucks were being used to transport forest and agriculture products [18]. The trucks were considered economically viable for large shipment volumes and where loading and unloading would not become considerably harder or more time-consuming. The report concluded that a shift from rail freight traffic was likely to be most substantial for forestry products and GHG emissions from road traffic would be reduced by 77,000 tonnes annually if the estimated changes were realized.

In the Netherlands, several trials were conducted during the 2000s in which trucks with a capacity of 60 tonnes were allowed temporarily. Kindt et al. [19] evaluated the third trial by means of a stakeholder survey of terminal operators, shipping companies and transporters. They found that the modal split in terminals had been unchanged compared to the situation before the trials started. Jonkeren and Aarts [20] concluded that the number of LHVs operating in the Netherlands had increased substantially from 2001 to 2016 but that the shift of freight to LHVs solely was derived from regular trucks and not from rail and inland waterway transportation. Results from the trials showed that although fuel consumption per vehicle-km increased, the fuel efficiency (measured in relation to load) rose which led to lower emissions of GHG, NO_x, and PM [21].

An evaluation of the Norwegian trial where trucks up to 60 tonnes and 25.25 m were allowed found that the actual usage of the LHVs during the period of study (2008–2013) had been relatively low. The firms that were using the LHVs did indeed experience large reductions in costs, largely because they could move the same amount of goods using fewer vehicles [22]. The results pointed towards lower emissions per tonne-km and/or cubic meter-km for LHVs compared to regular trucks, which suggested that these vehicles would bring about reductions in GHG, NO_x, and PM.

Trials have also been conducted in Denmark, where vehicle dimension up to 60 tonnes and 25.25 m were temporarily allowed in 2008. The Danish Road Directorate [23] found in their analysis that after two years of trials, around 400 LHVs contributed to 3.6% of annual transport performance and were mainly being used on trip distances of around 200–300 km. Comparing road freight transportation in 2007 and 2010, the report concluded that using LHVs would not have changed the emission GHG significantly.

In Germany, permission was granted to vehicles with a maximum length of 25.25 m, compared to the conventional length of 18.75 m. An evaluation of the trials between 2012 and 2014 found that forwarders reported LHVs to bring about a cost advantage of 16% compared to conventional vehicles, given that they managed a utilization rate above 83% [24]. Although the LHVs required more fuel, the fact that fewer vehicles were required to move the same amount of freight yielded the cost reductions for firms. The report also showed that LHVs were used as a replacement for conventional trucks and none of the companies in the study adopted the LHV in favor of rail freight transportation.

Evaluations of the Swedish weight reforms in the 1990s have focused on modal shift effects and cost–benefit analysis. Nelldal [25] reviews the development in the freight market in the 1990s and argues that rail transportation stagnated during this period due to the truck weight reforms in 1990 and 1993. He estimates that the reforms jointly entailed a price reduction for road freight of 22% at full capacity utilization which is in line with subsequent estimates [7,26]. Nelldal et al. [27] argue that the decline of rail transportation also depended on the extension of the road network, lack of industrial railway tracks, and increasing foreign trade.

Vierth et al. [7] make use of the Swedish national freight model Samgods to analyze how LHVs affect the Swedish freight market. They show that allowing trucks of trucks of 60 tonnes and 25.25 m reduce road vehicle-km substantially and leads to a higher tonne-km road share. Haraldsson et al. [28] use the Samgods model to conduct a cost benefit analysis of an increase in maximum weight from 60 tonnes to 90 tonnes. They find that such measure reduces the amount of road vehicle-km by 21%. Vierth and Karlsson [29] study the effects of allowing road vehicle combinations up to 25.25 m and 60 tonnes on a designated freight corridor between Sweden and Germany. They find that this increases the tonne-kilometre road freight transportation by 0.5% and decreases rail freight by 0.7%.

3. Methodology and Data

3.1. Methodology

Our analysis of the increase in road vehicle weight limits starts off by examining the uptake of LHVs in the vehicle fleet following the reforms in 1990 and 1993. We subsequently investigate the development along three other dimensions of the freight transportation market: the modal split, the efficiency of road freight and the emissions of GHG, NO_x, PM, NMVOC, and SO_2. Our methodological approach is to first compile the statistics of freight transportation activity and emissions per year and generate a set of time series of these outcomes. We subsequently plot each time series in a graph and rely on visual inspection to compare the outcome of interest before, during, and after the weight reforms in 1990 and 1993. We expect any effects of the reforms to be the most visible from the year of the first reform up to some years after the last reform and focus our analysis on this time horizon. It is more difficult to attribute changes over a 10–15-year interval to the weight reforms.

The statistics allow us to identify time trends in the outcomes and try determining whether the reforms had any impact on the development over time. This is an important distinction from simply assessing the post-reform development if there were underlying trends in the outcomes of interest before the reforms. We therefore put more weight on findings showing that the development of a particular outcome changed substantially during or after the weight reforms. This kind of investigation could also be implemented in a time series regression analysis, but we opt for graphing the outcomes and using ocular inspection instead. In our view, this approach increases the transparency by allowing the reader to assess the development of the outcomes. The limited number of observations also makes it difficult to implement test of structural breaks at the weight reforms in a regression analysis with robustness and precision.

Our methodological approach lets us analyze the overall development in the freight market but does not allow an identification of the isolated impact of higher vehicle weight limits. Other events in the transportation sector coinciding with the weight reforms are likely to also have mattered for the development, as are other structural changes during this period. We return to this point in the analysis of the results.

3.2. Data

We combine data from several sources to conduct our analysis. Our measures of the use of LHVs in Sweden, the road freight efficiency and the modal split are based on annual freight transportation statistics for domestic transportation. The statistics cover road, rail, and waterborne transportation (the latter includes only short-sea shipping and not inland waterway transportation). The figures come from various governmental authorities which have been or are responsible for collecting the data during our sample period (State Railways, Statistics Sweden, SIKA and Transport Analysis). The modal split is measured in tonne-kms and covers domestic transportation because the weight reforms only applied in Sweden. Road freight surveys have been conducted annually in Sweden except for the period 1987–1993 when only three surveys were made. Statistics Sweden therefore imputed aggregate road freight statistics for the years 1988, 1989, 1991, and 1992, which we use in our analysis. However, disaggregated statistics are completely missing for these years which causes a break in the time series of the use of LHVs.

Our analysis of emissions is based on information about the emission of GHG and air pollution from domestic transportation on Swedish territory and is compiled by the Swedish Environmental Protection Agency. Vehicle-km performed by non-Swedish trucks on Swedish roads are excluded based on the statistics authority's data for 2000–2013. The share ranged from 11% in 2000 to 17% in 2013. We extrapolate this trend from 1990 to 2000.

We assign monetary values on the yearly emissions based on the European handbook on external costs of transport and show the cost of emissions from each mode and by pollutant [30]. This exercise requires some assumptions about the transportation sector in Sweden. First, we assume that freight transportation constitutes 70% of the transport performance of rail (and that the rest is passenger transportation). Second, based on studies finding that the fuel consumption of water transportation is twice the size of that reported in the sales statistics, we multiply the emissions from water transportation by two [31]. These assumptions change the level of emissions from water transportation but affects emissions by rail only marginally as this mode is powered by electricity. Table 2 shows the value per tonne and pollutant based on our assumptions (in 2010 price levels). Each figure shows the external cost of a tonne emitted of the corresponding pollutant. The external cost refers to the cost to society due to freight transport (e.g., health cost, material and biosphere damages) that is not borne by the transport user.

Table 2. Euro per tonne and pollutant.

Mode	NO_x	NMVOC	SO_2	PM2.5	GHG
Rail	5247	974	5389	29,208	90
Water	4700	1100	5250	13,800	90
Road	5247	974	5389	42,009	90

Source: Korzhenevych et al. [30] and own calculations.

Table 3 shows summary statistics for the outcomes of interest as well as the data coverage over time. Our analysis of the use of LHVs, modal split development and road freight efficiency covers the time 1985–2013 and the analysis of road freight emissions range from 1990 to 2013. See Vierth et al. [32] for further details about the data.

Table 3. Summary statistics.

Variable	Unit	Mean	Min	Max	Coverage
Transportation activity					
Rail freight	Million tonne-km	10,515	8463	13,450	1985–2013
Water freight	Million tonne-km	7755	6504	9447	1985–2013
Road freight	Million tonne-km	32,167	22,611	38,807	1985–2013
Road freight (0–30 t)	Million tonne-km	5365	2805	8506	1987–2013
Road freight (30–40 t)	Million tonne-km	12,695	5904	16,771	1987–2013
Road freight (40+ t)	Million tonne-km	14,106	1347	28,164	1987–2013
Road efficiency (0–30 t)	Tonne-km per vehicle-km	0.12	0.10	0.16	1985–2013
Road efficiency (30–40 t)	Tonne-km per vehicle-km	0.14	0.12	0.16	1985–2013
Road efficiency (40 t)	Tonne-km per vehicle-km	0.17	0.13	0.20	1985–2013
Emissions					
GHG by road	Kilo-tonne	4337	3379	5196	1990–2013
GHG by rail	Kilo-tonne	50	37	71	1990–2013
GHG by water	Kilo-tonne	541	296	812	1990–2013
$PM_{2.5}$ by road	Tonne	1264	487	1840	1990–2013
$PM_{2.5}$ by rail	Tonne	22	16	30	1990–2013
$PM_{2.5}$ by water	Tonne	494	194	854	1990–2013
NO_x by road	Tonne	38,972	21,825	48,445	1990–2013
NO_x by rail	Tonne	888	467	1326	1990–2013
NO_x by water	Tonne	7849	4557	11,443	1990–2013
SO_2 by road	Tonne	465	12	3294	1990–2013
SO_2 by rail	Tonne	10	0	81	1990–2013
SO_2 by water	Tonne	3666	1173	7013	1990–2013
NMVOC by road	Tonne	4194	783	7800	1990–2013
NMVOC by rail	Tonne	74	39	107	1990–2013
NMVOC by water	Tonne	142	76	212	1990–2013
Cost of road emissions	Million euros	741.5	647.5	831.9	1990–2013
Cost of rail emissions	Million euros	9.9	6.3	14.8	1990–2013
Cost of water emissions	Million euros	111.8	60.2	168.4	1990–2013

4. Results

In this section, we present the development of the four outcomes of interest. We start by the use of LHVs and subsequently investigate the modal split development, road freight efficiency, and emissions.

4.1. Use of LHVs

To track the uptake of LHVs in Sweden over time we use road freight statistics segmented by maximum payload of the road vehicle combinations, defined as the sum of the load capacity of the truck and the load capacity of trailers. We divide the vehicle combinations into three categories based on their maximum payload (0–30, 30–40, and 40+ tonnes) and show the tonne-km for each segment in Figure 1. As a rule of thumb, vehicles of length 24–25.25 m and weight of 60–64 tonnes, are assumed to have a maximum payload between 30–42 tonnes [33]. The absence of road freight surveys in the late 1980s and early 1990s causes breaks in the time series.

Figure 1 shows that at the time of the weight limit increase in 1990, trucks with a maximum load capacity above 30 tonnes already accounted for more than 60% of the road transport performance. This reflects the fact the maximum permissible weight had been high for a long period of time in Sweden. The share of tonne-km performed by trucks with a load capacity above 40 tonnes increased substantially in the 1990s, from 10% in 1990 to 45% in 2000, which shows the high degree of incorporation of LHVs in the Swedish vehicle fleet. The development mainly came at the expense of the vehicles with the lowest capacity, which saw its share go from 35% in 1990 to 15% in 2000.

During the 2000s, the tonne-km by trucks with a load capacity above 30 tonnes expanded moderately and peaked just before the economic recession in 2009. The activity by trucks in the smallest capacity segment declined somewhat during this period. Since 2009, the largest trucks have increased their share of road transport performance considerably and reached their peak in 2013.

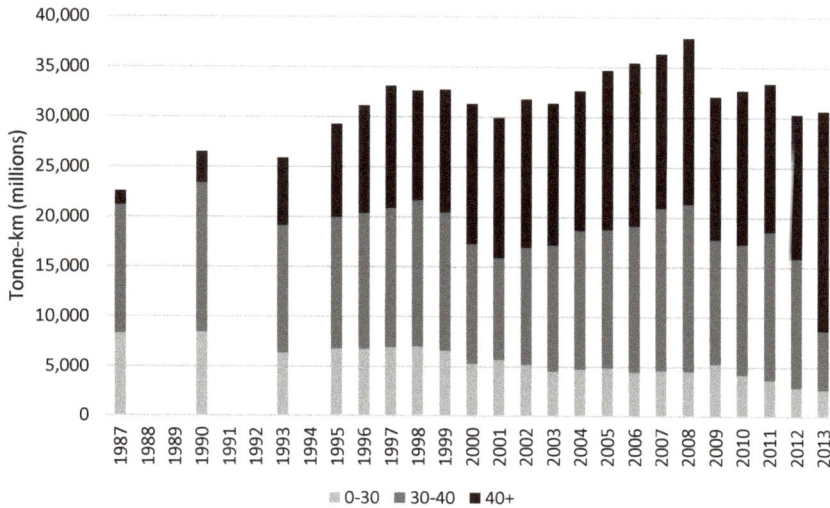

Figure 1. Domestic road transport performance by maximum load capacity.

4.2. Modal Split

In Figure 2 we have stapled the share of each mode (road, rail, and water) of the total domestic transport performance for each year. We have also created an index of the share for road and rail, which shows the change in the share of each mode relative to the baseline year 1990. An index above 100 corresponds to a higher modal share for that particular year compared to the share in 1990. The figure shows how the rail share was decreasing from 1985 up until 1995 when the trend reversed. In 2000, the rail share was back to its pre-reform value and continued to rise in the 2000s. Waterborne transport was consistently losing market shares from 1985 and onwards. It went from having 24% of the market in 1985, to 19% in 1990, and then to 14% in 2013. Road freight developed in the opposite way. The road share increased steadily between 1985–1990 and continued this way during most of the 1990s, until it stabilized around 60–65%. What is noticeable is the lack of break in the modal split trends at the time of the increasing in maximum weights in 1990 and 1993. On the contrary, the share for each mode is continuing its long-term development.

Figure 3 shows the level of tonne-km for each mode. It also includes the tonne-km index for road and rail, which shows the percentage change in transport performance for each mode relative to the baseline year 1990. From this figure, it is apparent that the tonne-km was growing both for road and rail in the 1990s. The increase in the road share that was documented in the previous figure therefore seems to be driven by the fact that the tonne-km growth rate was higher for road than for rail. Overall, it is difficult to trace out substitution patterns based on the aggregate statistics. In addition, the weight reforms coincided with the 1990–1993 economic recession in Sweden [34], deregulation of the Swedish railway freight sector in 1996 [35], and the replacement of the distance-based road tax by a tax on diesel fuels in 1995, which may have influenced the development.

In Vierth et al. [30] we use the weight reforms to estimate short- and long-run demand elasticities of road and rail with respect to road freight cost. We also show that the modal split developed differently for various commodity groups, which may reflect that LHVs are more suitable for commodities with certain volume and weight characteristics.

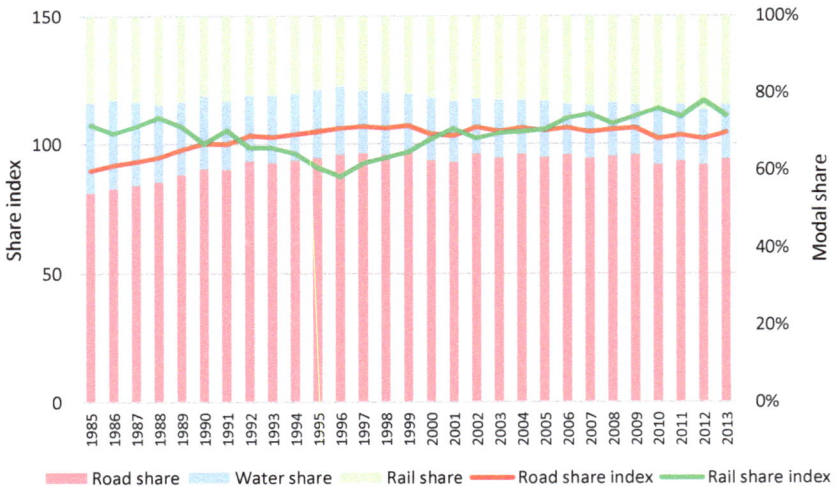

Figure 2. Domestic modal shares (in tonne-km) and share index (1990 = 100).

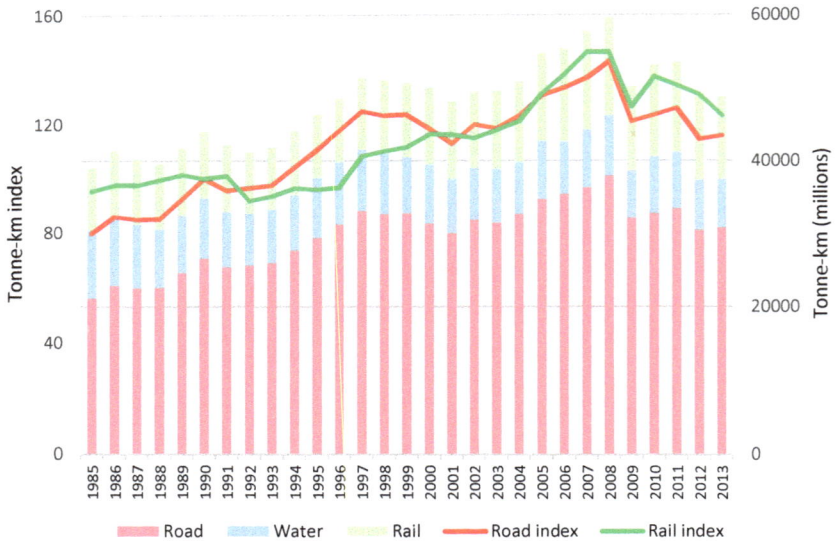

Figure 3. Domestic transport performance and tonne-km index (1990 = 100).

4.3. Road Freight Efficiency

Figure 4 shows the amount of tonne-km per vehicle kilometre for the three road vehicle categories based on maximum payload (0–30, 30–40, and 40+ tonnes). This gives a crude measure of the transportation efficiency of each class but highlights important dimensions of the development in the road freight sector. As expected, vehicles with higher load capacity have a higher transport performance per vehicle-km driven. What is noticeable is the increase in the ratio for vehicles in the larger capacity class between 1990 and 1997, compared to the development in the other classes. Their ratio of tonne-km to vehicle-km increase by 23% between 1990 and 1997 whereas the corresponding change was 4% and 10% for vehicles in the smallest and medium capacity classes respectively. This illustrates the differential change in road freight efficiency for the heaviest vehicles following the weight reforms.

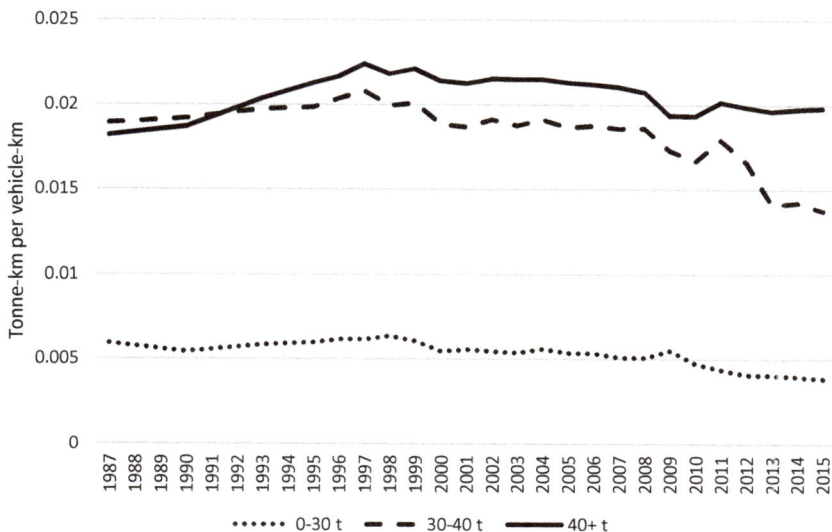

Figure 4. Domestic road freight efficiency (tonne-km per vehicle-km) by max load capacity.

4.4. Emissions

Our analysis relates the level of emission of each pollutant to the road freight transport performance between 1990–2013. Figure 5 shows the road freight emissions of GHG and its ratio to transport performance by road. Both the emissions and transport performance rose in the years following the weight reforms. The large drop in the emission factor (GHG per tonne-km) during this period shows that the surge in tonne-km surpassed that of emissions. However, from 1997 and onwards, the GHG per tonne-km has risen consistently.

Figures 6 and 7 show the amount of emissions of PM, NO_x, SO_2, and NMVOC and their relationship to the transport performance by road. There is a clear downward trend in the emission factor (emissions per tonne-km) for all pollutants, in particular for SO_2 and NMVOC. The drop in PM and NO_x appears to wear off in the end of the 1990s. The development of the emissions of pollutants in the 1990s is also likely to affected by the Swedish and European environmental and energy policies during this period. This includes the introduction of an energy- and carbon dioxide tax on fuels [36], restrictions on the sulfur content of fuel [37], a NO_x charge on energy producers [38], as well as the European Union emissions standards.

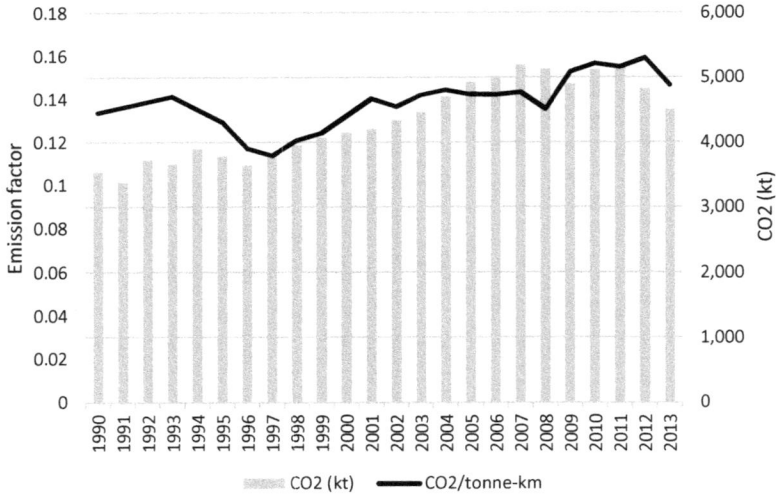

Figure 5. GHG emissions in levels and per domestic road tonne-km.

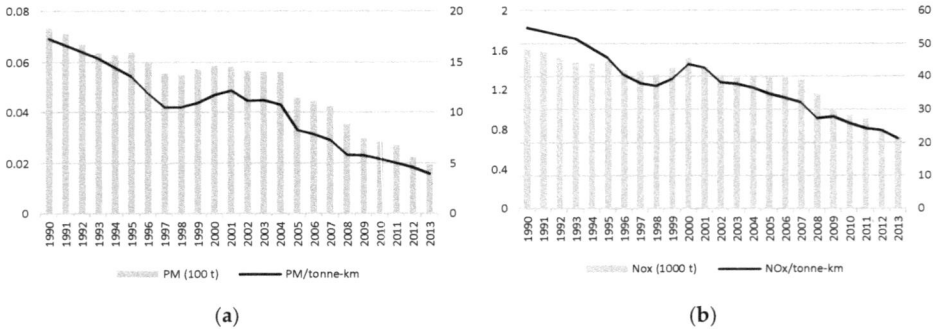

(a)

(b)

Figure 6. (a) $PM_{2.5}$ emissions in tonnes (right axis) and per domestic tonne-km by road (left axis); (b) NO_x emissions in levels (right axis) and per domestic tonne-km by road (left axis).

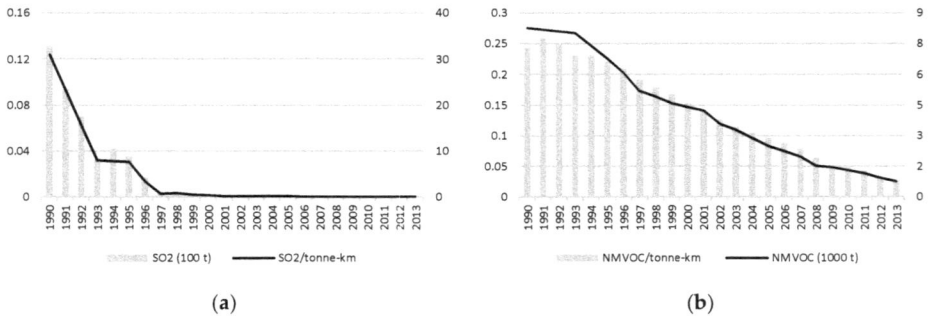

(a)

(b)

Figure 7. (a) SO_2 emissions in 100s tonnes (right axis) and per domestic tonne-km by road (left axis); (b) NMVOC emissions in levels and per tonne-km by road.

We conclude our analysis of the emissions by assigning a monetary value on the yearly emissions based on the Ricardo valuation and show the cost of emissions from each mode and by pollutant.

Figure 8 shows the development of the cost of GHG and air pollution between 1990–2013. Road transportation contributes by far the most to emission costs, followed by water transportation. The reduction in the cost of emissions from freight transportation in the early 1990s is likely to be driven by the drop in transport performance coinciding with the economic recession in Sweden during this time. The annual cost of road freight emissions is relatively constant in the 1990s and there is no noticeable change coinciding with the weight reforms. One explanation for this is two counteracting forces reviewed in the previous sections. As documented in Section 4.2, the road transport performance increased in the 1990s which would increase emissions, all else equal. On the other hand, road freight appears to have increased its efficiency judging by the results in Section 4.3. This means that fewer vehicle-km were needed for a given amount of tonne-km. The reductions in the emission factors also suggest that trucks became more fuel efficient and/or that fuel have become cleaner during this period. A zero-net effect of these forces could therefore explain the constant cost of road freight emissions in the 1990s. To further investigate this hypothesis, Figure 9 shows the value of the road freight emission of each pollutant. The composition of road freight emissions has changed significantly over time, with the cost of GHG rising consistently, both in levels and as a share of the total cost of emissions. This development is line with that in the rest of the EU [39].

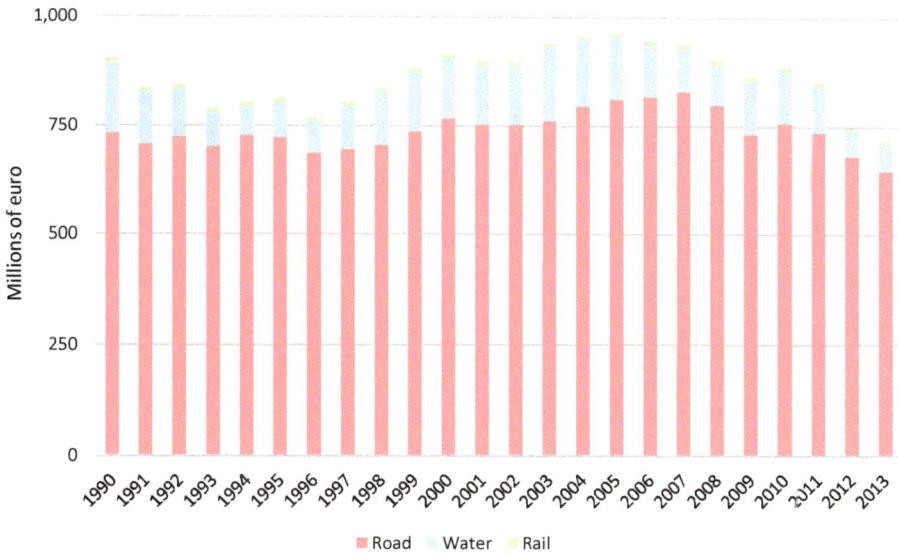

Figure 8. Valuation of all emissions from freight transportation by mode (2010 price level).

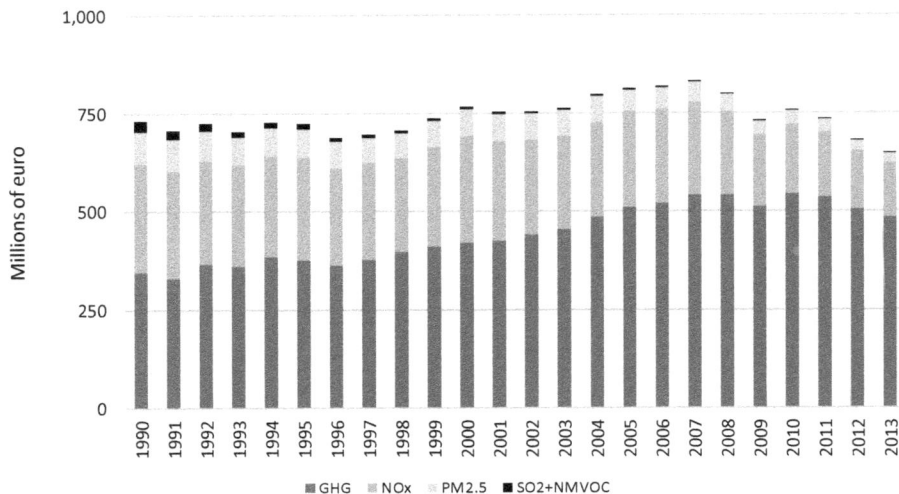

Figure 9. Valuation of emissions from road freight by source (2010 price level).

5. Discussion

In this study, we have compiled domestic freight statistics and emissions data to investigate how the use of LHVs, modal split, road freight efficiency, GHG emissions, and air pollution have developed before, during, and after the increases in maximum permissible weight for vehicle combinations in Sweden. We have focused on the increase from 51.4 tonnes to 56 tonnes in 1990 and to 60 tonnes in 1993 and the period 1985–2013.

We find that the share of tonne-km by trucks with a load capacity above 40 tonnes increased substantially in the 1990s, which mainly came at the expense of the vehicles with the lowest capacity. This shows the high degree of incorporation of LHVs in the Swedish vehicle fleet and is similar to the full-scale introduction of LHVs in Australia during the same time period.

Our analysis of the modal split shows that the road share in Sweden increased steadily before, during, and after the weight reforms. Rail and water transportation on the other hand were decreasing between 1985 and 1995. We document increases in the levels of tonne-km by both road and rail during this period, which implies that the falling rail share was driven by relatively higher tonne-km growth for road. What is noticeable is the lack of break in the modal split trends at the time of the weight reforms. On the contrary, the share for each mode is continued on its long-term development. It is difficult to trace out substitution patterns between the modes. Whether the freight that was shifted away from waterborne transportation benefitted road or rail the most is not apparent. Neither is any shift of freight between road and rail transportation.

We document an increase in road freight efficiency after the weight reforms, meaning that fewer vehicle-km were needed for a given amount of tonne-km. This increase was particularly noticeable among vehicles of the highest maximum load capacity. This suggests that carriers managed to increase the use of LHVs while being able to utilize the extra load capacity.

Finally, we show that road transportation contributes by far the most to emission costs, followed by water transportation. The composition of emissions from road freight changed in the years following the weight reforms. While GHG emissions increased, emissions of PM, NO_x, SO_2, and NMVOC dropped significantly since the end of the 1990s. The consistent increase in road transport performance during this period lead to falling emission factors (emissions per tonne-km) for each pollutant. We synthesize these findings by showing that the external costs of road freight emissions remained

Sustainability **2018**, *10*, 1731

constant in the 1990s. We hypothesize that the observed development is due to higher road freight efficiency off-setting the increase in road transport performance.

In conclusion, our results suggest that allowing higher weight dimensions on a large scale would increase the use of those vehicles fairly quickly but not lead to considerably adverse environmental impacts, at least not in terms of the outcomes considered in this study. These findings may be limited to the specific country and time period of study and future research should conduct ex-post analyses of the implementation of LHVs on various scales to assess whether these results hold in the current freight transportation market and in other countries.

Author Contributions: H.L. and S.L. collected the data, all authors analyzed and interpreted the data and wrote the paper.

Funding: This research was funded by Centre for Transport Studies, Sweden.

Acknowledgments: We would like to thank Thomas Asp, Swedish Road Administration, and Tom Andersson, Transport Analysis for useful comments on an earlier version of this study. We are also grateful to Mohammad-Yeza Yahya, IVL Swedish Environmental Research Institute and Tobias Lindé, VTI, for helpful assistance with the data.

Conflicts of Interest: The authors declare no conflict of interest.

References

1. Ortega, A.; Vassallo, J.M.; Guzmán, A.F.; Pérez-Martínez, P.J.; Ortega, A.; Vassallo, J.M.; Guzmán, A.F.; Pérez-Martínez, P.J. Are longer and heavier vehicles (LHVs) beneficial for society? A cost benefit analysis to evaluate their potential implementation in Spain. *Transp. Rev.* **2014**, *34*, 150–168. [CrossRef]
2. Sanchez-Rodrigues, V.; Piecyk, M.; Mason, R.; Boenders, T. The longer and heavier vehicle debate: A review of empirical evidence from Germany. *Transp. Res. Part D* **2015**, *40*, 114–131. [CrossRef]
3. Steer, J.; Dionori, F.; Casullo, L.; Vollath, C.; Frisoni, R.; Carippo, F.; Ranghetti, D. *Review of Megatrucks. Major Issues and Case Studies*; European Commission Directorate General for Internal Policies: Brussels, Belgium, 2013.
4. Ericsson, J.; Lindberg, G.; Mellin, A.; Vierth, I. Co-modality—The socio-economic effects of longer and/or heavier vehicles for land-based freight transport. In Proceedings of the 12th WCTR, Lisbon, Portugal, 11–15 July 2010.
5. McKinnon, A. The economic and environmental benefits of increasing maximum truck weight: The British experience. *Transp. Res. Part D* **2005**, *10*, 77–95. [CrossRef]
6. De Ceuster, G.; Breemersch, T.; Van Herbruggen, B.; Verweij, K.; Davydenko, I.; Klingender, M.; Jacob, B.; Arki, H.; Bereni, M. *Effects of Adapting the Rules on Weight and Dimensions of Heavy Commercial Vehicles as Established within Directive 96/53/EC*; European Commission Directorate General Energy and Transport: Brussels, Belgium, 2008.
7. Vierth, I.; Berell, H.; McDaniel, J.; Haraldsson, M.; Hammarström, U.; Yahya, M.-R.; Lindberg, G.; Carlson, A.; Ögren, M.; Björketun, U. *The Effects of Long and Heavy Trucks on the Transport System*; VTI: Stockholm, Sweden, 2008.
8. Leach, D.Z.; Savage, C.J. *Impact Assessment: High Capacity Vehicles*; University of Huddersfield: Huddersfield, UK, 2012.
9. Gutberlet, T.; Kienzler, H.P.; Labinsky, A.; Eckert, S.; Faltenbacher, M. Longer Heavy Goods Vehicles in Germany—"Ecocombis" or "Climate Killers"? In Proceedings of the European Transport Conference, Barcelona, Spain, 4–6 October 2017.
10. McKinnon, A. Improving the Sustainability of Road Freight Transport by Relaxing Truck Size and Weight Restrictions. In *Supply Chain Innovation for Competing in Highly Dynamic Markets*; Evangelista, P., McKinnon, A., Sweeney, E., Esposito, E., Eds.; IGI Global: Hershey, PA, USA, 2012.
11. Doll, C.; Fiorello, D.; Pastori, E.; Reynaud, C.; Klaus, P.; Lückman, P.; Hesse, K.; Kochsiek, J. *Long-Term Climate Impacts of the Introduction of Mega-Trucks*; Study for the Community of European Railways and Infrastructure Companies; CER: Brussels, Belgium, 2008.
12. Knight, I.; Newton, W.; McKinnon, A.; Palmer, A.; Barlow, T.; McCrae, I.; Dodd, M.; Couper, G.; Davies, H.; Daly, A.; et al. *Longer and/or Longer and Heavier Goods Vehicles (LHVs)—A Study of the Likely Effects if Permitted in the UK: Final Report*; DfT: London, UK, 2008.

13. Salet, M.; Aarts, L.; Honer, M.; Davydenko, I.; Quak, H.; de Bes van Staalduinen, J.; Verweij, K. *Longer and Heavier Vehicles in The Netherlands. Facts, Figures and Experiences in the Period 1995–2010*; Rijkswaterstaat: The Hague, The Netherlands, 2010.

14. OECD. *Moving Freight with Better Trucks: Improving Safety, Productivity and Sustainability*; ITF Research Reports; OECD Publishing: Paris, France, 2011.

15. Bureau of Infrastructure, Transport and Regional Economics. *Truck Productivity: Sources, Trends and Future Prospects*; Report 123; Bureau of Infrastructure, Transport and Regional Economics: Canberra, Australia, 2011.

16. Mitchell, D. Heavy Vehicle Productivity Trends and Road Freight Regulation in Australia. In Proceedings of the Australasian Transport Research Forum 2010 Proceedings, Canberra, Australia, 29 September–1 October 2010.

17. State of Victoria. *Victorian Freight Network Strategy*; Victoria Dept. of Transport: Melbourne, Australia, 2008.

18. Lapp, T.; Ikkanen, P. *Transport System Impacts of HCT Vehicles*; Research Reports of the Finnish Transport Agency 57/2017; Finnish Transport Agency: Helsinki, Finland, 2017.

19. Kindt, M.; Burgess, A.; Quispel, M.; van der Meulen, S.; Bus, M. *Monitoring Modal Shift. Longer and Heavier Vehicles. The Follow-Up Measurement*; Rijkswaterstaat: The Hague, The Netherlands, 2011.

20. Jonkeren, O.; Aarts, L. Dutch Experience of Modal Split. In Proceedings of the ITF Cambridge Workshop on Modal Shift, Cambridge, UK, 7 December 2016.

21. Arcadis. *Monitoringsonderzoek Vervolgproef LZV: Resultaten van de Vervolgproef Met Langere en Zwaardere Voertuigcombinaties op de Nederlandse Wegen*; Rijkswaterstaat: The Hague, The Netherlands, 2006.

22. Brevik-Wangsness, P.; Bjørnskau, P.; Hovi, I.B.; Madslien, A.; Hagman, A. *Evaluation of Norwegian Trials with European Modular System (EMS) Vehicles*; Report 1319/2014; Institute of Transport Economics: Oslo, Norway, 2014.

23. Danish Road Directorate. *Evaluation of Trial with European Modul System*; Final Report; Danish Road Directorate: Copenhagen, Denmark, 2011.

24. Limbeck, S.; Gail, J.; Schwedhelm, H.; Jungfeld, I. Configurable and Adaptable Trucks and Trailers for Optimal Transport Efficiency. Appendix A of Deliverable D5.5. Available online: http://modularsystem. odeum.com/download/bast_report.pdf (accessed on 23 May 2018).

25. Nelldal, B.L. *Järnvägssektorn efter Järnvägsreformen 1988—Förändringar i Omvärlden, Trafikpolitiken och Järnvägsbranschen och i Järnvägens Marknad 1990–2000*; Royal Institute of Technology: Stockholm, Sweden, 2002.

26. Pålsson, H.; Winslott, L.; Wandel, S.; Khan, J.; Adell, E. Longer and heavier road freight vehicles in Sweden: Effects on tonne- and vehicle-kilometres, CO_2 and socio-economics. *Int. J. Phys. Distrib. Logist. Manag.* **2017**, *47*, 603–622. [CrossRef]

27. Nelldal, B.L.; Troche, G.; Wajsman, J. *Effekter av Lastbilasavgifter*; Royal Institute of Technology: Stockholm, Sweden, 2009.

28. Haraldsson, M.; Jonsson, L.; Karlsson, R.; Vierth, I.; Yahya, M.; Ögren, M. *Cost Benefit Analysis of Round Wood Transports Using 90-Tonne Vehicles*; Report 758/2014; VTI: Stockholm, Sweden, 2008.

29. Vierth, I.; Karlsson, R. Effects of Longer Lorries and Freight Trains in an International Corridor between Sweden and Germany. *Transp. Res. Procedia* **2014**, *1*, 188–196. [CrossRef]

30. Korzhenevych, A.; Dehnen, N.; Bröcker, J.; Holtkamp, M.; Meier, H.; Gibson, G.; Varma, A.; Cox, V. *Update of the Handbook on External Costs of Transport*; Final Report; MOVE/D3/2011/571; Ricardo-AEA: London, UK, 2011.

31. Windmark, F.; Jakobsson, M.; Segersson, D. *Modellering av Sjöfartens Bränslestatistik Med Shipair*; Report 2017-10; Swedish Meteorological and Hydrological Institute: Norrköping, Sweden, 2017.

32. Vierth, I.; Lindgren, S.; Lindgren, H. *Impact of Higher Road Vehicle Dimensions on Modal Split: An Ex-Post Analysis for Sweden*; Report 34A-2017; VTI: Stockholm, Sweden, 2018.

33. Nelldal, B.L. Competition and co-operation between railways and trucking in long distance freight transport—An economic analysis. In Proceedings of the 3rd KFB-Research Conference "Transport Systems—Organisation and Planning", Stockholm, Sweden, 13–14 June 2000.

34. Englund, P. The Swedish Banking Crisis. Roots and Consequences. *Oxf. Rev. Econ. Policy* **1999**, *15*, 80–97. [CrossRef]

35. Nilsson, J.E. Restructuring Sweden's railways: The unintentional deregulation. *Swed. Econ. Policy Rev.* **2002**, *9*, 229–254.

36. Swedish Code of Statutes. General Energy Tax Act, 1994: 1776. Available online: https: //www.riksdagen.se/sv/dokument-lagar/dokument/svensk-forfattningssamling/lag-1994:776-om-skatt-pa-energi_sfs-1994-1776 (accessed on 23 May 2018).

37. Swedish Code of Statutes. Sulfur Tax Act, 1990: 587. Available online: http://riksdagen.se/sv/dokument-lagar/dokument/svensk-forfattningssamling/lag-1990587-om-svavelskatt_sfs-1990-587 (accessed on 23 May 2018).

38. Swedish Code of Statutes. Act on an Environmental Charge on Emissions of Nitrogen Oxides in Energy Production, 1990: 613. Available online: https://www.riksdagen.se/sv/dokument-lagar/dokument/svensk-forfattningssamling/lag-1990613-om-miljoavgift-pa-utslapp-av_sfs-1990-613 (accessed on 23 May 2018).

39. European Commission. *Strategy for Reducing Heavy-Duty Vehicles' Fuel Consumption and CO_2 Emissions*; COM 2014:285 Final; European Commission: Brussels, Belgium, 2014.

![sustainability logo] *sustainability*

MDPI

Concept Paper

The Prism of Elasticity in Rebound Effect Modelling: An Insight from the Freight Transport Sector

Franco Ruzzenenti [1,2]

1   Energy and Sustainability Research Institute, University of Groningen, 9747 AG Groningen,
     The Netherlands; f.ruzzenenti@rug.nl; Tel.: +31-5036-34-691
2   Advanced Systems Analysis, International Institute for Applied Systems Analysis, 2361 Laxenburg, Austria

Received: 17 July 2018; Accepted: 7 August 2018; Published: 13 August 2018

Abstract: If the rebound effect is to be considered a major obstacle to sustainable freight transport, then action and timely policy must be made in advance. This, however, requires a theoretical understanding of the nature of the rebound effect and an empirical grasp of its underlying mechanism. Elasticity is the centrepiece of current models on the rebound effect (or Jevons paradox). Although elasticity is a metric of indisputable usefulness for empirical purposes, it may be misleading when applied to the complex rebound effect. Drawing on the parallel case of the 'distance puzzle' in international economics, it will be shown how elasticity can be misinterpreted or how it can misdirect an investigation of the phenomenon by following a predetermined mindset. This particular bias is shown to widen in the long term and evolving systems in which the elasticity metric continues to output a constant number, eliciting a persistent effect. Drawing on previous research, an alternative approach to studying the rebound effect based on complex network theory and statistical mechanics of networks will be described. It will be shown how the interplay between spatial and non-spatial effects in freight transport networks can inform us about the evolution of the effect of distances on trade relationships, upon which a new metric for the rebound effect can be built.

Keywords: freight transport; rebound effect; Jevons paradox; gravity models; distance puzzle; network theory; statistical mechanics of networks; complexity

---

1. Introduction

In 1865, Stanley Jevons published his famous pamphlet 'The Coal Question', in which, for the first time in history, it was suggested that the consequences of energy efficiency might be very different from their intended, conservative goals. In fact, 'the very contrary is true': a more efficient technology delivers a more economical use of energy, thereby encouraging 'more and new applications' [1]. The Jevons paradox has surfaced several times throughout the history of energy and environmental studies and in such different circumstances and guises that in an editorial of a landmark special issue of Energy Policy, Lee Schipper referred to it as the 'Loch Ness monster' of energy efficiency [2]. The first occurrence was after the oil shocks of the 1970s in the guise of the 'Khazzoom-Brookes postulate', named as such after two economists who independently questioned the effectiveness of pursuing energy conservation by imposing efficiency mandates. Rejuvenated by a new climate change awareness, the second appearance came at the end of the century, bringing with it a new wave of energy efficiency announcements. This time it arrived in its definitive form, the rebound effect (RE), which remains the most common description of the paradoxical connection between energy efficiency and energy consumption. According to a widely used definition, the rebound effect is a 'behavioural or systemic response' to a new, more efficient technology, which can lead to partially or totally offsetting expected savings [3]. Despite it still being a controversial topic, often engendering exorbitant reactions, the rebound effect has attracted considerable scientific interest, as demonstrated by the volume of articles published in international journals every year, which increased more than six times in the last decade. Between 1998 and 2008, less than 10 articles per year addressing or mentioning the rebound (in abstract or text) could be found on Scopus, whereas, in the last five years, the number went from 50 to 60 articles, many of the them in the field of transport research.

The transport sector has always been a favoured subject for investigating the rebound effect. There are several reasons for this. It epitomizes the transformational power of energy efficiency more than any other field. It is often blessed with abundant data and offers a real world subject for conceptualizing RE [4]. This is why a car is very often used as a tangible example of the rebound effect. The thrill you experience if you buy a more efficient vehicle means you are likely to drive for leisure more often and choose driving a car in place of public transport. It also enables you to commute longer distances and choose a more distant workplace. The problem presented here is an emblematic case of behavioural response, but what is the analogous problem for freight transport? There is probably none to be found in cargo or railway freight, where behavioural changes are difficult to envisage, and few in the case of road freight. More efficient trucks could, feasibly, encourage more aggressive driving behaviour, but this is arguably only a secondary factor in road freight transport energy consumption. Rather than a behaviour response, energy efficiency has always had seen a more profound response in transport costs and what those are inimitably intertwined with: the spatial and time distribution of the value chain, both up-stream (supply chain) and down-stream (distribution and retail). Market integration of factors and goods is just one of the positive, unintended consequences of more efficient transportation that has always had an arguably deep impact on energy demand [5]. Indeed, in this field, the balance between behavioural and systemic responses seems to lean towards the latter and makes the analysis of the rebound effect far more complex and with much wider implications. The transport-intensive economy which arose after Fordism was initially supported by the road transport sector (in the words of Baldwin [6] it was the first 'unbundling') and later by the cargo and aviation sector (the second 'unbundling')—it is not possible to disentangle the evolution of global value chains from that of transport costs whose energy efficiency is a non-marginal component [7]. In this article, after briefly reviewing the literature about the rebound effect in freight transport (method and findings), what could be considered the prism of almost all current models on RE will be considered—the energy service to energy efficiency elasticity—to show how a rebound effect metric based on elasticity can be very misleading in the long term when there are structural effects in the system. To do this, it will be drawn on the experience of gravity theory in trade economics, where the inability of models to capture declining transport costs sparked a long-standing debate (the 'distance puzzle'). A completely different approach will be proposed based on (complex) network theory and statistical network mechanics to model the rebound effect in freight transport, followed by contemplation of how and where future research should proceed.

2. Rebound Effect in Freight Transport: The Role of Elasticity

Although the number of studies on private mobility is much higher, the increasing interest over the past decade on the rebound effect in freight transport (mostly road freight) has partially closed the gap. Table 1 offers a snapshot of the literature. The work of Walnum et al. [8] provides a more complete and in-depth review of the subject and, despite being current only to 2014, frames the issue clearly in the context of rebound effect studies and sustainability in transportation. The percentage shown in the last column of Table 1 indicates how much of the expected savings rebounds within a specific time-span. A 100% rebound value means that all expected savings are absorbed by increased demand. A value greater than 100% indicates that consumption grew higher than before the new technology was introduced. Table 1 reports authors, case study (country and time span) and the metric used. With a few exceptions, most are econometric regressions [9,10]. The variables used to assess the rebound effect are: energy service (tonne-kilometres or vehicle-kilometres) or energy consumption (diesel) versus energy efficiency (generally fuel economy, per vehicle or tonne-kilometres) or energy prices (taken as a proxy of energy efficiency). When possible, regressions are performed on all the variables above, plus some additional explanatory variables (such as economic output, population, load factors, etc).

Table 1. Rebound effect (RE) in road freight transport, estimates review: the range of values refer to different regions, countries or type of vehicles; values in parenthesis refer to short-term estimates.

Study	Case Study	Metric	Estimate
Anson and Turner [9]	Scotland, 1999	Service, efficiency	36–39%
Matos and Silva [11]	Portugal, 1987–2006	Service, efficiency	24%
Borger and Mulalic [12]	Denmark, 1980–2007	Service, efficiency	17% (10%)
Wang and Lu [13]	China, 1999–2011	Service, price	52–84 %
Leard et al. [14]	USA, 1977–2002	Service, price	9–30%
Winebrake et al. [15,16]	USA, 1997–2006	Service, price and efficiency	0%
Llorca and Jamasb [10]	EU(15), 1992–2012	Energy, price	0–68% *
Sorrell and Stapleton [17]	UK, 1970–2014	Service, efficiency and price	21–137%

(*) indicates a 0–100% RE range imposed by the model.

The most accepted measure [18] of the rebound effect is the elasticity of demand for energy services S with respect to energy efficiency ϵ. However, when (physical) efficiency data is lacking, the elasticity of energy services with respect to energy price p, namely $\eta_\epsilon(S)$ and $\eta_p(S)$, is often used. When there is no data on energy services, the following relation holds between the elasticity of energy consumption E and energy services S over efficiency (price): $\eta_{\epsilon(p)}(E) = \eta_{\epsilon(p)}(S) - 1$ [18]. Regardless of the approach used, elasticity has become a popular metric to assess the rebound effect, though there are some exceptions (for example, the model by [10]). In most cases (including beyond the domain of transportation), it acts as a prism by which we look at the paradoxical phenomenon of greater energy efficiency leading to higher energy consumption.

The present article does not discuss the limitations of other approaches, nor provide a comprehensive survey of those few that do not use elasticity as a metric for the rebound effect (several studies have used Life Cycle Assesment(LCA), a good review is given by Font Vivanco [19] and there is also the work of Freeman et al. who estimate RE using system dynamics modelling [20]. Others have used decomposition analysis, which is ultimately still based on elasticity). Instead, this article aims to show how elasticity has ultimately shaped our perception of the rebound effect. Indeed, elasticity has taken a paradox (a thing existing in the domain of words) and turned it into a number (a thing we can measure and count). The reasons for its success are not only empirical but also axiomatic. The concept of elasticity combines well with the rebound effect because both indicate a state of quietness or balance, which is suggestive of a particular level of energy savings without RE. This is an important analogy because, once we have decided to measure the rebound effect, we must first be able to set the expected consumption level without the rebound effect. Elasticity provides us with a suitable solution for this problem with the value 0 (perfectly inelastic). This level need only be an estimate. The elasticity concept presents the idea that RE is inherent to the system, like a durable property, in the same way certain materials are more or less flexible. This is consistent with the idea that elasticity is immanent and persistent. Nearly all rebound effect studies assume that this does not change over time. In some cases, the estimations given for the rebound effect span three decades or more, but can we reasonably assume that the 'behavioural or systemic response' of the system remains the same over such a long period of time? Are behaviours today the same as they were in the 1970s? Is the transport system the same? Does it use the same means and infrastructure? The next chapter will show how similar questions have puzzled trade economics scholars for years and specifically in terms of elasticity and longer timescales.

3. The 'Distance Puzzle'

In 1954, using a remarkably simple and elegant idea, the economist Walter Isard proposed a model emulating gravity theory to predict trade flows (Some give credit to Pöyhönen [21] and others to Tinbergen [22] for the introduction of gravity models in economics, but this dispute is beyond the scope of the present article). The concept follows that the amount of trade between two countries should be directly proportional to a measure of size (typically gross domestic product (GDP)) and inversely proportional to distance [23]. The simplest formulation of this relationship is:

$$w_{ij} = G \frac{x_i x_j}{d_{ij}^{\gamma}}, \tag{1}$$

where w_{ij} is the flow between country i and country j, G is an empirical constant (the 'gravitational constant'), x is the GDP and γ is the power factor of distance d (a quadratic law, for mass–mass interaction).

Gravity models (GM) are a whole family of spatial interaction models rather than a single model. Wilson's pioneering work provides a clear categorization of gravity models according to parameter specifications: production, attraction and attraction–production constrained models. In the first case, the model is parameterized according to observed flows, in the second to the observed mass term and in the last case the model is constrained to both. Remarkably, Wilson also proposes a maximizing entrop gravity model where the general function is replaced by an exponential function and the resulting value indicates the expected flow given the observed constraints. The functional form, obtained by maximizing the log (likelihood), is the same as the exponential random graph models, but in the case proposed here, expected values are intended as a null model rather than predictions of future flows. Furthermore, Wilson's approach is dyadic (bilateral) and does not consider the topology of the whole transport system as we do in our previous works. Nevertheless, the similarity is striking and Wilson's intuition remarkable [24]. Gravity models proved to be good predictors of trade volumes but exhibited major shortcomings [25]. As a result, they have seen a long, uneven success and found as many advocates as detractors prompting as much research to improve them as to overcome them. For example, a limitation is their inability to predict zero-flows in a world where not all countries are linked by trading relationships (i.e., the network of trade is not fully connected). A second major flaw is that GM predict perfectly symmetrical flows between each pair of countries: this is also empirically untrue [26]. Mutual relationships are indeed relevant. However, bilateral trade is never perfectly balanced, nor is the trade matrix perfectly symmetrical [27]. Both findings are not surprising given that gravity force presents a rotational symmetry, i.e., particles along the same radius experience the same force. In other words, in Newton's third law (action–reaction), two particles at a given distance experience the same force (obviously, the same relationship does not apply to the volume of trade between two countries). In essence, the problem is that GM assume trade occurs in a homogeneous space, but, in reality, trade occurs in an oriented, structured space [28].

Nevertheless, scholars have been mostly bewildered by the fact that empirical estimation of 'distance coefficients' (measuring the effect of distance on trade) were practically constant over time [29]. In a widely cited book, Frances Cairncross [30], senior fellow at UCLA (University of California, Los Angeles), proclaimed the 'death of distances' as a result of advancements in communication and transport technologies. However, as remarked by Lin and Sim, 'while the death of distance seems sensible in light of globalization, the task of establishing this empirically has proven to be challenging' [31]. This 'task' has triggered a vast and varied scientific endeavour to find declining distance coefficients in time, mostly in the field of economics whose reach does not extend to this article (nonetheless, it is worth recalling three works which notably address the 'distance puzzle' by modifying its most crucial hypothesis: the homogeneity of space. The first article describes a model to study the spatial homogeneity in trade by assessing the autocorrelation in trade volumes with Hurst exponent [28]. The second introduced fractality to interpret the distance coefficients [32] and the third introduced topological constraints to a radiation model [33]). Hence, where trade economists were desperately searching for a pattern of change, scholars interested in RE were looking for immutability. The goal of these was to show that elasticity between trade and distance was changing over time, whereas the others aimed to measure constant elasticity (of transport service to energy efficiency). Both were motivated by an unspoken vision of the system which in one case struggled to correspond with the paradigm expressed by model and metric, and in the other case aligned perfectly with the model. This was a model whose idea of hypostasis was reassuring or, in other words, had an underlying substance: a fundamental, immutable reality behind it. The concept of elasticity in economics resembles that in materials sciences where the relationship between stress δ and strain ϵ (or deformation) is proportional to a constant E:

$$\delta = E\epsilon \tag{2}$$

Deformations (ϵ) generated by external forces (δ) can vary according to boundary conditions (i.e., temperature), but are essentially pre-determined by some fundamental structural properties of the molecules' matrix measured by E. In solid-state physics, the molecular lattice changes size and shape when forces are applied (which means that the energy of a system increases) and when these cease the lattice returns to its original lower energy state. As for the 'lower energy state', in the RE model, this is analogous with the sought-after 'state of normality' which is essential to establish the hypothetical energy consumption in which energy efficiency has expressed all its potential. However, the structure of interactions in social and economic systems is more like a complex network than a regular matrix and trade (or freight transport) is not an exception [34]. This is not a trivial question because the dynamics of processes (such as shock propagation or epidemic phenomena) on complex networks can be remarkably different from that on a lattice or lattice-like systems [35]. The space of interactions in a complex network is not homogeneous and the incumbent structure (topology) of agents' interactions shape the speed, frequency and intensity [28,32]. Complex network theory proved to be a highly informative approach to understand trade, its connection with economy [36] and how this evolved over time [37]. Can network theory help investigate RE in freight transport using an approach which is evolutionary and takes into account the complexity of the system? In the next section, an attempt recently taken in this direction [38] will be illustrated.

4. Modelling Rebound Effect with Network Theory

Network theory primarily focuses on the study of interactions, in contrast to most scientific approaches which are concerned with the study of their constituents. It is also a holistic approach because it studies the structure emerging from the entire organization of nodes. Specifically, network theory focuses on how the combination of many local interactions, generally formed following a decentralized and non-engineered process, can give rise to unexpected structures at a global level. In turn, these non-designed global (macroscopic) properties can affect the properties of individual nodes or edges and, consequently, the local (microscopic) structure. Interactions between intermediate (mesoscopic) levels are also possible, for instance when the entities under study are so-called communities, i.e., sets of vertices more densely connected internally than with the rest of the network.

In recent years, the use of complex network theory in the field of transport studies has gained momentum mostly, but not exclusively, in air transport [39]. Two articles that provide an extensive and detailed review of the subject are [40,41]. A recent attempt by Calatayud et al. used a network (and multi-network) approach to study global trade as a transport network, though it must be said that this study lacks analysis of any real spatial input (that is, its spatial embedding) as it only concerns topological properties [42]. Network theory has also gained the attention of supply chain analysis scholars as a way of overcoming the established linear and static view of production processes using 'complex adaptive systems' [43]. Similar to the idea presented here, but at a lower scale of analysis, supply chains are approached as evolving networks exchanging 'materials and information' [44]. Few studies, however, take a strictly quantitative approach to describe the structure and properties of such networks [45]. Interestingly, according to a wide and accurate review dated 2013, none of the 126 articles assessed consider supply chain networks as spatial networks, where spatial effects (or transports) would be a factor shaping their topology [46]. Even when logistics are considered in the analysis, they appear to be adaptive to the network structure rather than the opposite. The spatial fix of the network's nodes is given.

Needless to say, to establish the connection between energy efficiency and a transport network, the spatial embedding (also known as geographical or Euclidean embedding) of the network and the spatial information it provides is of paramount importance. It is worth noting that a network is an abstract object described by a matrix whose entries are specified by the existence (or the magnitude, in the case of weighted rather than binary graphs) of a link between two elements of a set. Mathematically, the embedding of the network in space is irrelevant. In principle, different spatial configurations are just different arbitrary representations of the same graph. However, if we are looking for explanations of the structure of a particular network or predictions of its future evolution, then it might well turn out that the spatial information is relevant. Networks that are subject to strong spatial constraints generally display a high degree of regularity and so are more predictable (depending on the spatial information) than other graphs. For example, in a lattice, each atom has the same number of connections (degree). Conversely, complex networks show a much more complex architecture with a degree distribution following a (non-regular) power law. In a complex transport network, the transport system embodies the spatial information concerning the topological configuration which we expect to be affected by the efficiency of transport. In conclusion, we have on the one hand the topological information delivered by the structure of the complex network (for example, the trade network), and on the other hand the spatial information delivered by the transport system (i.e., the spatial network), and we want to combine the two to obtain information about the mutual influence.

Following previous work on spatial networks [47], it will be now shown how to gauge spatial effects on complex networks aimed at developing a spatial embeddin metric of a network upon which, as will become clearer later, it is possible to estimate RE. The trade network of 27 European countries (EU27) will be used as a case study. The two dimensions of the analysis, topological and spatial, are condensed into two matrices, the matrix **A** of trade relationships and the matrix **D** of distances (Figure 1).

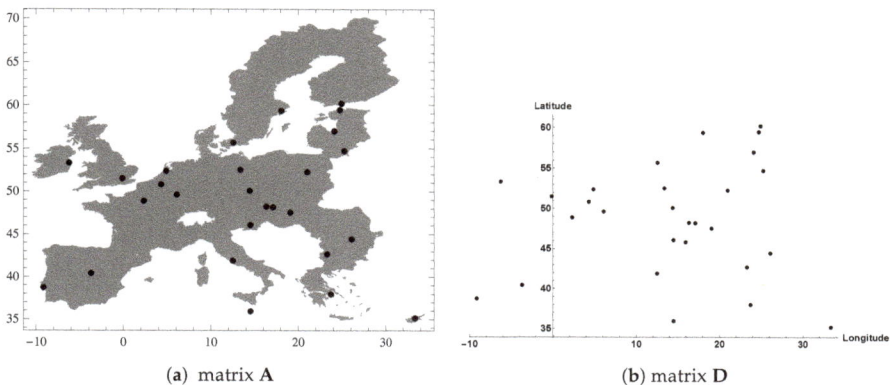

(a) matrix **A**   (b) matrix **D**

Figure 1. The cross-border transport network of Europe condensed into its (**a**) topological dimensions (bilateral trading relationships, whose entries are trade flows in mass units (tonne)) and (**b**) spatial dimensions (binary distances between capitals, whose entries are geodesic distances measured on longitudes and latitudes (deg)).

The simplest, linear choice for a measure that exploits the entire topological and spatial information provided by the network is:

$$F \equiv \sum_{i=1}^{n} \sum_{j \neq i} a_{ij} d_{ij}, \tag{3}$$

where a and d are the entries of the binary matrix **A** (a matrix with entries 0 or 1 if a link exists between tow nodes) and **D** respectively and n is the number of nodes of the network (27 in our case). The second step is normalization because the value F expressed in Equation (3) is not a suitable measure for comparing different networks or the evolution of a single network in time as it is size-dependent. Networks can grow in terms of number of links (edges) or nodes (vertexes) and in the latter case also the embedding space. Therefore, we consider a slightly more complicated definition for the single purpose of having a normalized quantity ranging between 0 and 1:

$$f \equiv \frac{\sum_i \sum_{j \neq i} a_{ij} d_{ij} - F_{min}}{F_{max} - F_{min}}, \tag{4}$$

where F_{min} and F_{max} are the minimum and maximum values that F can take, given the distance matrix **D** and the total number of links L. The vector $V^{\uparrow} = (d_1^{\uparrow}, \ldots, d_n^{\uparrow}, \ldots, d_{N(N-1)}^{\uparrow})$ denotes the list of all (off-diagonal) distances ordered from the smallest to the largest where $d_n^{\uparrow} \leq d_{n+1}^{\uparrow}$ and symmetrically the vector $V^{\downarrow} = (d_1^{\downarrow}, \ldots, d_n^{\downarrow}, \ldots, d_{N(N-1)}^{\downarrow})$, considers from the largest to the smallest distances, where $d_n^{\downarrow} \geq d_{n+1}^{\downarrow}$. Hence, in terms of the two lists $V^{\uparrow}$ and $V^{\downarrow}$, the maximum and minimum values for F read $F_{min} = \sum_{n=1}^{L} d_n^{\uparrow}$ and $F_{max} = \sum_{n=1}^{L} d_n^{\downarrow}$. It is worth noting that the matrix A is non-relevant to the computation of $F_{min(max)}$ as it only depends on the existing number of links L and it is not influenced by the observed topology. The value ranges between 0 and 1. The former extreme ($f = 0$) represents the case where the L links are placed among the closest pairs of nodes (the maximum shrunk network). The latter extreme ($f = 1$) instead represents the case where the L links are placed among the most distant pairs of nodes (the maximum stretched network). In our EU27 trading network example, these two extremes are shown in Figure 2a,b respectively, where, for visualization purposes, we have actually chosen a value of L equal to $n = 27$, much less than the real value (which would fill the plot with links). This would have been the case, for instance, if in the original network each node had exactly one outgoing link. Networks between the two extremes would have a value $0 < f < 1$. As rendered intuitively by the figures, a larger value of f implies a more pronounced filling of the available space. Therefore, we denote f as the (spatial) filling of the network represented by the matrix **A**.

So far, we have considered the binary network, but an extension of f to the weighted network, which is more suitable for assessing transport flows, reads as follows:

$$f_w \equiv \frac{\sum_i \sum_{j \neq i} w_{ij} d_{ij} - F_{min}}{F_{max} - F_{min}}, \tag{5}$$

where w is the entry of the weighted matrix and F_{min} and F_{max} are chosen as the two extreme values that F can take in a network with the same total weight W ($\sum_i \sum_{j \neq i} w_{ij}$) as the original network, i.e., $F_{min} = W d_1^{\uparrow}$ and $F_{max} = W d_1^{\downarrow}$ ($d_1^{\uparrow}$ and $d_1^{\downarrow}$ are the smallest and largest distance between vertices, respectively). The concept here is that from the binary to the weighted representation, unitary links can cumulate between dyads, forming a (weighted) flow as a discrete summation of unitary links (similarly, particles that follow the Bose–Einstein distribution can occupy the same energy level). Therefore, the maximally shrunk(expanded) configuration is when all links cumulate between the closets(farthest) couple of nodes, that is, when the total weight is placed on the shortest(longest) distance. A different normalization could be done preserving the local weight distribution. In this case, the $F_{min(max)}$ would result as the summation of the total strength of the node i times the distance with the closest (farthest) j, for al the i nodes. This would change the weighting factor, but leave the trend unaffected, which is the focus of the present analysis. This measure, like its binary equivalent, will vary between 0 and 1 (though the visual representation is more complex here) and informs us about the spatial embedding of the network: the lower the value, the higher the embedding, that is, distances are more important in determining the network's interactions (topology). Can we use the weighted filling (Equation (5)) to assess the rebound effect? This measure indeed gives us an indication of the role of distance in shaping the network topology, but we know that topology is never entirely, if only ever marginally, explained by its spatial constraints [48]. We expect, for example, the Netherlands to trade abundantly with Germany not just because of their geographic proximity, but because Germany is a hub in the European trading network. Therefore, to carve out the purely spatial information found in trading relationships, we need to disentangle spatial and non-spatial (topological effects) using a null model based on exponential random graph theory and maximum entropy graph ensembles [49]. A maximum entropy graph ensemble is, in essence, a sample of randomized networks that conserve some desired properties similar to the network under investigation, say the in and out degree sequence (or in and out strength sequence for weighted networks). The functional form of maximum entropy ensemble can be solved analytically or performed by means of an iterative process. In this ensemble, we can assess the expected (average) value of the investigated measure (for example, the filling) in order to gauge if this is trivially explained by the constraints we imposed on the ensemble (say, the sequence of imports and exports observed in the network). We are thus able to implement the expected filling value of the spatial filling and control for non-spatial effects embodied in a null model via the filtered filling ϕ:

$$\phi \equiv \frac{f - \langle f \rangle}{1 - \langle f \rangle}, \tag{6}$$

where $\langle f \rangle$ is now obtained by replacing w_{ij} with $\langle w_{ij} \rangle$ (the flows generated by the null model) in Equation (5). Positive (negative) values of ϕ indicate a network which is more stretched (shrunk) than expected by the null model (hence, by imposing some topological constraints). Notably, for a space-independent network, $\phi = 0$ sets the level of spatial neutrality, that is, the level at which the spatial embedding of the network is fully (and trivially) explained by its topological properties. The network is therefore indifferent to distance, meaning that it is neither positively nor negatively affected by distance. Arguably, a more efficient technology will reduce the weight of distance in trade relationships and drive the system towards spatial neutrality. In Figure 3, we show how the decade-long trend of filtered spatial filling ϕ of Europe compared to two measures of energy efficiency [38]. Amid a fast increase in energy efficiency, as marked by both the energy intensity of freight transports (toe/tkm) and fuel economy of trucks (L/km), the filtered filling becomes more negative, indicating that the system increased its spatial embedding (given its topology). In other words, distances have become more binding, even if only marginally (a variation of the order of 10^{-2}). Overall, the value of ϕ of Europe shows us that the European transport network is very close to spatial neutrality and thus is not significantly affected by efficiency variations and we should not see a rebound effect. This is peculiar to Europe. Other spatial networks generally show a marked negative value, decreasing with time (showing that distances become less binding). For example, the world as a whole displays a negative ϕ of -0.2 [38]. This result is in line with that obtained by [10] who also found a close-to-zero rebound effect for Europe during the same period. Nevertheless, broadening the time scope of the analysis suggests that this result might only be true for Europe in the last few decades.

Figure 3b shows the long-term trend of spatial filling for Europe measured in monetary units instead of mass units. The long-term trend of the filling value shows that the European trade network has undergone two marked phases: an initial phase of spatial contraction which ended in the mid-1970s, and a following phase of spatial expansion which ended in the mid-1990s. It is plausible that, after this second wave of spatial expansion, Europe entered a stable, spatially-neutral phase as portrayed by the analysis of the last decades (Figure 3a). Consequently, RE was more marked and effective between the 1970s and the 1990s than in the last two decades. However, this is just a tentative analysis. More research is needed, supported by reliable and fine-grained data on mass flows and efficiency in freight transports.

(a)

(b)

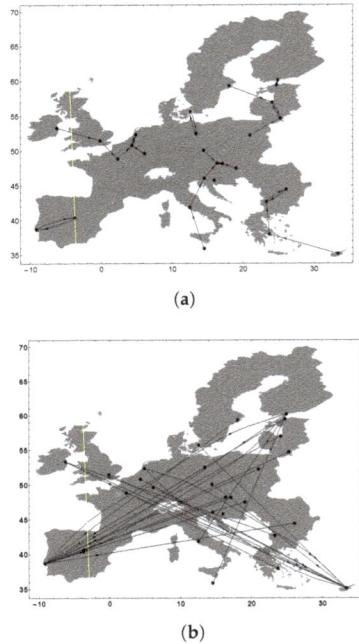

Figure 2. The two extreme degrees of spatial embeddedness (filling) for the binary trading network of Europe. (**a**) Maximum shrunk configuration for the EU27 trading network with a constraint on the number of links: $f = 0$ ($N = 27, L = 27$); (**b**) Maximum stretched configuration for the EU27 trading network with a constraint on the number of links: $f = 1$ ($N = 27, L = 27$).

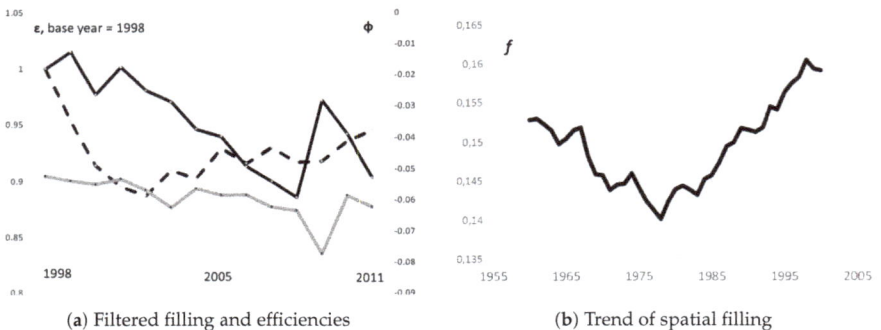

(a) Filtered filling and efficiencies

(b) Trend of spatial filling

Figure 3. Trends in European (EU27) spatial embedding: (**a**) Filtered filling ϕ_w in (gray line); fuel economy (dashed black line) and energy intensity (black line). Data source: CEPII www.cepii.fr/CEPII/en/bdd_modele/bdd.asp and Odysee http://www.indicators.odyssee-mure.eu/energy-efficiency-database.html. (**b**) Unfiltered filling f_w (Equation (4)) assessed on trade relationships in monetary units (black line). Data source: Gleditsch http://journals.sagepub.com/doi/abs/10.1177/0022002702046005006.

Except in the last two decades where results are difficult to interpret, the model incontrovertibly shows that, after the late 1970s, spatial embedding of the European trade network fell dramatically (and the network expanded). The oil shocks triggered a drive for higher efficiency across all sectors of the economy but foremost the automotive sector [50]. As a result, road freight transport became the most economical mode of transport for goods and trucks became the conduit for the onset of the new transport intensive, post-Fordian economy. Between 1970 and 1995, the tKm of the road freight transport sector in Europe (EU15) grew more than 130% compared to a growth rate of GDP and industrial production of 53% and 56%, respectively. In Italy, between 1973 and 1990, the traffic density of goods vehicles grew 132% (that of semi-trailers alone accounted for a whopping 172%) as opposed to a growth rate of 25% for all types of road vehicles [5]. The environmental burden of the transition towards a transport intensive economy (a transition that has never reversed, but only attenuated or transferred to greater spatial scales) is twofold: local air quality deteriorated and global greenhouse gas emissions rose. Measures to counteract this burden are doomed by the rise of the diesel engine, whose unquestionable dominance appears now to be unchallenged. If energy efficiency triggered this transition, can it continue to be considered a potential solution or obstacle for achieving more sustainable transport?

5. Conclusions

Can rebound effects explain why sustainable mobility has not been achieved? This was the difficult question Walnum, Aall and Løkke addressed in an article published in this journal four years ago [8]. Far from giving a definitive answer, the authors came to the conclusion that under 'certain circumstances', energy efficiency improvements could lead to an overall increase in transport volume and completely offset energy savings, concluding with a measure of resignation that the rebound effect, 'will be evident as long as the economy keeps growing'. If RE is a serious threat to any energy-efficiency based strategy aiming for a more sustainable freight transport system, understanding its phenomenology is of paramount importance. Understanding the phenomenology means being able to decipher its true, complex nature and ideally its causes with a clearer insight into the scope, timescale and all its tangled interactions. Reducing RE to a number might be of practical use for gauging, for example, short-term, counteractive measures or estimating immediate response to a new, more efficient process, but, unfortunately, it does not add much to our fundamental comprehension. On the contrary, it may persuade us that the true nature of RE is a number. A number, and even more so an unequivocal one, can be reassuring, but it can also be soporific, especially when it is applied as a verdict or definitive response to a complex subject. With very few exceptions, elasticity is the prism through which we currently look at RE, regardless of the model used. In this article, an attempt was made to define elasticity as a metric for RE. Epistemological scrutiny showed that elasticity shapes our perception of RE by eliciting the idea that this is an inherent and permanent phenomenon rather than an evolutionary process. While studies on RE aim to measure the constant elasticity of energy service to energy efficiency, trade economics scholars are puzzled by the unchanged elasticity of trade to distance. Sometimes, we are misled by our expectations, even more so if the model we use is reassuring. Studies on RE should explore new avenues of research and join with different disciplines to gain new, different perspectives on the topic (For further reading on new approaches and models on RE: https://www.frontiersin.org/research-topics/6598/the-rebound-effect-and-the-jevons-paradox-beyond-the-conventional-wisdom). In this article, network theory was proposed as a new method for approaching the rebound effect in transport and the use of spatial (filtered) filling as an alternative metric to elasticity. Structural change and complexity are endemic to the network theory paradigm and the large amount of data the network provides annually (702 data points for the EU27, which is one order of magnitude greater than the annual data used for regression) allows us to perform sound statistical analysis every year and observe the evolution of the system as time elapses. Therefore, by inspecting the topology of interactions and their relation to spatial constraints (distances), a complexity and evolutionary perspective of the system is achieved. With spatial filling, we can observe spatial embedding of the network (contraction or expansion) and with filtered filling we can set the level of spatial neutrality (indifference) and the extent to which the observed network extends. By comparing these two measures with the evolution of energy efficiency, we have an indication of how the rebound effect unfolds over time, that is, how the spatial embedding responds to increased transport efficiency. Lastly, a question which should be addressed is whether a network approach to RE can be extended to sectors other than transportation. Our approach takes advantage of the concept of spatial embedding [38] and studies the interplay between the adjacency matrix and the distance matrix (Figure 1). A straightforward extension of this conceptualization to non-metric distances (non-spatial networks), although appealing, is unconvincing. It is also difficult to apply this framework to private mobility where distance is only one factor in decision-making. However, with increasing attention and research on Jevons paradox, it is expected that more scholars and scientists will follow the path of complex network theory, leading to new models and metrics, and new insights and puzzles.

Funding: This research received no external funding.

Acknowledgments: I would like to thanks Mir Mohammadi Kooshknow for the insightful comments and attend revision of the paper and two unknown reviewers for helping me in discovering a new ambit of knowledge of interest and enhancing my, indeed poor, dissemination skill.

Conflicts of Interest: The author declares no conflict of interest.

Abbreviations

The following abbreviations are used in this manuscript:

RE	Rebound Effect
GDP	Gross Domestic Product
GM	Gravity Models
LCA	Life Cycle Assessment
EU27	European Union, 27 countries.
toe	Tonnes of Oil Equivalent
tKm	Tonnes Kilometres
η	elasticity
E	energy
S	energy service
ϵ	energy efficiency
p	energy prices

References

1. Madureira, N.L. The anxiety of abundance: William Stanley Jevons and coal scarcity in the nineteenth century. *Environ. Hist.* **2012**, *18*, 395–421. [CrossRef]
2. Schipper, L. On the rebound: The interaction of energy efficiency, energy use and economic activity. An introduction. *Energy Policy* **2000**, *28*, 351–353.
3. Maxwell, D.; Owen, P.; McAndrew, L.; Muehmel, K.; Neubauer, A. *Addressing the Rebound Effect, a Report for the European Commission DG Environment*; European Commission DG ENV: Bruxelles, Belgium, 2011.
4. Greening, L.A.; Greene, D.L.; Difiglio, C. Energy efficiency and consumption—The rebound effect—A survey. *Energy Policy* **2000**, *28*, 389–401. [CrossRef]
5. Ruzzenenti, F.; Basosi, R. The rebound effect: An evolutionary perspective. *Ecol. Econ.* **2008**, *67*, 526–537. [CrossRef]
6. Baldwin, R. *Trade and Industrialisation after Globalisation's 2nd Unbundling: How Building and Joining a Supply Chain Are Different and Why It Matters (No. W17716)*; National Bureau of Economic Research: Cambridge, MA, USA, 2011.
7. Picciolo, F.; Papandreou, A.; Hubacek, K.; Ruzzenenti, F. How crude oil prices shape the global division of labor. *Appl. Energy* **2017**, *189*, 753–761. [CrossRef]
8. Walnum, H.J.; Aall, C.; Løkke, S. Can rebound effects explain why sustainable mobility has not been achieved? *Sustainability* **2014**, *6*, 9510–9537. [CrossRef]
9. Anson, S.; Turner, K. Rebound and disinvestment effects in refined oil consumption and supply resulting from an increase in energy efficiency in the Scottish commercial transport sector. *Energy Policy* **2009**, *37*, 3608–3620. [CrossRef]

10. Llorca, M.; Jamasb, T. Energy efficiency and rebound effect in European road freight transport. *Transp. Res. Part A Policy Pract.* **2017**, *101*, 98–110. [CrossRef]

11. Matos, F.J.; Silva, F.J. The rebound effect on road freight transport: Empirical evidence from Portugal. *Energy Policy* **2011**, *39*, 2833–2841. [CrossRef]

12. De Borger, B.; Mulalic, I. The determinants of fuel use in the trucking industry—Volume, fleet characteristics and the rebound effect. *Transp. Policy* **2012**, *24*, 284–295. [CrossRef]

13. Wang, Z.; Lu, M. An empirical study of direct rebound effect for road freight transport in China. *Appl. Energy* **2014**, *133*, 274–281. [CrossRef]

14. Leard, B.; Linn, J.; McConnell, V.; Raich, W. *Fuel Costs, Economic Activity, and the Rebound Effect for Heavy-Duty Trucks*; Discussion Paper; Resources for the Future: Washington, DC, USA, 2015.

15. Winebrake, J.J.; Green, E.H.; Comer, B.; Corbett, J.J.; Froman, S. Estimating the direct rebound effect for on-road freight transportation. *Energy Policy* **2012**, *48*, 252–259. [CrossRef]

16. Winebrake, J.J.; Green, E.H.; Comer, B.; Li, C.; Froman, S.; Shelby, M. Fuel price elasticities for single-unit truck operations in the United States. *Transp. Res. Part D Transp. Environ.* **2015**, *38*, 178–187. [CrossRef]

17. Sorrell, S.; Stapleton, L. Rebound effects in UK road freight transport. *Transp. Res. Part D Transp. Environ.* **2018**, *63*, 156–174. [CrossRef]

18. Sorrell, S.; Dimitropoulos, J. The rebound effect: Microeconomic definitions, limitations and extensions. *Ecol. Econ.* **2008**, *65*, 636–649. [CrossRef]

19. Vivanco, D.F.; van der Voet, E. The rebound effect through industrial ecology's eyes: A review of LCA-based studies. *Int. J. Life Cycle Assess.* **2014**, *19*, 1933–1947. [CrossRef]

20. Freeman, R.; Yearworth, M.; Preist, C. Revisiting Jevons' paradox with system dynamics: Systemic causes and potential cures. *J. Ind. Ecol.* **2016**, *20*, 341–353. [CrossRef]

21. Pöyhönen, P. A tentative model for the volume of trade between countries. *Weltwirtschaftliches Arch.* **1963**, *90*, 93–100.

22. Tinbergen, J.; Hekscher, A. *Shaping the World Economy: Suggestions for an International Economic Policy*; Twentieth Century Fund: New York, NY, USA, 1962.

23. Walter, I. Location Theory and Trade Theory: Short-Run Analysis. *Q. J. Econ.* **1954**, *68*, 305–320. [CrossRef]

24. Wilson, A.G. A family of spatial interaction models, and associated developments. *Environ. Plan. A* **1971**, *3*, 1–32. [CrossRef]

25. Duenas, M.; Fagiolo, G. Modeling the international-trade network: A gravity approach. *J. Econ. Interact. Coord.* **2013**, *8*, 155–178. [CrossRef]

26. Squartini, T.; Garlaschelli, D. Jan Tinbergen's legacy for economic networks: From the gravity model to quantum statistics. In *Econophysics of Agent-Based Models*; Abergel, F., Aoyama, H., Chakrabarti, B., Chakraborti, A., Ghosh, A., Eds.; Springer: Chambridge, UK, 2013; pp. 161–186, ISBN 978-3-319-00022-0.

27. Squartini, T.; Picciolo, F.; Ruzzenenti, F.; Garlaschelli, D. Reciprocity of weighted networks. *Sci. Rep.* **2013**, *3*, 2729. [CrossRef] [PubMed]

28. Chiarucci, R.; Ruzzenenti, F.; Loffredo, M.I. Detecting spatial homogeneity in the world trade web with Detrended Fluctuation Analysis. *Phys. A Stat. Mech. Its Appl.* **2014**, *401*, 1–7. [CrossRef]

29. Buch, C.M.; Kleinert, J.; Toubal, F. The distance puzzle: On the interpretation of the distance coefficient in gravity equations. *Econ. Lett.* **2004**, *83*, 293–298. [CrossRef]

30. Cairncross, F. *The Death of Distance: How the Communications Revolution Will Change Our Lives*; Harvard Business School Press: Cambridge, MA, USA, 1997.

31. Lin, F. Are distance effects really a puzzle? *Econ. Model.* **2013**, *31*, 684–689. [CrossRef]

32. Karpiarz, M.; Fronczak, P.; Fronczak, A. International trade network: Fractal properties and globalization puzzle. *Phys. Rev. Lett.* **2014**, *113*, 248701. [CrossRef] [PubMed]

33. Masucci, A.P.; Serras, J.; Johansson, A.; Batty, M. Gravity versus radiation models: On the importance of scale and heterogeneity in commuting flows. *Phys. Rev. E* **2013**, *88*, 022812. [CrossRef] [PubMed]

34. Ruzzenenti, F.; Garlaschelli, D.; Basosi, R. Complex networks and symmetry II: Reciprocity and evolution of world trade. *Symmetry* **2010**, *2*, 1710–1744. [CrossRef]

35. Haldane, A.G.; May, R.M. Systemic risk in banking ecosystems. *Nature* **2011**, *469*, 351–355. [CrossRef] [PubMed]

36. Garlaschelli, D.; Loffredo, M.I. Fitness-dependent topological properties of the world trade web. *Phys. Rev. Lett.* **2004**, *93*, 188701. [CrossRef] [PubMed]

37. Fagiolo, G.; Reyes, J.; Schiavo, S. The evolution of the world trade web: A weighted-network analysis. *J. Evolut. Econ.* **2010**, *20*, 479–514. [CrossRef]
38. Ruzzenenti, F.; Basosi, R. Modelling the rebound effect with network theory: An insight into the European freight transport sector. *Energy* **2017**, *118*, 272–283. [CrossRef]
39. Wang, J.; Mo, H.; Wang, F.; Jin, F. Exploring the network structure and nodal centrality of China's air transport network: A complex network approach. *J. Transp. Geogr.* **2011**, *19*, 712–721. [CrossRef]
40. Ducruet, C.; Beauguitte, L. Spatial science and network science: Review and outcomes of a complex relationship. *Netw. Spat. Econ.* **2014**, *14*, 297–316. [CrossRef]
41. Rocha, L.E. Dynamics of air transport networks: A review from a complex systems perspective. *Chin. J. Aeronaut.* **2017**, *30*, 469–478. [CrossRef]
42. Calatayud, A.; Mangan, J.; Palacin, R. Connectivity to international markets: A multi-layered network approach. *J. Transp. Geogr.* **2017**, *61*, 61–71. [CrossRef]
43. Surana, A.; Kumara, S.; Greaves, M.; Raghavan, U.N. Supply-chain networks: A complex adaptive systems perspective. *Int. J. Prod. Res.* **2005**, *43*, 4235–4265. [CrossRef]
44. Pathak, S.D.; Day, J.M.; Nair, A.; Sawaya, W.J.; Kristal, M.M. Complexity and adaptivity in supply networks: Building supply network theory using a complex adaptive systems perspective. *Decis. Sci.* **2013**, *38*, 547–580. [CrossRef]
45. Hearnshaw, E.J.; Wilson, M.M. A complex network approach to supply chain network theory. *Int. J. Oper. Prod. Manag.* **2013**, *33*, 442–469. [CrossRef]
46. Bellamy, M.A.; Basole, R.C. Network analysis of supply chain systems: A systematic review and future research. *Syst. Eng.* **2013**, *16*, 235–249. [CrossRef]
47. Ruzzenenti, F.; Picciolo, F.; Basosi, R.; Garlaschelli, D. Spatial effects in real networks: Measures, null models, and applications. *Phys. Rev. E* **2012**, *86*, 066110. [CrossRef] [PubMed]
48. Squartini, T.; Picciolo, F.; Ruzzenenti, F.; Basosi, R.; Garlaschelli, D. Disentangling spatial and non-spatial effects in real networks. In *Complex Networks and their Applications*; Cherifi, H., Ed.; Cambridge Scholars Publishing: Newcastle upon Tyne, UK, 2014; pp. 1–28, ISBN 978-1-4438-5370-5.
49. Squartini, T.; Garlaschelli, D. Analytical maximum-likelihood method to detect patterns in real networks. *New J. Phys.* **2011**, *13*, 083001. [CrossRef]
50. Ruzzenenti, F.; Basosi, R. Evaluation of the energy efficiency evolution in the European road freight transport sector. *Energy Policy* **2009**, *37*, 4079–4085. [CrossRef]

Article

Greenhouse Gas Emissions from Intra-National Freight Transport: Measurement and Scenarios for Greater Sustainability in Spain

Carlos Llano [1,3,*], Santiago Pérez-Balsalobre [2,3] and Julian Pérez-García [2,3]

1   Departamento de Análisis Económico, Teoría Económica e Historia Económica, Facultad de Ciencias
    Económicas y Empresariales, Universidad Autónoma de Madrid, Campus Cantoblanco,
    28049 Madrid, Spain
2   Departamento de Economía Aplicada, Facultad de Ciencias Económicas y Empresariales,
    Universidad Autónoma de Madrid, Campus Cantoblanco, 28049 Madrid, Spain;
    santiagoj.perez@uam.es (S.P.-B.); julian.perez@uam.es (J.P.-G.)
3   L. R. Klein Institute, Facultad de Ciencias Económicas y Empresariales, Universidad Autónoma de Madrid,
    Campus Cantoblanco, 28049 Madrid, Spain
*   Correspondence: carlos.llano@uam.es; Tel.: +34-914-972-910

Received: 11 May 2018; Accepted: 5 July 2018; Published: 13 July 2018

Abstract: Greenhouse Gas (GHG) emissions is a topic of major concern worldwide. Following previous articles which provide a methodology for estimating GHG emissions associated with international trade by transport mode at the world level, in this paper, we estimate an equivalent database of GHG emissions for inter-regional trade flows within a country (Spain). To this end, we built a new database of GHG emissions for origin–destination flows between Spanish provinces during 1995–2015. For each year, we combine industry-specific flows by four transport modes (road, train, ship and aircraft) with the corresponding GHG emissions factor for each mode in tons*km, drawn from the specialized literature. With this dataset of GHG emissions, we generate and analyze the temporal, sectoral and spatial pattern of Spanish inter-regional GHG flows. We then forecast emissions for 2016–2030 and consider how transport mode shifts might produce a more sustainable freight system within the country through the substitution of environmentally friendly alternatives (railway) for specific origin–destination–product flows in high-polluting modes (road).

Keywords: greenhouse gas emissions; national freight transport emissions; interregional trade by transport mode; modal shift

---

1. Introduction

In December 2015, at the Paris climate conference (COP21), 195 countries adopted the first universal, legally binding global climate deal, with the aim of keeping global warming below 2 °C. All signatories were to turn their commitments into concrete policy actions after COP21 and report periodically on their progress. The European Union (EU) was the first major economy to submit its intended contribution to the new agreement in March 2015, pledging an ambitious 40% reduction of greenhouse gas emissions by 2030 from 1990 levels [1,2]. This target was in line with the EU's previous "2030 climate and energy framework" [3] and with the European Commission's "White Paper of Transport" from 2011 [4].

The EU's 40% reduction of GHG emissions by 2030 has two parts: On the one hand, sectors covered by the emissions trading system (ETS) will have to lower emissions by 43% from 2005 levels. Sectors outside the ETS, which include the *"transport sector"*, will need to reduce them by 30% from 2005 levels. For these sectors, the Effort Sharing Decision (ESD) sets the maximum annual tonnage of GHG emissions for each EU member state based on its relative wealth (GDP per capita).

As suggested in several official documents [2,4,5], transport generates about a quarter of EU GHG emissions and is the second-most-polluting sector, after energy. However, while other sectors have seen their GHG emissions decrease, transport has seen them rise. Moreover, the transport modes with the sharpest increase in traffic have also had the largest increase in GHG emissions. From 1990 to 2012, international aviation, international shipping and road transport saw increases of 93%, 32% and 17%, respectively.

The EC's 2011 "White Paper of Transport" [4] put forward several non-binding longer-term targets for the transport sector, with an overall goal to cut transport GHG emissions by at least 60% by 2050 (with respect to 1990 levels). Some reductions have been achieved since 2008, and transport GHG emissions fell by 3.3% in 2012, with the biggest reduction in road (3.6%) and aviation (1.3%). However, in 2012, EU transport emissions remained 20.5% above 1990 levels and will need to fall 67% by 2050.

According to a recent report [5], heavy duty vehicles (HDVs) were responsible for around 30% of road transport emissions, that is, more than 5% of EU GHG emissions and around 10% of total non-ETS emissions. This implies that less than 5% of all vehicles on the road emit around 30% of road transport CO_2 emissions. Moreover, forecasts for this highly polluting mode are negative: HDV emissions are projected to rise 22% by 2030.

In light of these trends, the EC has adopted a new strategy to promote low-emissions mobility, for both passengers and freight [2], proposing several measures to curtail excessive use of the road mode. In parallel, all member states, including Spain, are increasing their efforts to meet the general commitment and hit each specific target. In all cases, accurate measurements of and follow-up on emissions are critical.

As Davydenko et al. suggested [6], if we are to see gains in transport efficiency, we will need to establish a certain basis of comparison between the different methods of calculating emissions. This basis should be set at both the national and the international level, and cover the full extent of the complex logistical chain, door to door. As these authors reported, there is to date "no single globally-recognized and accepted standard for the calculation of the carbon footprint that covers the entire freight transport supply chain". They used as their benchmark the EN-16258 methodology for the calculation of transport-service GHG emissions and laid out the criteria for an accurate methodology. Note that, in 2012, the European Committee for Standardisation (CEN) published European norm EN 16258 "Methodology for calculation and declaration of energy consumption and GHG emissions of transport services (freight and passengers)" (CEN standard EN 16258), which is the only official international—though European—standard aiming at the specific topic of transport supply chains. Davydenko et al. [6], also set forth three levels of aggregation at which transport-sector GHG emissions can be obtained: "micro", "meso" and "macro".

Keeping Davydenko et al. [6] and their categories in mind, we now turn to the Spanish case. At the "macro" level, the main official effort to compute GHG emissions is the Informative Inventory Report (IIR), produced by the Spanish National Inventory System (SEI) within the Ministry of Agriculture and Fishing, Food and Environment [7]. The 2018 IIR report was compiled in the context of the United Nations Economic Commission for Europe (UNECE) Convention on Long-Range Transboundary Air Pollution (CLRTAP), and contains detailed information on annual emissions estimates of air quality pollutants by source in Spain for the EMEP domain (excluding the Canary Islands) from 1990 onwards. According to the last IIR report, the "energy sector" generates more than 50% of the Inventory's emitted pollutants. Within the "energy sector", transport accounts for a large share of current emissions, with road transport being the worst offender. This subcategory encompasses pollutant emissions from vehicular traffic, including both passengers and freight.

The IIR's methodology is thorough, involving the use of hundreds of variables at the production and consumption levels. Transport emissions are estimated through a detailed process, mainly based on the national figures for energy use by transport mode. Although the number of statistics is large, the estimation essentially follows a *top-down approach*, where the specific origin–destination–product–mode for each flow is given scant attention. Moreover, the Spanish IIR does not offer a sub-national allocation of GHG emissions by sector, but just a raw top-down aggregate imputation with no detail for the "transport sector". Exact allocation of responsibility for polluting activities within the country is thus not possible [8]. This becomes critical when we consider sub-national entities within a highly decentralized country such as Spain, where not just the national government but also the regions (Nuts 2), provinces (Nuts 3) and municipalities (Nuts 5) are co-responsible for moderating GHG emissions. To this regard, it is interesting to consider, for example, how regions and cities are responsible for the development and control of transport infrastructures and services within urban areas, that is, the areas of densest congestion and pollution. Similarly, in Spain, Municipalities, Diputaciones Provinciales and Comunidades Autónomas all share with the national government various responsibilities with respect to follow-up on the quality of fresh water for human consumption. Similar arrangements exist for the management of waste, residuals, etc. Both COP21 and the "European Strategy for Low-Emission Mobility" [1,2] make a point of recognizing the role that sub-national entities (cities and regions) must play in any policy aiming to foster a more sustainable economy.

As Cristae et al. [9] noted, the situation within Spain is similar to the one observed for the international freight flows worldwide. While the International Transportation Forum uses a top-down approach to generate aggregate estimates of emissions from international transport, Cristea et al. suggested an alternative bottom-up procedure that more clearly allocates responsibility for pollution across countries and sectors. In fact, they highlighted the convenience of bottom-up approaches, where data on GHG emissions by mode are combined with data on traffic (tons*km) by mode, and detailed information is provided on the origin–destination and product type for each delivery.

In line with the methodology of Cristea et al. [9] for international freight flows and the recommendations of Davydenko et al. [6] for GHG estimation at the shipment level, this paper aims to estimate GHG emissions for intra- and inter-provincial freight flows within a country (Spain). GHG emissions for freight flows within a country are commonly estimated with input–output frameworks and CGE modeling [1,7,10–12]. However, with the exception of inter-regional input–output tables [13,14], it is impossible using this method to allocate emissions by specific origin–destination–product flows. The main reason, as Cristea et al. pointed out [9], is scarcity of data on origin–destination flows at the sectoral level in most countries. To the best of our knowledge, there has been no previous attempt with our methodology to cover origin–destination emissions for freight flows in Spain or any other EU country.

Drawing from a previous investigation [15], which develops and applies a detailed inter-provincial trade dataset to analyze transport-mode competition within Spain for a given year (2007), we build an extended database on intra- and inter-provincial freight flows and use it to obtain GHG emissions for origin–destination flows between Spanish provinces during 1995–2015. Note that, by adopting the province (Nuts 3) as the spatial unit of reference, we approached as close as possible to city-level figures, since for most Spanish provinces the capital city agglomerates the bulk of the population and economic activity. The flow data are based on a permanent dataset that was collected and prepared by the C-intereg Project (www.c-intereg.es) and based on the country's most detailed available data on origin–destination–product statistics for freight flows by transport mode (*road*, *train*, *ship* and *aircraft*). The C-intereg project generates alternative figures covering intra-national Spanish trade at different spatial and sectoral levels. The data available to the public on the website are censored to some degree, while the sponsoring institutions and the research group in CEPREDE have access to the full detailed data. We used the full detailed data, especially "*raw freight flows*" measured in tons. By *raw* we mean closest to official figures on freight flows reported independently by the institution responsible for each mode in Spain (e.g., Ministerio de Fomento for road, RENFE for railway, Puertos del Estado for ship, and AENA for aircraft). We use this dataset to forecast the origin–destination–product–mode flows for 2015–2030 by means of gravity models, using intra- and inter-provincial origin–destination distance by mode, as well as the predictions described before about the evolution of provincial GDP in Spain for the same period.

In addition, for each transport mode and year, we build a corresponding dataset for GHG emissions, measured in gCO_2 per tons*km. These indicators cover 1995–2015 and are drawn from estimates already published by official institutions and other sound academic publications in the field. We then generate forecasts for 2015–2030 for each mode, extrapolating observed time trends.

Next, in line with the EN-16258 methodology [6], for each year, we combined industry-specific flows for each of the four transport modes, measured in tons*km, with the corresponding GHG emissions factors. Once the dataset of GHG emissions is built, we generate and analyze the temporal, sectoral and spatial patterns of Spanish inter-provincial GHG flows, and compare them with official national figures. Then, to search for a more sustainable freight system within the country, we address the possibility of promoting transport mode shifts from high-polluting modes (road) to more environmentally friendly alternatives (railway) for specific origin–destination–product flows. Two scenarios for transport mode shifts are considered, both inspired by targets suggested by the EC's "White Paper of Transport" [4] and striving to achieve railway's desirable future share.

To return to the conceptual framework suggested by Davydenko et al. [6], the methodology developed herein does not fulfil all of the authors' recommendations. It fits in with their "meso" level and should be considered complementary to the "macro" official estimates published by the Spanish IIR. It includes at least two relevant aspects explicitly considered by the authors: the estimation of emissions at the shipment level, with the origin–destination–product–mode for each delivery; and the use of actual freight flows in volume and actual distance traveled by each mode for each origin–destination delivery. Because of data constraints, the main limitations of our methodology by comparison with the holistic approach described by Davydenko et al. [6] are: (i) lack of information on multimodal deliveries (our methodology assumes that points of origin and destination for each delivery correspond to points of production and consumption and does not consider multimodality); (ii) lack of information on GHG emissions associated with product handling at the origin and destination or with intermodal-connections; (iii) lack of information on the vehicle type used in each delivery; and (iv) lack of information on specific routes taken by freight haulers on their deliveries.

The rest of the paper is structured as follows: Section 2 reviews recent literature on the measurement and reduction of GHG emissions for freight flows at the international, European and country level, with a final focus on Spain. Section 3 describes our empirical estimation strategy for GHG emissions within Spain. Subsections lay out our two parallel datasets (origin–destination freight flows vs. GHG emissions indicators) and the two periods considered (1995–2015 vs. 2016–2030). Section 4 is an empirical analysis of trade and emissions patterns for each region, product type and transportation mode, and concludes with a description of the main results for the suggested scenarios.

2. GHG Emissions and Freight Flows

In the Introduction, we cite Davydenko et al. [6], who stressed the need for a standard measurement of the transport sector's GHG emissions. However, there have been other approaches to the topic for given countries and other attempts to deal with the usual data constraints. An interesting paper in this regard was presented by McKinnon and Piecyk [16], and reviews several ways to estimate CO_2 emissions from freight transport flows. They claimed that, despite the interest of alternative estimation methods, the variability in the figures from official sources to academic approaches can erode the confidence of industry stakeholders in the validity of the estimates. McKinnon and Piecyk [16] used UK data, focused on the road mode, and evaluated various estimation methods for national emissions in a given year. More specifically, they considered four alternative methodologies used in the UK in 2006. Two are taken directly from official government sources, while the others are calculated by the authors: (i) National Environmental Accounts estimate for the "road transport of freight"; (ii) HGV-activity of British-registered haulers on UK roads using survey-based fuel efficiency estimates; (iii) all HGV-activity using survey-based fuel efficiency estimates; and (iv) all HGV-activity in the UK using test-cycle fuel efficiency estimates. These four approaches differ mainly in the scope of their calculations, their methodology and their alignment of vehicle classifications. The two lowest estimates relate solely to British-registered operators and therefore provide only a partial view of road freight activity in the UK. As this reference illustrates, an academic estimate such as ours can differ from the alternative official one, if only because of the use of alternative road data. McKinnon and Piecyk [16] based their calculations on heavy truck surveys or on traffic stations. One limitation of their approach is the lack of detail at the delivery level, which makes it difficult to assign responsibility for emissions at the sub-national level.

As noted in the Introduction, another interesting reference among the short literature analyzing GHG emissions associated with international trade using origin–destination flows by mode was presented by Cristae et al. [9], who collected extensive data on worldwide trade by transport mode and used it to provide detailed comparisons of GHG emissions associated with output versus international transport of traded goods. According to their analysis, international transport is responsible for 33% of worldwide trade-related emissions and over 75% of emissions for major manufacturing categories. Their approach covers emissions associated with both the production (output) and the transport of goods to destinations abroad, and allows them to distinguish between the two. Moreover, for the latter, they also considered the *scale effect* (i.e., changes in emissions due to changes in demand for international transport) and the *composition effect* (i.e., changes in the mode mix). They concluded that including transport dramatically changes the ranking of countries by emissions per dollar of trade. They also investigated whether trade inclusive of transport can lower emissions. In one quarter of cases, the difference in output emissions is more than enough to compensate for the emissions cost of transport. More interestingly for us, they also tested how likely patterns of global trade growth could affect modal use and emissions. According to their results, full liberalization of tariffs and GDP growth concentrated in China and India should lead to much faster growth in transport emissions than in the value of trade, because of shifts toward distant trading partners. However, the main limitation of their approach is to consider international trade in isolation from internal freight flows, which in most countries [17,18] account for a larger share of economic activity.

Whereas McKinnon and Piecyk [16] covered the entire UK, Zanni and Bristow [19] analyzed CO_2 emissions for freight flows in London, using historical and projected road freight CO_2 emissions. They also explored the potentially mitigating effect of a set of freight transport policies and logistical solutions for the period up to 2050. Despite the effectiveness of such measures, the resulting reduction would, it seems, only partly counterbalance the projected increase in freight traffic. Profound behavioral measures are need if London wishes to hit its CO_2 emissions reduction targets. The main interest of Zanni and Bristow [19] was that it opens the way to future alternative scenarios for emissions in specific sub-national entities, such as London, but it fails to provide a general perspective on the whole country or on alternative transport modes.

There are also several interesting papers on Spain. For example, Sánchez-Choliz and Duarte [12] used an input–output model to analyze the sectoral impacts of Spanish international trade on atmospheric pollution. They analyzed direct and indirect CO_2 emissions generated in Spain and abroad by Spanish exports and imports. Their results show that the sectors of transport material, mining and energy, non-metallic industries, chemicals and metals are the most relevant CO_2 exporters, while other services, construction, transport material and food are the biggest CO_2 importers. In addition, Cadarso et al. [20] examined the growth in offshoring as a result of production chain fragmentation and measures CO_2 emissions due to increases in final and intermediate imports. Their main contribution is a new methodology (also input–output) for quantifying the impact of international freight transport by sector, which serves to assign responsibility to consumers. As expected, industries with the most intense offshoring show the greatest increases in carbon emissions related to international transport. These are significantly higher than emissions from domestic inputs in certain industries with significant and increasing international fragmentation of production. As already noted, these two papers present two main drawbacks for our purposes: first, their approach allocates emissions by sector but fails to address the regional dimension in a country where regions and cities are developing their own political strategies towards sustainability; and, second, similar to Cristea et al. [9], they focused on the effect of international trade to the exclusion of GHG emissions from internal freight.

Finally, we found some additional papers discussing potential measures to curb GHG emissions within Spain. On the one hand, López-Navarro [21], given the EC's urging of intermodal transport through, for example, the "motorways of the sea", reviewed the existing literature to examine the relevance of environmental considerations to modal choice in the case of short sea shipping and the motorways of the sea. He also used EC-provided values to calculate the external costs of Marco Polo freight transport project proposals to estimate the environmental costs for several routes, comparing the use of road haulage with the intermodal option that incorporates Spanish motorways of the sea. The results of this comparative analysis show that intermodality is not always the best choice in environmental terms. Its main limitation is not to consider inter-regional flows within the country, while focusing on modal shifts for Spanish international deliveries. The same topic was analyzed by Pérez-Mesa et al. [22].

3. Empirical Strategy

Let us begin by considering a country with I provinces (for Spain, $I = 52$, given its 50 provinces and 2 autonomous cities in Africa). Intra- and inter-provincial transport (freight) flows are registered in volume (tons) separately for each transport mode (m), $\{F_{ij}^R; F_{ij}^T; F_{ij}^S; F_{ij}^A\}$, namely: road (R), train (T), ship (S) and aircraft (A). In the absence of intermediation (re-exportation schemes), the aggregate of all deliveries is obtained by adding together the corresponding mode-specific flows, $F_{ij} = F_{ij}^R + F_{ij}^T + F_{ij}^S + F_{ij}^A$. The same can be said regarding product k specific flows using each of these m modes. An additional t suffix for time serves to consider the panel data configuration of the dataset described here.

Now, following Cristae et al. [9], Equation (1) defines the general expression for estimating GHG emissions for each freight flow within a country:

$$E_{ijt} = \sum_k \sum_m F_{ijt}^{mk} * Dist_{ij}^{mk} * e_t^{mk}. \tag{1}$$

where E_{ijt} denotes GHG emissions for all freight flows from origin i to destination j in year t. Emissions are determined by adding across modes m and products k, for every given i-j trip within the country, considering the weight (tons) of the corresponding flows by mode (F_{ijt}^{mk}), the distance in km traveled by each mode for each delivery $(Dist_{ij}^m)$, and a set of vectors e_t^{mk}, with GHG emissions factors produced by mode m for product k when providing one $ton*km$ of transport services. Note that this final element also has a subscript t, to incorporate efficiency gains in terms of emissions factors for each mode by year. Moreover, a suffix k is also added, to indicate that in some cases it is possible to introduce certain heterogeneity within each transport mode, because the use of specific types of vehicles might induce different emissions levels. Although we do not put much emphasis on this component (suffix k drops from term e_t^m onwards), it is interesting to include for further extensions. To this regard, Demir et al. [23] reviewed several variables to determine emissions for the road mode alone. These can be divided into five categories, vehicle, environment, traffic, driver and operations, and include variables such as speed, acceleration, congestion, road gradient, pavement type, ambient temperature, altitude, wind conditions, fuel type, vehicle weight, vehicle shape, engine size, transmission, fuel type and oil viscosity. Similar variables can be applied for railway, ship and airplane, which suggests that the average vectors used here describe only the most likely general trends. We assume distance to be constant over time between any i-j dyad for each mode.

3.1. Estimating GHG Emissions by Mode for 1995–2015

Using Equation (1) and considering the case of Spain, we estimate GHG emissions per mode for 1995–2015 by combining two parallel datasets: (i) that containing intra-nd inter-provincial freight flows by year, product and mode, which contributes with the elements (F_{ijt}^{mk}) and $Dist_{ij}^{m}$; and (ii) that containing GHG indicators by mode (e_t^m).

3.1.1. Inter-Provincial Freight Flows by Mode

The freight flow data used in this paper are based on the most accurate data on Spanish bilateral transport flows of goods by transport mode (road, train, ship, aircraft). This rich dataset was collected and filtered in accordance with the methodology described in Llano et al. [18] and published as part of the C-intereg project (www.c-intereg.es). It includes refinements and extensions with respect to the dataset published in previous papers [15,24]. It is analyzed at the province level (Nuts 3), using the largest possible sectoral detail (29 products) compatible with the four transport modes. No alternative dataset with equivalent amount of data is available for Spain.

The dataset is built on a set of origin–destination transport statistics, such as roads (Permanent Survey on Road Transport of Goods by the Ministerio de Fomento), railways (Complete Wagon and Containers flows, RENFE), ship (Spanish Ports Statistics, Puertos del Estado) and aircraft (O/D Matrices of Domestic Flows of Goods by Airport of Origin and Destination, AENA). Since the original purpose of this dataset in the C-intereg project was to serve as a basis for obtaining monetary flows between regions in the country, flows not associated with economic transactions were eliminated (i.e., it does not include empty trips, removals, military or fair materials moved within the country, etc.). This fact introduces differences in the levels of tons and tons*km for each mode with respect to the general statistics used by the official top-down estimates (e.g., Spanish IIR). This can be a drawback for accurate estimation of final emissions levels, but it is also a virtue in that it is more directly connected with the real economic activity capture in the National Accounts—usually an obligatory reference for any environmental analysis. In addition, the dataset has been subject to a debugging process with the aim of removing potential inter-national hub-spoke and re-exportation structures hidden within the intra-national freight flows [24]. This prevents the double counting of transit flows, mainly by road and railway, from *hinterlands* to ports before/after their loading/unloading for exporting to/importing from foreign markets.

3.1.2. GHG Emissions Indicator by Mode

In addition, we have built a dataset for GHG emissions indicators per transport mode with the information from the specialized literature. Several official documents offer environmental indicators that are useful for our purposes:

The Spanish Ministry of Public Works (SMoPW = Ministerio de Fomento) regularly publishes different indicators for the transport sector's overall GHG emissions. Although none of them fully meet the requirements of the analysis conducted here, they provide a good basis for estimation. The SMoPW publishes the following indicators:

- Total GHG emissions generated by the Spanish transport sector, including passengers and freight movements, by mode. More specifically, the ministry publishes the annual ktCO$_2$ equivalent generated separately by road, railway and air over a long period: 1990–2015. Ship is omitted. These figures appear in the National Inventory of GHG Emissions, produced by the SMoPW in coordination with the MAPAMA and in accordance with the international methodology established by the European Environmental Agency (EEA). The estimates follow a top-down approach and are based on consumption data. Unfortunately, emissions for passenger and freight transport are not systematically distinguished or determined for all modes. We therefore cannot compare our estimates with official estimates.
- GHG emissions factors for freight deliveries within Spain by just three transport modes (road, railway, and air), measured in gCO$_2$ equivalent per tons*km. These figures are reported for 2005–2015. The emissions factor for railway does not include indirect emissions for electric power.

- Aggregate figures for internal freight flows in Spain (measured in tons*km) by transport mode (road, railway, ship and air). The largest statistical series for this indicator corresponds to 1996–2015, but it is not always fully compatible with the emissions indicators noted above.

In addition to this official data, we have found interesting references with which to build a set of alternative scenarios regarding average GHG emissions factors per transport mode and year within Spain. The main sources considered are Cristea et al. [9], Ministerio de Fomento (several years http://observatoriotransporte.fomento.es/OTLE/LANG_CASTELLANO/ BASEDATOS/), and Monzón et al. [25]. Table 1 summarizes alternative GHG emissions indexes by mode reviewed in the literature. For each mode, the three main references considered in this paper appear in pale grey. Note that, in general, these estimates are prudent by comparison with the higher factors in the literature, mainly for *aircraft*.

Table 1. Review of alternative values for GHG emissions factors by mode.

	CO_2 Emissions Intensity (gCO$_2$/T*Km)	Energy Type	Scope	Source
Ship	**10.1**	**Total**	**World**	**Cristea et al. [9]**
	21.37 *	**Total**	**Spain**	**Spanish Ministry of Public Works (Fomento), 2016**
	30.9	**Total**	**Spain**	**Monzón et al. [25].**
	18.9	Fuel oil	Global	Kristensen [26] cited in Monzón et al. [25].
	20	Fuel oil	Australia	Lenzen [10] cited in Monzón et al. [25].
	23.4	Fuel oil	EEUU	Kamakaté y Schipper [27] cited in Monzón et al. [25].
	32.8	Fuel oil	Canada	Steenhof et al. [28] cited in Monzón et al. [25].
	44	Fuel oil	Holanda	Wee et al. [29] cited in Monzón et al. [25].
Railway	**22.7**	**Total**	**World**	**Cristea et al. [9]**
	6.94 (for 2015)	**Total**	**Spain**	**Spanish Ministry of Public Works (Fomento), 2016**
	22.8 *	**Mix**	**UE 15**	**ECMT [30], TRENDS [31] cited in Monzón et al. [25].**
	17.7	Mix	Canada	Steenhof et al. [28] cited in Monzón et al. [25].
	19.4	Mix	EEUU	Kamakaté y Schipper [27] cited in Monzón et al. [25].
	40	Mix	Australia	Lenzen [10] cited in Monzón et al. [25].
	44	Mix	Holanda	Wee et al. [29] cited in Monzón et al. [25].
	45	Diesel and Mix	Holanda	Wee et al. [29] cited in Monzón et al. [25]
Road	**119.7**	**Total**	**World**	**Cristea et al. [9]**
	83.93 *	**Total**	**Spain**	**Spanish Ministry of Public Works (Fomento), 2016**
	123.1	**Mix**	**UE 15**	**ECMT [30], TRENDS [31]**
	110	Diesel. Art. Trucks.	Australia	Lenzen [10] cited in Monzón et al. [25].
	160.7	Diesel. Art. Trucks.	Canada	Steenhof et al. [28] cited in Monzón et al. [25].
	226.5	Diesel. Road total	France	Kamakaté y Schipper [27] cited in Monzón et al. [25].
	260	Diesel. Rigid. Trucks.	Australia	Lenzen [10] cited in Monzón et al. [25].
	490.2	Diesel. Rigid. Trucks.	Canada	Steenhof et al. [28] cited in Monzón et al. [25].
Air	**809.2**	**Total**	**World**	**Cristea et al. [9]**
	139.72 *	**Total**	**Spain**	**Spanish Ministry of Public Works (Fomento), 2016**
	358.6	**Kerosene**	**UE 15**	**ECMT [30], TRENDS [31]**

Note: figures marked with an * are the benchmarks for their respective modes in the reference year. Own elaboration on basis of information published by several sources: Cristea et al. [9]; Ministerio de Fomento (www.fomento.es). See Monzón et al. (2009) for a longer list of references.

The GHG emissions factors used in this paper's baseline scenario are the following:

- For *road* (79.88 gCO$_2$ per tons*km in 2005) and *aircraft* (149.64 gCO$_2$ per tons*km in 2005): The emissions factors are taken from the SMoPW for the period in which they are available (2005–2015). For the remaining years, 2005–1995, the time series are obtained by combining the information in tons*km and total emissions by mode published by the SMoPW.
- For *ship* (22.15 gCO$_2$ per tons*km in 2005): Since the SMoPW does not publish them, we have estimated emissions factors for internal freight flows by ship by considering the relative intensity of this mode with respect to the other three, as reported by Monzón et al. [25], ECMT [30], and TRENDS [31].
- For *railway*: To include direct plus indirect emissions for electric power (the SMoPW does not include them in its estimates), we consider emissions reported by Monzón et al. [25], ECMT [30], and TRENDS [31] for the year 2000 (22.8 gCO$_2$ per tons*km). The change in this level over the rest of the period is obtained in the same way as for *road* and *aircraft*; we combine the information in tons*km for this mode with total emissions published by the SMoPW for *railway* (direct emissions only).

3.2. Predicting GHG Emissions by Mode for 2016–2030

As in the previous section, estimating GHG emissions per mode for 2016–2030 requires the combination of different forecasts able to produce the equivalent elements F_{ijt}^{mk} and e_t^m. For each step:

- We start by estimating F_{ijt}^{mk}, the intra- and inter-provincial trade flows for 2016–2030. This step uses the gravity equation, which entails estimating the GDP for each Spanish province over the period, assuming a time-invariant vector for distance $Dist_{ij}^m$ and a set of control dummy variables.
- We then obtain corresponding predictions for GHG emissions factor e_t^m.

3.2.1. Forecasting Provincial GDPs for 2016–2030

The aim of this section is to obtain a *GDP* prediction for each Spanish province in the forecasting period. It should be noted at the outset that the Spanish Institute of Statistics (INE) publishes *GDP* and Value Added (*VA*) figures on yearly basis, for both regions (Nuts 2) and provinces (Nus 3). However, the level of disaggregation is lower for provinces, the reference spatial unit for this paper. For this reason, it is convenient to predict the change in *GDPs* at the regional and provincial level simultaneously to take advantage of the richer information available for Spain at the Nuts 2 level.

Thus, the point of departure is the forecasts provided by CEPREDE (www.ceprede.es) at the national and regional level (Nuts 2) in Spain, with a breakdown of 23 activity branches covering the needed forecasting horizon. We obtain these forecasts through different linked models developed by the "Lawrence R. Klein" Institute at the Universidad Autónoma de Madrid. Their general structure is described in Figure 1. A more detailed description of the main macro-econometric model (Wharton-UAM) can be found in Pulido and Perez [32], whereas the detailed methodology for the long-term international scenario is found in Moral and Pérez [33]. The Wharton-UAM model provides forecasting trends for the Spanish economy from Project Link, an international collaborative research group for econometric modeling, coordinated jointly by the Development Policy and Analysis Division of United Nations/DESA and the University of Toronto (https://www.un.org/development/desa/dpad/project-link.html).

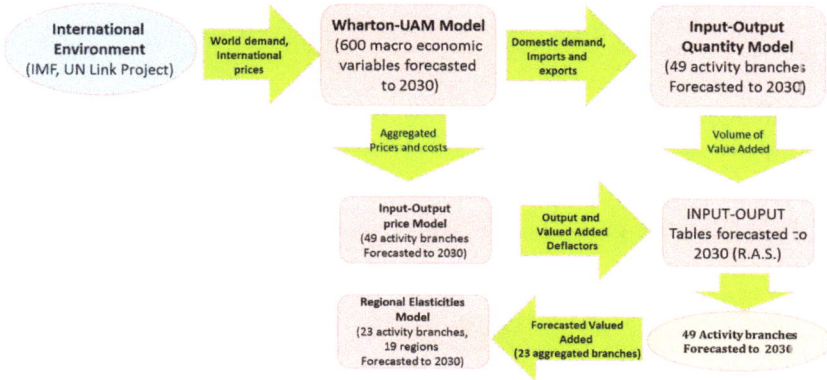

Figure 1. General outline of long-term forecasting model: regional level (Nuts 2). Source: Own elaboration.

The scheme for regional disaggregation of *GDP* is quite similar to the one described below for provinces, and it is based on the sectoral structure of each region in terms of *VA*, and the corresponding elasticities between regional and national performance by sector.

Drawing from the regional figures provided by the Wharton-UAM model, we obtain our predictions for each province by considering their sectoral mix and changes in the regions they belong to. With this aim, we use data from the Spanish Regional Accounts published by National Statistical Institute (INE), which include provincial *GDP* and *VA* figures for 2000–2015, broken down into seven activity branches. For each province *i*, total GDP in each period *t* can be obtained by the aggregation of the *VA* for the seven activity branches *b*, plus net production taxes *I*.

$$GDP_{i,t} = \sum_{b=1}^{7} VA_{b,i,t} + I_{i,t} \tag{2}$$

For each branch and province (Nuts 3), we compute historical elasticity between the growth rate of the *VA* in volume (Chained Linked Volume Index, *IVA*) for provinces *i* (Nuts 3) and regions *r* (Nuts 2).

$$\varepsilon_{b,i}^{z} = \frac{1}{T} \sum_{t} \frac{\Delta IVA_{b,i,t}}{\Delta IVA_{b,r,t}}, \ \forall \ 2016 < z < 2030 \tag{3}$$

where $VA_{b,r,t}$ is the volume index of *VA* in branch *b* from region *r* where province *i* is located.

These initial elasticities are harmonized to guarantee the stability of predictions, making unitary the weighted average of the elasticities in the different provinces of each region.

$$\varepsilon_{b,i} = \varepsilon_{b,i}^{z} * \frac{1}{\sum_{i} \omega_{b,i} * \varepsilon_{b,i}^{z}}, \ \forall \ 2016 < z < 2030 \tag{4}$$

where $\omega_{b,i}$ is the weight of province *i* over the regional *VA* in branch *b*.

Once these elasticities have been computed, we can obtain an initial *GDP* for each province *i* by multiplying them by regional forecasts in each branch. Note that net taxes are treated as an additional activity branch.

$$GDP^{0}_{i,t+z} = \sum_{b=1}^{7} VA_{b,i,t+z-1} * (1 + \varepsilon_{z,i} * \Delta IVA_{b,r,t+z}), \ \forall \ 2016 < z < 2030 \tag{5}$$

Afterwards, these initial values are corrected to match regional *GDP* with the aggregation of provincial *GDPs*.

$$GDP_{i,t+z} = GDP^0_{i,t+z} * \frac{GDP_{r,t+z}}{\sum_p GDP^0_{i,t+z}}, \quad \forall\, 2016 < z < 2030 \tag{6}$$

Thus, we obtain *GDP* figures for each province in the forecasting period that match the regional and national predictions produced by CEPREDE. Although provincial *GDP* can be broken down into the *VA* of seven branches, in this paper, we use only aggregate provincial *GDP* figures.

3.2.2. Forecasting Inter-Provincial Flows in 2016–2030

Using the provincial *GDPs* predicted in the previous section for 2016–2030, we now estimate, for the same period, the corresponding intra- and inter-provincial flows by each of the four modes and by each of the 29 product types in the historical sample. For this, we use the gravity equation, the most standard methodology to model international and interregional trade flows [34,35]. This approach is rooted in previous articles modeling equivalent flows in Spain [15,18,24,34]. The baseline model is described by Equation (7):

$$F^{mk}_{ijt} = \beta_0 + \beta_1 lnY_{it} + \beta_2 lnY_{jt} + \beta_3 Own\_pro + \beta_4 lnDist^m_{ij} + \beta_5 Contig + X_i + X_j \\ + \mu_{i,mk} + \mu_{j,mk} + \varepsilon_{ijmkt}. \tag{7}$$

F^{mk}_{ijt} is the volume (*tons*) of freight flows of product *k* transported in year *t* by mode *m* from province *i* to province *j*. Note that *i* and *j* are two of the 52 Spanish provinces, so any flow where $i = j$ is intra-provincial, while any flow where $i \neq j$ is inter-provincial. Suffix *m* indicates the transport mode used in the delivery, which can take four values (R = road, S = ship, A = aircraft, and T = railway). Suffix *k* can take 29 values, corresponding to the 29 product types described in Table A4 in the Appendix A. Note that as a robustness check, equivalent flows have been obtained for intra- an inter-provincial trade flows measured in current euros.

Variables lnY_{it} and lnY_{jt} are the logarithms of nominal *GDP* for exporting and importing provinces, respectively. Note that *GDPs* in the historical period correspond to official figures published by the INE, whereas in the forecasting period they correspond to the figures obtained in the previous section, which are compatible with the national and regional predictions provided by CEPREDE. This is the only set of time-variant variables specific to each *i-j* pair that are taken into account when we forecast intra- and inter-provincial flows for 2016–2030.

In addition to these time-variant variables for the forecasting period, we also consider a number of time-invariant ones. First, as is standard in this type of modeling, a dummy variable, *Own_Prov*, is included to control for the different nature of flows within and between Spanish provinces. This dummy variable takes the value 1 if the flow's origin and destination are the same province and 0 otherwise. The anti-log of this dummy is the *own-province effect* or *home bias* extensively discussed in the literature of international trade [15,24,32,36,37].

The variable $lnDist^m_{ij}$ is the logarithm of the distance between province *i* and province *j* for each mode *m*. Note that, in line with Gallego et al. [24], we have used alternative distance measures per transport mode. Each of these alternative distances is obtained as follows:

$Dist^R_{ij}$ represents the most likely bilateral distance (in *km*) for deliveries by road:

(i) For all bilateral deliveries within the peninsula (47 inner provinces), we follow Zofío et al. [38], where GIS software determines the shortest trip distance between any two places based on the actual network of roads and highways (including such parameters as slope, quality and maximum legal speed). We thus obtain raw bilateral distances for a detailed picture of the Iberian Peninsula, split into more than 800 areas. These raw distances are aggregated, with averages weighted by the various populations of these areas, to produce a province-to-province matrix of inter-provincial distances.

(ii) For the three island provinces, we obtain bilateral distances between them and the inner provinces by taking the official distance traveled by ship between the islands and the main maritime ports

(Cádiz for the Canary Islands and Barcelona for the Balearic Islands) and adding it to the road distance from these two main ports to each inner province. This road distance exactly corresponds to the distance described above. This treatment is justified because deliveries between inner regions and the islands are in fact made by *Ro-and-Ro* and similar strategies, with trucks loaded onto ships.

We have checked the results against alternative *distance* measures, such as actual distances reported by trucks upon their deliveries, and found them to be robust. However, we have decided to use GIS distances, as they avoid problems related to the computation of intra-provincial distances traveled by trucks within each province. GIS distances are simply not affected by the huge number of short trips entailed by capillary distribution from wholesalers to retailers (see Díaz-Lanchas et al. [39]).

$Dist^T_{ij}$ represents the bilateral distance (in *km*) traveled by *railway*, as reported by RENFE (the former Spanish rail monopoly). RENFE expresses data on bilateral flows between any two provinces in tons*km and tons. By dividing the first measure by the second, we obtain a fairly precise average distance traveled by trains in a given year (2007) for the main inter-provincial pairs. When a specific bilateral distance is not available, we substitute road distance.

$Dist^S_{ij}$ represents bilateral distance (in *km*) by *ship*. The distance between ports (coastal and islands provinces) is reported by the official Spanish port authority, Puertos del Estado. Again, to fill gaps in the data and in the unlikely event that an island reports flows by ship with inner regions (multimodal), we substitute road distance.

$Dist^A_{ij}$ represents the most likely bilateral distance (in *km*) for air transport. This is the straight-line distance computed by GIS between airport locations in Spain. For provinces with no airports, we substitute road distance.

To capture the positive effect of adjacency between provinces, we introduce the dummy variable *Contig*, which takes the value one when trading provinces *i* and *j* are contiguous and zero otherwise. This variable conveniently controls for higher inter-provincial trade flows between contiguous Spanish provinces. ε_{ijmkt} denotes the classical disturbance term.

The specification also includes several time-invariant variables to control for different factors that may affect the magnitude of the flows across provinces. Such variables are summarized in $X_i; X_j$:

Coastal$_i$; Coastal$_j$ A dummy indicating whether the exporting or importing province is coastal or land-locked. These dummies are considered for *ship*, while for the other modes they become non-significant.

Island$_i$; Island$_j$ A dummy variable identifying the three island provinces of Spain (Islas Baleares, Las Palmas and Santa Cruz de Tenerife) as exporting regions.

Finally, three additional dummy variables have been added for the *road* mode, with the aim of controlling for the special case of trade between the Canary Islands and the Balearic Islands with the provinces in the Iberian Peninsula, which relates to the "*Ro-and-Ro*" logistic strategy. Failure to include these resulted in excessively high predictions for flows in the forecasting period.

The estimation of the equation adopts a pooled regression format with several fixed effects, following the standard approach in the literature as an alternative to pure panel data specifications, which will absorb the time-invariant dyadic variables such as distance. The terms μ_{imk} and u_{jmk} correspond to multilateral-resistance fixed effects for the origin–mode–product and the destination–mode–product, respectively. Their inclusion follows Anderson and van Wincoop [17] and Feenstra [37] and is meant to control for competitive effects exerted by the non-observable price index of partner provinces and by other competitors. They are also meant to capture other particular characteristics of the provinces in question. It is worth mentioning that, because of their cross-section dataset, the origin and destination fixed effects in Anderson and van Wincoop [17] and Feenstra [37] do not consider their interaction with time. We also cluster the residuals by α_{ijmk}. Following the most recent literature on the estimation of gravity models in presence of many zero flows, we use the Poisson pseudo-maximum likelihood technique (PPML). This approach was proposed by Silva and Tenreyro [40], which sorts out Jensen's inequality (note that the endogenous variable is in levels) and produces unbiased estimates of the coefficients by solving the heteroskedasticity problem. Note that, with PPML estimation, it is recommended that the endogenous variable be included in levels rather than in logs (F_{ijt}^{mk})). Time-fixed effects are not considered here because of the problems that arise in the forecasting exercise.

The results obtained in the estimation of the gravity model for each mode are reported in Table A1 in the Appendix A. Regressions are based only on the period 2013–2015, as an attempt to avoid the recent economic crisis. Therefore, the elasticities obtained are based on the most recent relationship between internal trade, provincial *GDPs*, distance and the rest of the time-invariant controls. Alternative samples and specifications have also been tested, while the one reported here offers the best results in the forecasting exercise. However, it is important to stress the results' great sensitivity to levels, something that will affect final GHG emission estimates. Note that, even if we limit the sample to this short window of time, and remove zero flows to increase forecasting accuracy, the regressions consider between 5103 (railway) and 64,191 (road) observations. Despite these long panels, the R^2 obtained are reasonably high in all sectors but ship, ranging from 0.6 in railway to 0.89 in aircraft. In general, the coefficients are significant and the signs match expectations. However, interesting variability is found, in line with certain previous analyses [24], for each transport mode.

Starting with *road*, whose results are the most standard within the literature, both *GDPs* for the exporting and importing province are significant and positive, with values close to unity. The coefficient for the $lnDist^m_{ij}$ is negative and close to -1, which is in line with standard values for international and interregional deliveries [24,36,37]. The coefficients for contiguity and *OwnProv* are also positive and significant, and dummies related to the *Islands* are always negative and significant.

The results for the other modes are more surprising but also easy to account for: non-significant coefficients for *GDP* in *ship* and *railway* indicate that certain provinces more specialized in these two modes are associated with heavy industries and bulk freight movements. These, with some exceptions (País Vasco), do not correspond to the richest regions in the country. The opposite happens with the positive and high coefficient of GDP_i of the exporting province, and the negative and significant coefficient for the GDP_j of the importing province, found in *aircraft*. This result indicates that the main exporting provinces using this mode are Madrid and Barcelona, while the main importers are the Canary and the Balearic Islands. This is also reflected in the *Island* dummies, as well as in the positive and significant coefficient found for the log of *distance*. This result, singular for a gravity model, perfectly matches our intuition that aircraft is more efficient for the furthest destinations.

3.2.3. Forecasting GHG Emissions Factors by Mode for 2016–2030

Finally, it is now time to predict the evolution of GHG emissions factors e_t^m for each mode m in the forecasting window 2016–2030. There is a complex and interesting technical literature on the prognosis of efficiency gains affecting the emissions of each transport mode [23,24,41]. The analysis of this literature is beyond the scope of this article. Instead, we opt for a more automatic approach, with the projection of observed trends over a recent 20-year period (1995–2015). To select the best time-trend option, we have tested four alternatives for each mode, with following specifications:

$$e_t^m = \alpha^m + \beta^m * t \tag{8}$$

$$e_t^m = \alpha^m + \beta^m * t + \gamma^m * t^2 \tag{9}$$

$$e_t^m = a^m * (b^m)^t = \alpha^m + \beta^m * ln(t) \tag{10}$$

$$e_t^m = a^m * t^{b^m} \leftrightarrow ln(e_t^m) = \alpha^m + \beta^m * ln(t) \tag{11}$$

where t is the time trend variable, and α^m, β^m and γ^m are the coefficients to be estimated. For each trend specification and transport mode m, the statistical significance of each trend is tested by means of the *t-statistics* associated with trend coefficients (β^m, γ^m). Finally, for each mode the best time trend option is selected by the *sum of squared errors*.

Once the best trend has been selected, and once the baseline trend (Baseline scenario) has been forecast, a sensitivity analysis is performed using the confidence statistical intervals estimated for trend coefficients β^m. Thus, for each mode, we have computed an upper and lower bound by moving the trend coefficient between the 95% confidence interval estimated through the standard error $Sd_\beta{}^m$ of the trend coefficients.

$$Max. \rightarrow \beta^m + 2 * Sd_{\beta^m} \tag{12}$$

$$Min. \rightarrow \beta^m - 2 * Sd_{\beta^m} \tag{13}$$

As we explain in the next section, these terms serve to define alternative scenarios. The main econometric results of this section are shown in Table A2 in the Appendix A.

To illustrate the variability of the emissions factors obtained, Figure 2 plots the evolution of the *Observed*, *Baseline*, *Max* (upper bound) and *Min.* (lower bound) factors for each mode. Note that in all cases the change points to clear gains in efficiency, which can be explained by the development of greener technology and its progressive adoption within each mode. No exogenous shocks regarding policy actions are considered here.

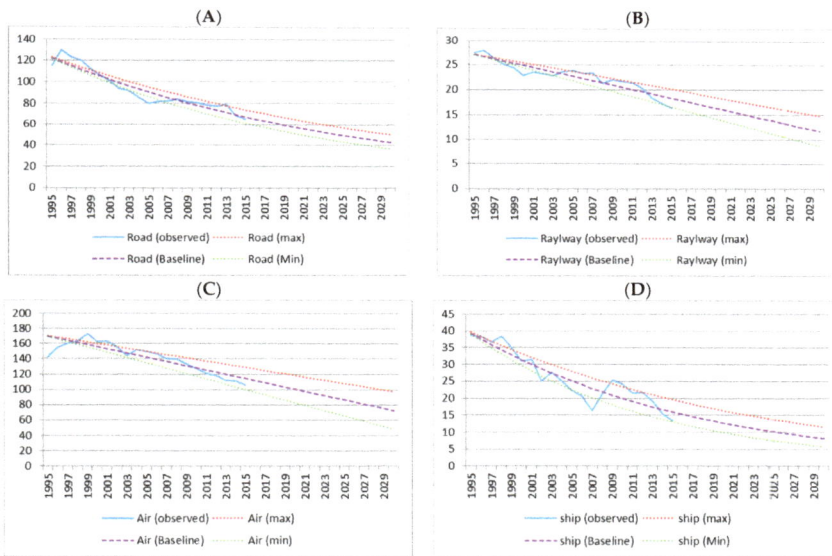

Figure 2. GHG emissions factors by mode for 1995–2030 (gCO$_2$/t*km): (**A**) Road; (**B**) Railway; (**C**) Air; and (**D**) Ship. Source: Own elaboration based on several sources (mainly Spanish Ministry of Public Works).

3.3. Scenarios for Reducing GHG Emissions by 2030

Once the entire dataset is obtained, we define the main modal choice scenarios for the flows. We then consider alternative sub-scenarios for the GHG emissions factors defined by Equations (8)–(11). In summary, each scenario is described as follows:

Scenario 1: Baseline

For the Baseline scenario, we obtain GHG emissions by combining the actual-flow dataset (observed + predicted) with the emissions factors described as benchmarks in Section 3.2, that is, the factors taken from the SMoPW and combined with Monzón et al. [25]. We then consider two alternative sub-scenarios:

(a) **Baseline-Max.:** Using 2016–2030 GHG emissions factors considering the upper bound (Max.) for each year and mode, as defined by Equation (12).
(b) **Baseline-Min.:** Using 2016–2030 GHG emissions factors considering the lower bound (Min.) for each year and mode, as defined by Equation (13).

Scenario 2: Moderate modal shift from road to railway

In **Scenario 2**, the objective is to analyze additional emissions decreases due to hypothetical shifts from road to railway in certain flows. Based on findings from previous analyses [11,15,42], and the descriptive analysis reported in Figure 5, this scenario uses the following criteria:

(i) First, for the entire historical sample (1995–2015), we compute the share in tons for railway out of total tons moved for each i-j-k-t. Flows loaded/unloaded in the Islands and Ceuta and Melilla are excluded.
(ii) Then, for each i-j-k-t we identify the maximum share of railway, considering only trips with distances above 600 km in the entire historical sample (1995–2015). If this share for flows by railway for i-j-k-t is above 40%, it is truncated. We therefore assume the maximum share for each to be 40%. Again, flows from/to the Islands and Ceuta and Melilla are excluded.
(iii) Next, for every i-j-k triad with distance over 600 km, we compute load transfers from road to railway until the maximum share identified in Point (ii) is fulfilled for said triad. With this step, we impose a modal shift so that the maximum shares observed in the period for a given i-j-k are applied to every year in the forecasting period, within the limit of 40%.
(iv) Once the three previous steps have been applied to the whole dataset, we recalculate GHG emissions, considering the baseline emissions factors for each mode. We then also consider the two aforementioned alternative sub-scenarios, using the upper and lower bound from the emissions factors predicted for each mode. These sub-scenarios are labeled **Scenario 2-Max.** and **Scenario 2-Min**.

Scenario 2 adopts maximum shares by railway for a given product k as a benchmark for every flow of the same product k between any other i-j dyad whose bilateral distance is above 600 km. This protocol is applied throughout the forecasting period. The scenario is k specific to take into account the singular nature of each product: perishability, transportability, value/volume ratios, special infrastructures required for special k products (such as dangerous substances or refrigerated loads), etc. This may limit our ability to extrapolate a given share from any other product. In addition, the limit of 600 km is supported by previous analyses conducted in Spain [15,42,43], which suggest that railway is competitive with road beyond this threshold. Although, according to Figure 4, railway flows in *tons* agglomerate in short distances (<200 km), a detailed view of the distribution by product suggests that the agglomeration is driven by certain heavy products, whose performance is not comparable to that of products usually delivered by road. Moreover, considering the results in Figure 3, it seems reasonable to try to promote modal shifts in the longest-heaviest inter-provincial flows traveling by road between the farthest-most-populated provinces (Sevilla-Madrid-Valencia-Zaragoza-Barcelona), rather than to consider alternative transfers of short-distance deliveries by road. Without additional infrastructures, the latter are less likely to match with the current railway network or with absorption capacity. Moreover, the 40% maximum share imposed in Point (ii), although ad hoc, is rooted in the following facts: (i) currently, as reported in Figure 4, the product with the largest share in Spain holds 22%; and (ii) according to the EC's "White Paper of Transport" [4], railways will be handling 40% to 60% of EU traffic by 2050. It therefore seems highly optimistic to assume that a country that currently has only a tiny railway share (1.9%, according to official estimates) will attain a 40% railway share for the longest trips by 2030.

Scenario 3: Extreme modal shift from road to railway

We now consider a more radical alternative, where for each *i-j-k-t* in the forecasting period, all flows by road with a bilateral distance above 150 km are transferred to railway. We thus impose a prudential maximum railway share of 60% for each *i-j-k-t*, taking inspiration from the upper bound considered by the EC "White Paper in Transport" for the whole EU. Note, that by assuming this 60% maximum railway share for the longest *i-j-k-t* trips (>150 km), we end up obtaining an overall railway share of around 26% for the country's total freight traffic. Even assuming an extreme scenario like this, then, the overall railway share will remain around half of the reference one suggested by the EC for the entire EU transport sector. In this scenario, for brevity, we consider only the minimum emission factors reported in Figure 2 for each mode.

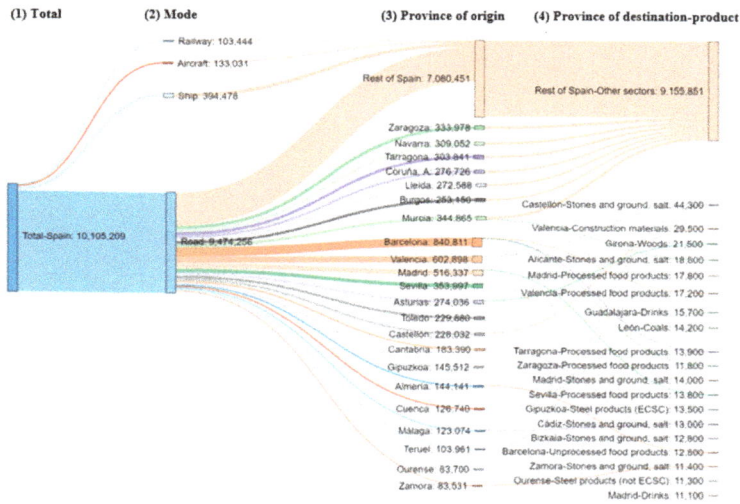

Figure 3. Main inter-provincial flows in terms of GHG emissions. 2015. tCO$_2$. Source: Own elaboration using http://sankeymatic.com/build/.

4. Results

The analysis starts with the main results reported in Table 2, which summarize the main GHG emissions for freight flows within Spain in the period 1995–2015 and their regional allocation (Nuts 2). Equivalent results are reported for provinces (Nuts 3) in Table A3 in the Appendix A. According to these results, GHG emissions reached a level of 10,105 ktCO$_2$ equivalent in 2015. To interpret this number, it is convenient to consider that, according to the official Spanish inventory IIS [7], total GHG emissions in Spain for 2015 were 335,662 ktCO$_2$, with 83,316 ktCO$_2$ attributed to the whole "*Transport sector*", mixing both passengers and freight intra-national flows.

It is important to remark that there are no official estimates of GHG emissions linked to freight flows with the same detail reported here. Moreover, the official estimates reported at the sub-national level are just top-down distributions at the regional level (Nuts 2) for all GHG emissions, with no detail by sector. They are published by the MAPAMA, which warns in a disclaimer that they are to be used "with caution". Thus, a full comparison is not possible. With this in mind, let us consider the SMoPW's report that GHG emissions attributed to freight flows in Spain (ship excluded) reached the value of 16,661 ktCO$_2$ for 2014, a figure 40% above the estimate obtained here (10,105 ktCO$_2$). Moreover, despite the difference, both estimates coincide in the contribution of each mode; both point to the road mode's huge share (95% in the official estimates, 94% in this paper). We obtain a 3% share for ship, while the official figure is 2%. Moreover, in our estimate, railway and aircraft each account for

1% in 2015, whereas in the official estimates *aircraft* reaches 3% and *railway* only 0.3% of total transport emissions for intra-national freight flows.

Table 2. GHG emissions by region (Nuts 2): Structure and change (ktCO$_2$ eq.).

	2015					Growth Rates (%)		
	Total	Intra-Regional	Inter-Regional	%	Total/GDP	2015–1995	2015–2009	2030–2015
	(1)	(2) in % of (1)	(3) in % of (1)	(4) = (1)/Spain	(5)	(6)	(7)	(8)
Andalucía	1378.78	52.3%	47.7%	13.6%	0.951	−11.6%	−30.4%	59.1%
Aragón	603.35	29.5%	70.5%	6.0%	1.804	4.4%	−19.3%	50.2%
Asturias	304.01	23.2%	76.8%	3.0%	1.433	−28.1%	−33.4%	43.3%
Baleares	30.73	70.1%	29.9%	0.3%	0.112	26.9%	−20.1%	76.8%
Canarias	105.93	46.1%	53.9%	1.0%	0.259	50.9%	−12.5%	49.2%
Cantabria	189.80	22.6%	77.4%	1.9%	1.556	−21.3%	−15.7%	51.3%
Castilla y León	1061.88	39.1%	60.9%	10.5%	1.979	4.2%	−27.4%	46.1%
C.-La Mancha	807.23	28.4%	71.6%	8.0%	2.156	2.5%	−33.2%	58.6%
Cataluña	1699.45	46.1%	53.9%	16.8%	0.827	−24.4%	−29.2%	52.5%
C. Valenciana	1063.11	41.1%	58.9%	10.5%	1.055	−11.8%	−26.4%	55.4%
Extremadura	225.97	42.5%	57.5%	2.2%	1.294	7.4%	−32.6%	81.0%
Galicia	700.12	42.5%	57.5%	6.9%	1.243	−5.5%	−24.7%	53.9%
Madrid	656.57	15.3%	84.7%	6.5%	0.322	0.9%	−29.7%	17.5%
Murcia	342.43	22.3%	77.7%	3.4%	1.214	1.0%	−13.6%	63.3%
Navarra	323.64	24.3%	75.7%	3.2%	1.743	−4.6%	−18.8%	59.6%
País Vasco	503.82	18.5%	81.5%	5.0%	0.758	−26.2%	−29.1%	36.2%
Rioja, La	107.75	10.9%	89.1%	1.1%	1.372	−3.2%	−18.7%	32.2%
Ceuta	0.39	0.0%	100.0%	0.0%	0.025	1045.1%	3950.7%	−36.0%
Melilla	0.24	0.0%	100.0%	0.0%	0.016	1931.9%	570.2%	76.0%
Total-Spain	10,1052.1	36.6%	63.4%	100.0%	0.936	−10.0%	−27.4%	51.1%

Pro memoria: Total official GHG emissions in Spain in 2015 = 335,662 ktCO$_2$; total official transport sector GHG emissions (passengers + freight flows) in Spain in 2015 = 83,316 ktCO$_2$; GHG emission attributed to freight flows in Spain (ship excluded) by SMoPW (year 2014) = 16,661 ktCO$_2$. Source: Own elaboration.

While there is indeed a lack of (full) comparability, the difference (40%) between our estimates and the official ones can be explained by a number of factors. First, the official estimates use top-down approaches and do not explicitly report on GHG emissions produced at the delivery level. Second, there is the matter of the statistics used for each mode, for both emissions factors and traffic in tons*km. For example, we use an emissions factor of 17.4 for railway in 2015; this includes direct and indirect emissions for the use of electric power. The SMoPW, although it does not publish an exact emissions factor or the tons*km traveled by railway for every year, uses a reference factor of 6.94; this does not include indirect emissions for the generation of electricity. Another source of difference is the use of different traffic values by transport mode, in tons*km. We borrow data, in tons, from C-intereg, and these differ from the official figures. The reason is that, in C-intereg, freight flows are used as proxies to obtain monetary flows between regions. Thus, freight flows by road that do not correspond to economic transactions, such as empty trips, movements of military materials, fairs, removals, etc., are eliminated. In the case of road, the largest mode by far in terms of tons*km and GHG emissions, the C-intereg data are based on the Spanish EPTMC survey on heavy truck road transportation. SMoPW estimates are based instead on a combination of this survey and the register of heavy-truck movements through Spanish networks (Aforos). As McKinnon and Piecyk [16] illustrate for the UK, the use of alternative sources like these leads to differences in tons*km and emissions. Moreover, according to the Spanish EPTMC, empty trips by road in Spain account for around 40% of total operations. The bare fact that the C-intereg data used here do not include empty-trips already explains much of the difference in estimates.

Despite the previous discussion about the aggregate levels, where further improvements in the methodology are possible, it is also interesting to compare our estimates with respect to the official ones in terms of growth rates. This analysis appears below, when we consider the forecasting exercise in Figure 6. For now, we turn back to Table 2, to explore the newest layer of the methodology developed therein: one that allows the allocation of GHG emissions to specific regions and provinces.

As reported in Column 4, the main polluting regions are those with the largest production capacity within Spain: that is, Cataluña, with almost 17% of GHG emissions, followed by Andalucía

(13.6%), Castilla-León (10.5%) and Comunidad Valenciana (10.5%). The Madrid region, surprisingly, accounts for just 6.5% of emissions, a lower level than its high GDP and its economic and geographic centrality would suggest. One reason could be a tendency towards trading over shorter distances than other, more extensive multi-provincial regions, or the delivery of products with higher value/volume ratios.

Indeed, in terms of *GDP* (Column 5), the geographical structure of polluters changes vividly: for example, Cataluña, the largest industrialized region, accounts for just 8.27% of emissions relative to its *GDP*, while Andalucía accounts for 9.5% and Madrid for 3.22%. To interpret this result, it is important to keep in mind that these calculations include only inter-regional exports of goods (no inter-regional imports) and exclude the service and construction sectors, which are more relevant in the richest regions.

It is also remarkable that 63.4% of all emissions are generated by inter-regional flows, and 36.6% by intra-provincial ones. The high value of the former, where short trips are prevalent, is explained by the most intensive use of *road* over the shortest distances. This is associated with products with low *value-to-volume ratio*, such as stones, minerals and construction materials.

Shares of inter-regional/intra-regional flows differ by region. For example, La Rioja and Madrid have the largest shares of emissions for inter-regional deliveries (89% and 81%), while Baleares (29.9%) and Andalucía (47.7%) have the smallest shares for inter-regional flows and the largest for intra-regional deliveries. Behind this heterogeneity lie the *product–mode mix* and the *geographical area* of each region. For example, Andalucía has an area of 87,599 km^2 and includes nine provinces, while Madrid is a single-province-region of 8028 km^2.

As for growth rates, the results suggest that GHG emissions for intra-national freight flows from 1995 to 2015 has decreased by 10%, the reduction being most intense after the economic downturn of 2008 (-27.4%). From 2009 to 2015, the Spanish economy suffered its worst crisis in recent history, with an intense decline of internal consumption and investment, and a clear re-orientation towards international trade. All these factors have greatly contracted freight deliveries within the country, and this has converged with political and individual measures towards sustainability. Finally, Column 8 shows the difference between GHG emissions in 2030 (baseline scenario) and in 2015. Here, we remark simply that, according to these figures, GHG emissions for internal freight flows in Spain are expected to rise 51.1% in the forecasting windows. These results are analyzed in more detail below.

To focus now on bilateral relationships between provinces (Nuts 3), Table 3 shows the ranking of the 20 highest flows in 1995 and 2015, reporting the point of origin and destination of the trip, as well as the product delivered and mode of delivery. Remarkably, as we see on the left panel, in 1995 only two out 20 flows were inter-provincial, that is, had an origin different from the destination: the 12th flow was from Valencia to neighboring Castellón, and the 19th from Tarragona to Barcelona. Moreover, all the main flows but four (15th, 16th, 17th, and 20th) correspond to just one sector, "*Rocks, sand and salt*", which has very low transportability (low *value/volume* ratio) and is highly linked to the building sector. The other three, corresponding to "*Cement and limestone*" (15th), "*Coal*" (16th), "*Chemical products*" (17th) and "*Construction materials*" (20th), are similar. In all cases, the mode used is road. The other main flows correspond to intra-provincial flows: 1st Barcelona–Barcelona; 2nd Valencia–Valencia; 3rd Navarra–Navarra; and 4th Madrid–Madrid. Note that in this analysis we use a fine spatial scale (provinces, Nuts 3). Had we used the region scale (Nuts 2), all flows would have appeared to be intra-regional. The conclusions derived from the other panel (flows in 2015) are very similar: the largest flows correspond to intra-provincial flows, while just three inter-provincial ones appear among the main flows. Moreover, these inter-provincial flows have gained positions in the ranking, rising from position 12th and 19th up to position 8th and 11th. The product types and modes are very similar, with a concentration on road, short distances and very heavy products. All these results reveal a great clustering of deliveries over short distances and suggest that road, one of the most polluting modes, is unbeatable when accessibility is crucial.

Table 3. Ranking of the 20 largest flows by GHG emissions within Spain: origin– destination–product–mode (1995 vs. 2015).

Rank	1995						2015					
	Origin	Destination	Product	Mode	Emis. ktCO$_2$	% (Overall in the Year)	Origin	Destination	Product	Mode	Emis. ktCO$_2$	% (Overall in the Year)
1°	Barcelona	Barcelona	Rocks, sand and salt	Road	218.0	1.94%	Barcelona	Barcelona	Rocks, sand and salt	Road	79.8	0.79%
2°	Valencia	Valencia	Rocks, sand and salt	Road	86.8	0.77%	Madrid	Madrid	Rocks, sand and salt	Road	43.2	0.43%
3°	Navarra	Navarra	Rocks, sand and salt	Road	73.8	0.66%	Lleida	Lleida	Rocks, sand and salt	Road	39.0	0.39%
4°	Madrid	Madrid	Rocks, sand and salt	Road	73.4	0.65%	Valencia	Valencia	Rocks, sand and salt	Road	36.4	0.36%
5°	Granada	Granada	Rocks, sand and salt	Road	56.7	0.50%	Castellón	Castellón	Rocks, sand and salt	Road	33.7	0.33%
6°	Murcia	Murcia	Rocks, sand and salt	Road	50.4	0.45%	Barcelona	Barcelona	Processed food products	Road	32.9	0.33%
7°	Girona	Girona	Rocks, sand and salt	Road	48.6	0.43%	Sevilla	Sevilla	Rocks, sand and salt	Road	32.5	0.32%
8°	Sevilla	Sevilla	Rocks, sand and salt	Road	41.4	0.37%	Castellón	Castellón	Rocks, sand and salt	Road	30.6	0.30%
9°	Lugo	Lugo	Rocks, sand and salt	Road	41.1	0.37%	Girona	Girona	Rocks, sand and salt	Road	30.4	0.30%
10°	Castellón	Castellón	Rocks, sand and salt	Road	40.6	0.36%	Almería	Almería	Cement and limestone	Road	29.9	0.30%
11°	Lleida	Lleida	Rocks, sand and salt	Road	40.0	0.36%	Castellón	Valencia	Construction materials	Road	29.5	0.29%
12°	Valencia	Castellón	Rocks, sand and salt	Road	36.9	0.33%	Burgos	Burgos	Rocks, sand and salt	Road	29.0	0.29%
13°	Asturias	Asturias	Rocks, sand and salt	Road	33.6	0.30%	Barcelona	Barcelona	Cement and limestone	Road	28.1	0.28%
14°	Malaga	Malaga	Rocks, sand and salt	Road	33.4	0.30%	Granada	Granada	Rocks, sand and salt	Road	25.1	0.25%
15°	Barcelona	Barcelona	Cement and limestone	Road	31.8	0.28%	Navarra	Navarra	Rocks, sand and salt	Road	24.1	0.24%
16°	Leon	Leon	Coal	Road	30.1	0.27%	Cantabria	Cantabria	Rocks, sand and salt	Road	22.7	0.22%
17°	Barcelona	Barcelona	Chemical products	Road	29.8	0.27%	Alicante	Alicante	Rocks, sand and salt	Road	22.3	0.22%
18°	Toledo	Toledo	Rocks, sand and salt	Road	29.8	0.27%	Coruña, A	Coruña, A	Woods	Road	22.1	0.22%
19°	Tarragona	Barcelona	Rocks, sand and salt	Road	29.2	0.26%	Asturias	Asturias	Woods	Road	21.5	0.21%
20°	Barcelona	Barcelona	Construction materials	Road	28.1	0.25%	Barcelona	Barcelona	Chemical products	Road	21.1	0.21%

Source: Own elaboration.

104

To dig deeper into our aforementioned results, Figure 3, using a multidimensional *Sankey diagram*, plots the main GHG emissions for inter-provincial freight flows in 2015, with the dense intra-provincial deliveries removed. The diagram should be read from left to right. The first subdivision (links between Columns 1 and 2) suggests that of the total 10,105 kt GHG emitted by internal freight flows within Spain in 2015 (Baseline scenario), 94% were produced by road, 4% by ship, and 1% by aircraft and railway. Within road (links between Columns 2 and 3), the main inter-provincial flows originate in Barcelona, Valencia, Madrid, Sevilla, Zaragoza, Navarra, Tarragona, A Coruña, Lleida, Asturias, etc. The first seven provinces (from Zaragoza to Burgos) are important exporters by road (origin provinces), but none of them are associated with the most polluting bilateral flows in the country, which appear in the links between Columns 3 and 4. Rather, the other provinces in Column 3 correspond to the origins of the most polluting flows shown in Column 4, where the province of destination and type of sector is shown. Column 4, after the general label "Rest of Spain-Other sectors", identifies the most polluting inter-provincial flows in the country in 2015 as follows:

(1)   Exports of "Rocks, sand and salt" (30,600 ktCO$_2$) by road from Valencia to Castellón.
(2)   Exports of "Construction materials" (29,500 ktCO$_2$) by road from Castellón to Valencia.
(3)   Exports of "Wood" (21,500 ktCO$_2$) by road from Asturias to Girona.
(4)   Exports of "Rocks, sand and salt" (18,800 ktCO$_2$) by road from Murcia to Alicante.

The rest of the list should be interpreted equivalently. Similar analysis could be done for the other modes, products and years.

Scenarios for Reducing Freight Emissions through Modal Shift and Efficiency Gains

As suggested in Section 3.1, the aim of this final section is to discuss alternative scenarios for the change in GHG emissions in the forecasting period 2016–2030. First, however, let us consider Figure 4, which offers relevant information in support of **Scenarios 2** and **3** (defined in Section 3.1), regarding the potential promotion of modal shifts from road to railway within Spain.

Figure 4 is complex but very informative. For clarity, the graph is split into two panels that can be analyzed in parallel, since the horizontal axis corresponds to the same variable, namely, the bilateral distance (*km*) between any given *i-j* pair of provinces. Figure 4A shows the scatter plot for railway's share of total flows for every *i-j-k* triad (*origin–destination–product*), versus bilateral distance. We use a different color for each *i-j* flow by product *k*, colored markers for the three sectors of interest, and hollow circles for the rest. Railway's specific share is reported on the left axis. Thus, for example, if railway is the only mode delivering product *k* from *i* to *j* (*k* = "*Chemical Products*" from *i* = "*Asturias*" to *j* = "*Cantabria*"), the share will reach a 100%; inversely, if zero flows are reported by railway for this *i-j-k* combination, a zero share will appear. Note that the number of 100% and 0% shares for *i-j-k* in the graphs is remarkable, since many dots of different colors appears in the 0% and 100% level for almost every distance and product. In many other cases, the railway share lies between these two extremes.

Figure 4B includes three bold-line graphs, measured on the right axis. The black one corresponds to the *kernel distribution* of *tons* delivered by railway against *bilateral distance (km)*, considering the whole historical sample. As can be easily seen, the shape of the distribution is clearly concentrated in the shortest distance (<200 km), with a *plateau* of constant intensity of deliveries at 200–500 km, and a *valley* thereafter, with some *bumps* at 700 km and 800 km. To illustrate the heterogeneity hidden within this aggregate distribution, we have also added the kernel distribution of tons of "*Paper*" (dark orange) and "*Transport material*" (pale blue), where the intensity of flows over longer distances is clearer.

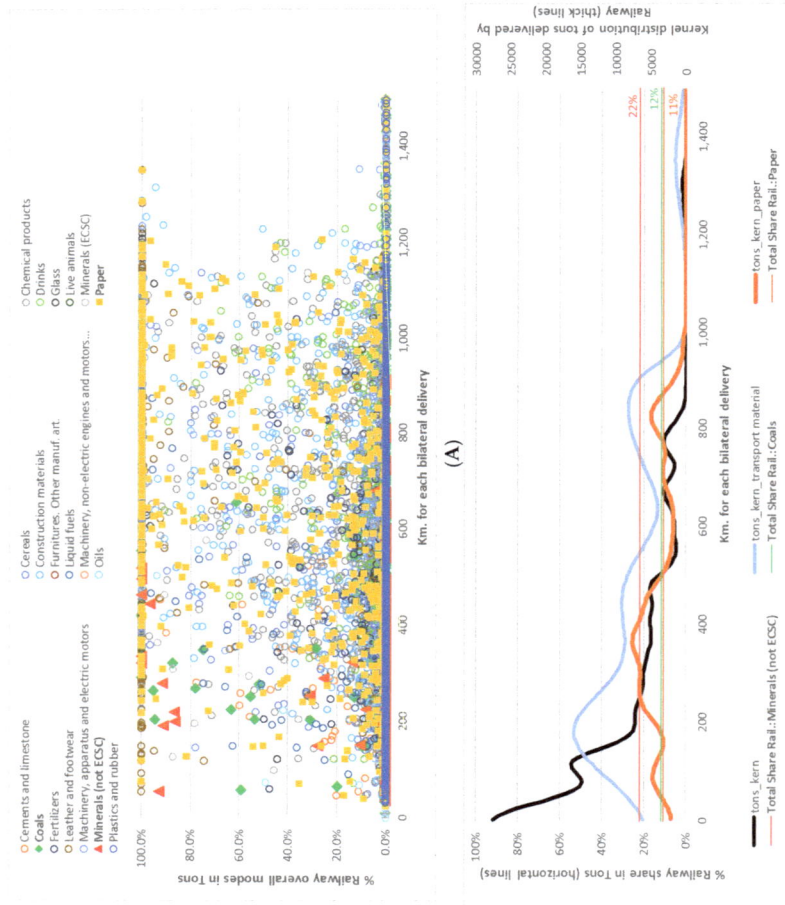

Figure 4. Share of Railway for each origin–destination–product triad (**A**); and selected kernel distributions (**B**) (1995–2015). Source: Own elaboration.

Moreover, also in Figure 4B, three horizontal thin lines in color indicate railway's total share of the aggregate flows for the three sectors with the largest shares. Note that the horizontal lines do not vary along bilateral distance, since they just represent scalars computed over the whole historical sample:

(i)   The thin red line shows railway's share in *"Minerals (not ECSC)"* in Spain as a whole, which accounts for 22% over the historic period. Note that although this 22% (Figure 4B) is the largest share that any of the 29 products has for railway in the entire country. for some specific *i-j* flows of this product (with distances shorter than 400 km), railway registers shares above 80% (red triangles ▲ in Figure 4A). We point out also that the red triangles for most of the bilateral distances (*i-j* pairs) in Figure 4A appear at 0% share, with very few at 100%. Probably, these short-distance trips by railway are explained by the existing interconnecting railway networks within industrial clusters of heavy industry, so the bulk of these products are moved very efficiently from maritime ports to factories, and from there to warehouses, storage infrastructures and other transformation plants, usually agglomerated within relatively short distances.

(ii) The second-largest average share is indicated with a green horizontal line (Figure 4B), which corresponds to *"Coal"* (12%). Our conclusions are similar to those for *"Minerals (not ECSC)"*, since in Figure 4A we also see green diamonds (◊) with shares above 12% for specific *i-j* (mainly located between 200 km and 400 km).

(iii) Finally, the orange horizontal line in Figure 4B indicates railway's total share for *"Paper"*, railway's third-largest share in Spain as a whole. In this case, we see many non-negative shares for specific *i-j*, marked with (■) in Figure 4A, over a wider range of distances. This may be a sign that railway is a more credible alternative to road over long distances for this product than for the other two. In their case, the higher general share is driven by few *i-j* specific pairs, which enjoy railway infrastructures of singular nature. More interestingly, for the *"Paper"* sector, railway will be a potential substitute for long trips by road. This can be clearly seen in the kernel distribution for this product in Figure 4B, plotted with a dark orange thick line, which has two humps for flows of 700 km and 800 km.

Given the previous analysis, it is worth remembering that railway accounts for a very small share of intra-national freight flows in Spain (just 1.97% in terms of *tons*, and 3% in tons*km, according to official figures). Note that this small share is compatible with a 100% share for specific *i-j-k* triads, and total shares of 22%, 12% or 11% for the three peculiar products commented before. However, it seems highly unlikely that a modal shift from road to railway would result in a total share of 40–60% for the entire country, as the EC's *"White Paper of Transport"* [4] has recommended.

It is, in fact, very difficult to change individual decisions about preferred mode use. In theory, railway can appear as the preferred option for some heavy products and certain *i-j* pairs. As previously suggested, however, current flows are associated with specific sectors over short-to-medium distances, and Spain's total share is still below both the shares of other countries and the emissions-reducing share recommended by the EC and strategized by Spain.

With this in mind, we turn to **Scenario 2**. Here, we take the maximum share by railway observed for a given product *k* for any possible *i-j* dyad and use it as our benchmark for every flow of the same product *k* for any other *i-j* dyad with a distance above 600 km. We use a limit of 40% for the sake of greater realism. Our more radical **Scenario 3** assumes a hypothetical modal shift from road to railway affecting all deliveries by road with a bilateral distance above 150 km and a maximum railway share of 60% for each *i-j-k-t*. This scenario is less realistic, since it includes trips of intermediate distance (150–600 km) and does not consider maximum historical shares for each product *k*. However, as we show, only **Scenario 3** leads to net reductions in GHG emissions in 2030. We analyze alternative distance segments and thresholds in future research.

Total *tons transferred* from road to railway for these two alternative scenarios is illustrated in Figure 5, where we have computed the *kernel regression* between the tons moved by road in **Scenarios 1** (no modal shift), **2** and **3**, with moderate and extreme modal shifts, and two alternative distance ranges. In Figure 5A for **Scenario 2**, total tons transferred is moderate, which indicates that, because of Spain's particular geographical features, the 600 km threshold may be too stringent. By contrast, total load transferred to railway in **Scenario 3** (Figure 5B) is quite large, given the strong concentration of flows over the shortest distance, as shown in Figure 4B.

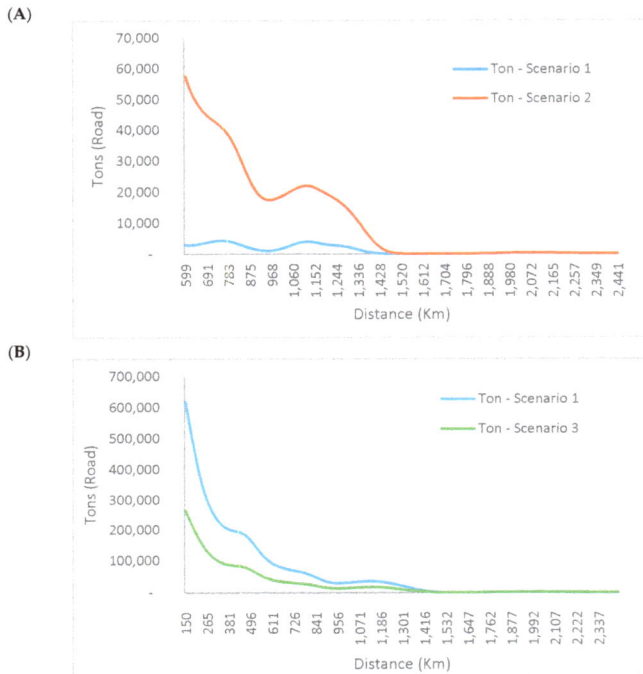

Figure 5. Kernel distribution of volume moved by road before/after the modal shift in Scenarios 2 and 3 vs. Scenario 1. Source: Own elaboration.

We now focus on changes in the main aggregate variables combined in the empirical exercise, which are plotted in Figure 6. The included time series (measured in growth rates) correspond to the GHG estimates (Scenario 1), the intra- and inter-provincial trade flows in volume (*Freight_Ton*) and euros (*Trade_Euros*), and their forecasts obtained with the gravity model. We also add the change in Spanish GDP, mixing official figures for the historical period with forecast (aggregated) figures, covering 2016–2030. Moreover, pro memoria, we add the change in GHG emissions predicted by the MAPAMA for the entire "Transport sector" in the latest Spanish Emissions Inventory [7].

First, it is interesting to compare the dynamics of each series before and after the crisis to interpret what is obtained for the forecasting period. In addition to the trends shown in the figure, we report average growth rates for each variable in three sub-periods:

(i) Before the crisis (1995–2007), the change in *freight flows in tons* (8.8%) was more dynamic than *GDP (7.4%)*, the GHG estimates obtained here (4.7%) and the official GHG emissions estimates (3.7%) from the MAPAMA for the entire transport sector (passengers + freight).

(ii) During the crisis and the take-off (2008–2015), GDP showed the slowest decline (0%), followed by the official GHG estimates (−3%) and the bottom-up GHG estimates described in this paper

(−8%). Changes in intra-national flows in tons and euros were more volatile than GDP in this period.

(iii) For the forecasting period (2015–2030), the models suggest a continuous positive trend for GDP in Spain (3.3%), followed, very closely, by intra-national trade in euros (3.7%). However, *freight flows in tons* (6%) appear to be more dynamic, following similar patterns to those observed during the pre-crisis period (8.8%). Note that the official forecast scenario for this period published by the MAPAMA uses a 1% average growth rate for GHG emissions for the entire transport sector (passengers + freight).

When all these trends are combined with the downward-trend GHG emissions factors plot in Figure 2, we obtain **Baseline-Scenario 1**, which shows permanent positive rates year by year, indicating higher levels of emissions than in 2015. These results suggest low levels of decoupling between freight traffic and *GDP* [13], and thus that the efficiency gains predicted in Figure 2 would be overwhelmed by the expected positive change in the economy. The blue line corresponds to the MAPAMA's official predictions for this period. Interestingly, these figures point to an increase in GHG emissions for the whole *"Transport sector"* in Spain, although with lower rates (1%) than those estimated here (6%).

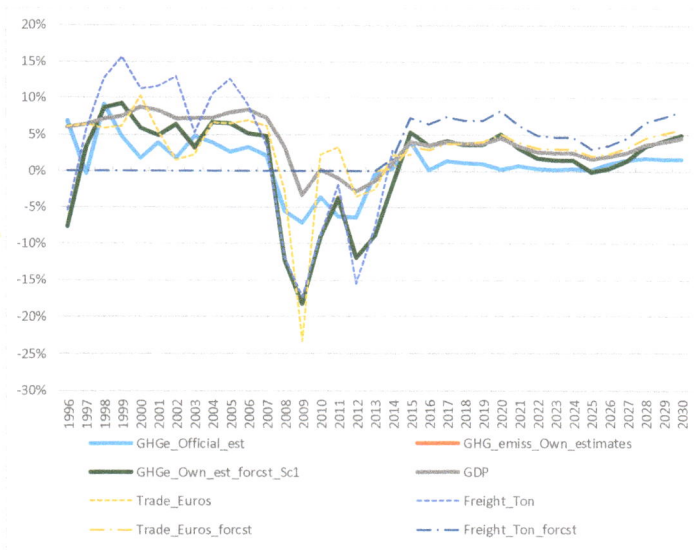

Figure 6. Observed and predicted evolution of main aggregates. Growth rates in percent. Source: Own elaboration.

Zooming in on these aggregates, and considering the alternative scenarios described before, Table 4 summarizes the main results obtained at the national and regional level (Nuts 2). Columns 1 and 2 report the levels for **Scenario 1-Baseline** for 2015 and 2030. Column 3 computes the difference in terms of $ktCO_2$ during the forecasting period. This difference is a consequence of the projected inertia for the emissions factors, which, as reflected in Figure 2, point to greater efficiency in all modes. As previously noted, the expected increase in freight traffic will more than compensate for gains in emissions efficiency, and result in an increase of 5167 $ktCO_2$. This corresponds to the 51.1% reported in column (8) of Table 2. In Table 4, Columns 4 and 5 report differences for **Scenario 1** using the **Max.** and **Min.** bounds. The results suggest that in both sub-scenarios increasing freight traffic will still drown out any efficiency gains, causing higher GHG emissions than in 2015. Additional emissions scale up to 2760 $ktCO_2$ for the Min. and 8018 $ktCO_2$ for the Max.

Table 4. GHG emissions scenarios for 2015–2030 in ktCO$_2$.

		Scenario 1				Scenario 2			Scenario 3
		Baseline	Difference	Max	Min	Baseline	Max	Min	Min
	2015	2030		2030	2030		2030	2030	2030
	(1)	(2)	(3) = (2) − (1)	(4) = Max − (1)	(5) = Min − (1)	(6) = Baseline_2 − (1)	(7) = Max_2 − (1)	(8) = Min_2 − (1)	(9) = Min_3 − (1)
Andalucía	1379	2193	815	1227	468	675	1068	343	−96.19
Aragón	603	906	303	467	164	270	430	134	−65.93
Asturias	304	436	132	215	61	102	181	34	−27.48
Baleares	31	54	24	35	14	24	35	14	13.79
Canarias	106	158	52	92	20	52	92	20	19.60
Cantabria	190	287	97	150	53	77	127	35	−16.90
Castilla y León	1062	1551	489	767	253	418	686	189	−125.29
C.-La Mancha	807	1280	473	702	279	441	665	250	−75.02
Cataluña	1699	2591	892	1370	487	725	1184	332	−45.57
C. Valenciana	1063	1652	589	895	332	514	809	263	35.18
Extremadura	226	409	183	256	121	159	229	100	15.61
Galicia	700	1077	377	576	209	270	454	113	−58.99
Madrid	657	772	115	282	−26	90	254	−50	−161.28
Murcia	342	559	217	322	129	177	278	91	3.15
Navarra	324	517	193	288	113	163	255	85	−23.03
País Vasco	504	686	182	315	72	134	260	29	−60.36
Rioja, La	108	142	35	62	12	30	56	8	−20.84
Ceuta	0	0	−0	−0	−0	−0	−0	−0	−0.22
Melilla	0	0	0	0	0	0	0	0	0.04
Total-Spain	10,105	15,272	5167	8018	2760	4322	7063	1989	−689.74
			51%	79%	27%	43%	70%	20%	−7%

Source: Own elaboration.

Results for **Scenario 2** are shown in Columns 6–8. Column 6 uses the same emissions factors as **Scenario 1-Baseline** (Column 1) but includes a moderate modal shift from road to railway. Results for 2030 suggest a total difference in emissions of 4322 ktCO_2 with respect to **Scenario 1**, which is slightly lower than that in Column 3. The equivalent increase in emissions for **Scenario 2-Max**, in Column 7, is 7063 ktCO_2. For **Scenario 2-Min**, in Column 8, it is 1989 ktCO_2.

In **Scenario 3** (Column 9), for brevity, we consider just the Min-GHG factor. Here, the suggested radical modal shift causes emissions to fall by 689 ktCO_2 from 2015 to 2030 (**Baseline-Scenario 1**). This represents a 7% reduction with respect to the last available figure in the historical period. Note that, before the extreme modal shift, road accounts for 43% of flows traveling less than 150 km and railway for just 1.13%. After the modal shift, road accounts for 18.5% of such deliveries, while railway scales up to 25.8% in 2030. There is, in other words, an increase of more than 24% in railway. Overall, even after this (hypothetical) huge structural change in the modal composition of Spanish freight, road still maintains a 70% share of all internal freight traffic, while railway maintains 26%.

5. Conclusions

The EU and its 28 member states (2017) have pledged to reduce domestic emissions by at least 40% between 1990 and 2030. In 2016, the EU's GHG emissions were already 23% below the 1990 level. According to the member states' most recent projections for existing measures, the 20% target for 2020 will be met. In 2030, emissions are expected to be 30% lower than in 1990 if no additional policies are implemented. Emissions not covered by the EU Emission Trade System (ETS) were 11% lower in 2016 than in 2005, exceeding the 2020 target of a 10% reduction. In addition, under the Effort Sharing Decision (ESD), EU member states must meet binding annual GHG emission targets for 2013–2020 in sectors not covered by the ETS, among them transport.

In this context, we have estimated a rich database of GHG emissions for intra- and inter-provincial flows within Spain for the period 1995–2015, considering 29 products and four transport modes (*road*, *train*, *ship* and *aircraft*). We have also projected origin–destination–product–mode specific flows and their corresponding emissions for the period 2016–2030. Having established the new dataset, we generate and analyze the temporal, sectoral and spatial pattern of Spanish inter-regional GHG flows. We then address the possibility of promoting transport mode shifts to achieve a more sustainable freight system within the country, transferring specific origin–destination–product flows from road to railway. The search, in other words, is for a better mode mix attainable within Spain, one that considers actual flows, product transportability and modal sustainability per distance category. In addition, we consider three alternative trends for the evolution of GHG emissions factors for each mode for the forecasting period.

The results suggest that Spain reduced GHG emissions by 10% from 1995 to 2015, with the larger reductions occurring in the most-polluting regions, with their denser industrial activity and higher trading volumes. The baseline scenario, however, suggests that this reduction might be a *mirage*, caused by the profound economic downturn of 2008–2012, which induced great reductions in GDP and freight deliveries. Even if there should be strong efficiency gains in the GHG emissions factors for each mode, the baseline scenario projects that positive GDP growth in the next 15 years will induce similar patterns of traffic within Spain to those observed before the crisis. Scenario 2, which assumes an important modal shift in long-distance trips (>600 km) from road to railway, shows no clear gains in GHG emissions for the forecasting period, even with optimistic changes in GHG emissions factors for each mode. However, the more radical alternative of Scenario 3, where modal shifts from road to railway also affect medium-distance trips (>150 km), generates a reduction of 7% from 2015 to 2030.

According to these results, if the predictions for GDP hold, and nothing else happens, the business-as-usual scenario predicts that GHG emissions will be higher in 2030 than in 2015. Furthermore, if modal shifts occur only for trips where railway is currently competitive (>600 km), GHG emissions will still rise above the current values, despite the international commitments of COP21 and the European Strategy for Lower Emission Mobility. Scenario 3, on the other hand, suggests that to change the trends Spain must implement a dual policy: first, it must promote a radical modal shift over long distances; second, it must induce a reduction in emissions over medium distances. In reality, this last assumption could be seen as a bidirectional vector of intervention, including both the promotion of railway for medium distances and/or radical efficiency gains in road freight deliveries, with a special focus on the capillarity and *last-mile* distribution. The former entails investments to enhance the competitiveness of railway over medium distances; the latter, a clearer promotion of efficiencies in the road mode for short distances.

Despite drastic changes from the automation and electrification of transport, as well as the development of other disruptive innovations, it seems more likely that government policy should move toward Scenario 3.

In this regard, it is interesting to reconsider one of the motivations behind this paper: the need to measure GHG emissions at the sub-national level, so that regions and cities can act more effectively to moderate current trends. This critical idea, suggested in COP21 and the European Strategy, is in line with our results, which highlight how Spain's main economic activity and densest freight interactions take place over the shortest distances and around metropolitan areas.

To conclude, we need more detailed measurements of national inventories, measurements that include sub-national entities and the emissions generated by each category of delivery. Better accountability will lead to better political coordination at all levels of government, which can then promote the more radical technological improvements in the road sector and sharp modal shifts for medium-long distances. For example, measures increasing the efficiency of inter-modal platforms at the local level can increase the competitiveness of combining railway and ship with road for short-medium distances. A study of "motorways of railways" in Spain [22,42], for example, argues that reducing transfer times and costs at loading/unloading spots can increase the efficiency of combining railway and sea with trucks for *last mile* delivery. Others [21,22] have made similar arguments for the promotion of *"motorways of the sea"*, not just for the most intense international deliveries, but also as a substitute for the longest trips within the country by road: for example, between Sevilla, Bilbao, Valencia or Barcelona.

Author Contributions: All authors contributed to every stage of the article, including data management, econometric analysis and final configuration of the manuscript, tables and figures. However, each one of them made special contributions in the following aspects. C.L. was coordinated the research strategy and the methodology, mainly with respect to the gravity equation. He produced most of the writing, tables and figures. He also contacted with the journal and the coordinators of the special issue. He is also responsible of the two projects where this research is developed. J.P.-G., developed all the work related to forecasting GDP figures at the province level, as well as GHG emission factors by mode. S.P.-B., developed the database on inter-provincial flows, applied the econometric apparatus and applied the methodology needed to estimate and predict inter-provincial flows using the gravity equation.

Funding: This paper was developed in the context of two research projects: (i) the C-intereg Project (www.c-intereg.es); (ii) The ECO2016-79650-P project from the Spanish Ministry of Economics and Innovation.

Conflicts of Interest: The authors declare no conflicts of interest.

Appendix A

Table A1. Econometric results for the gravity equation, used to forecast inter-provincial flows by mode. Estimation period, 2013–2015; Forecasting period, 2016–2030.

	M1	**M2**	**M3**	**M4**
Period	2013–2015	2013–2015	2013–2015	2013–2015
Transportation mode	Ship	Railway	Road	Aircraft
VARIABLES	Ton	Ton	Ton	Ton
Ln GDP origin	−0.499	1.385	0.931 **	5.328 ***
	(1.833)	(1.294)	(0.437)	(1.984)
Ln GDP destination	2.223	0.231	0.879*	−3.312 *
	(2.486)	(1.422)	(0.453)	(1.944)

Table A1. *Cont.*

	M1	M2	M3	M4
Ln Distance	−0.541 ***	−0.521 ***	−1.011 ***	0.902 ***
	(0.0874)	(0.129)	(0.0542)	(0.128)
Contiguity	−0.884 **	0.220	0.467 ***	−0.307
	(0.414)	(0.269)	(0.0694)	(0.815)
Own_Prov	−2.174 **	−0.924 **	1.732 ***	0.289
	(0.879)	(0.389)	(0.102)	(0.760)
Island origin	2.501	−1.081	−1.562 ***	−3.038 *
	(1.921)	(1.630)	(0.550)	(1.821)
Island destination	2.061	0.0559	0.806	3.734 **
	(2.814)	(1.391)	(0.602)	(1.883)
Coast origin	−1.993 *			
	(1.022)			
Coast destination	−0.889			
	(1.325)			
Canary Islands exports to Peninsula			−4.203 ***	
			(0.602)	
Canary Islands imports from Peninsula			−1.357 **	
			(0.588)	
Balearic Islands imports from Peninsula			−0.968 **	
			(0.492)	
Constant	−8.182 ***	−8.540 ***	−5.036 ***	−41.68 ***
	(3.176)	(2.147)	(0.809)	(13.31)
Observations	20,928	5,103	64,191	20,745
R-squared	0.262	0.600	0.728	0.895
Time FE	NO	NO	NO	NO
Sector FE	YES	YES	YES	YES
Region Origin FE	YES	YES	YES	YES
Region Destination FE	YES	YES	YES	YES

Robust standard errors in parentheses. *** $p < 0.01$, ** $p < 0.05$, * $p < 0.1$. Source: Own elaboration.

Table A2. Econometric results for forecast of GHG emissions factors.

Road							
Time-trend	β^m	Sd_{β^m}	Prob. T-Stat.	γ^m	Sd_{γ^m}	Prob. T-Stat.	Error Sum. of squares
Linear	−1.013	0.356	0.022				40.721
Quadratic	8.925	2.780	0.015	−0.321	0.089	0.009	25.780
Exponential	−0.013	0.005	0.023				32.730
Potential	−0.187	0.077	0.040				74.505

The Quadratic trend has been ruled out because of inconsistency in forecast. Exponential trend has been used instead.

Railway							
Time-trend	β^m	Sd_{β^m}	Prob. T-Stat.	γ^m	Sd_{γ^m}	Prob. T-Stat.	Error Sum. of squares
Linear	−0.661	0.086	0.000				1.157
Quadratic	1.247	0.876	0.198	−0.062	0.028	0.065	1.088
Exponential	−0.032	0.005	0.000				1.246
Potential	−0.471	0.079	0.000				2.204

The Quadratic trend has been ruled out because of inconsistency in forecast. Linear trend has been used instead.

Air							
Time-trend	β^m	Sd_{β^m}	Prob. T-Stat.	γ^m	Sd_{γ^m}	Prob. T-Stat.	Error Sum. of squares
Linear	−4.581	0.191	0.000				81.954
Quadratic	−4.520	2.515	0.115	−0.002	0.081	0.981	32.518
Exponential	−0.035	0.002	0.000				97.076
Potential	−0.532	0.033	0.000				228.323

The Quadratic trend has been ruled out because of statistical non-significance in trend coefficients. Linear trend has been used instead.

Ship							
Time-trend	β^m	Sd_{β^m}	Prob. T-Stat.	γ^m	Sd_{γ^m}	Prob. T-Stat.	Error Sum. of squares
Linear	−1.145	0.129	0.000				9.604
Quadratic	−2.353	0.472	0.000	0.058	0.022	0.017	7.782
Exponential	−0.043	0.005	0.000				8.013
Potential	−0.300	0.041	−7.242				17.516

The Quadratic trend has been ruled out because of inconsistency in forecast. Linear trend has been used instead.

Source: Own elaboration.

113

Table A3. GHG emissions by province (Nuts 3). Structure and evolution (ktCO$_2$ eq).

| | 2015 | | | | | Growth Rates | | |
| | Total | Intra-Provincial | Inter-Provincial | Percent | Total/GDP | 2015–1995 | 2015–2009 | 2030–2015 |
	(1)	(2) in % of (1)	(3) in % of (1)	(4) = (1)/Spain	(5)	(5)	(6)	(7)
Araba	97.5	5.8%	94.2%	0.1%	8.50	−8.6%	−32.0%	34.2%
Albacete	119.2	17.0%	83.0%	0.1%	16.50	−23.9%	−35.5%	53.6%
Alicante	190.1	25.3%	74.7%	0.2%	5.68	−11.2%	−26.4%	49.2%
Almería	149.8	39.1%	60.9%	0.1%	11.69	−2.0%	−12.4%	89.5%
Ávila	48.7	9.9%	90.1%	0.0%	16.24	40.3%	−31.0%	51.2%
Badajoz	166.1	26.6%	73.4%	0.2%	15.37	21.3%	−31.8%	76.5%
Balears, Illes	30.7	70.1%	29.9%	0.0%	1.12	26.9%	−20.1%	76.8%
Barcelona	923.3	28.9%	71.1%	0.9%	6.11	−23.6%	−30.2%	39.7%
Burgos	258.4	20.3%	79.7%	0.3%	27.67	−0.6%	−19.7%	35.1%
Cáceres	59.9	47.9%	52.1%	0.1%	9.00	−18.5%	−34.9%	93.4%
Cádiz	157.0	24.4%	75.6%	0.2%	7.88	−15.0%	−25.7%	58.0%
Castellón	243.3	26.5%	73.5%	0.2%	18.43	−21.9%	−20.0%	62.4%
Ciudad Real	170.9	11.4%	88.6%	0.2%	17.62	−3.4%	−15.7%	37.5%
Córdoba	129.8	23.8%	76.2%	0.1%	9.76	−14.3%	−24.8%	50.6%
Coruña, A	294.8	33.1%	66.9%	0.3%	12.06	6.1%	−20.3%	74.2%
Cuenca	127.1	17.5%	82.5%	0.1%	31.27	12.0%	−45.4%	49.3%
Girona	174.2	35.9%	64.1%	0.2%	8.74	−29.5%	−38.6%	72.0%
Granada	153.8	31.3%	68.7%	0.2%	9.88	−26.1%	−45.8%	80.8%
Guadalajara	157.1	8.5%	91.5%	0.2%	34.26	93.8%	−12.5%	51.9%
Gipuzkoa	158.9	9.6%	90.4%	0.2%	7.29	−10.5%	−31.8%	16.1%
Huelva	174.7	17.5%	82.5%	0.2%	19.35	−5.4%	−8.7%	41.6%
Huesca	149.3	20.6%	79.4%	0.1%	26.59	−2.8%	−19.4%	62.5%
Jaén	105.9	23.8%	76.2%	0.1%	9.76	−25.4%	−38.8%	56.1%
León	161.6	22.0%	78.0%	0.2%	17.22	−27.3%	−40.0%	74.1%
Lleida	274.5	33.5%	66.5%	0.3%	22.54	−26.4%	−24.3%	67.3%
Rioja, La	107.8	10.9%	89.1%	0.1%	13.72	−3.2%	−18.7%	32.2%
Lugo	173.1	16.5%	83.5%	0.2%	24.49	−13.4%	−13.0%	34.2%
Madrid	656.6	15.3%	84.7%	0.6%	3.22	0.9%	−29.7%	17.5%
Málaga	132.5	24.1%	75.9%	0.1%	4.78	10.4%	−40.7%	77.2%
Murcia	342.4	22.3%	77.7%	0.3%	12.14	1.0%	−13.6%	63.3%
Navarra	323.6	24.3%	75.7%	0.3%	17.44	−4.6%	−18.8%	59.6%
Ourense	85.0	18.6%	81.4%	0.1%	13.77	35.4%	−23.5%	9.8%
Asturias	304.0	23.2%	76.8%	0.3%	14.33	−28.1%	−33.4%	43.3%
Palencia	113.4	16.4%	83.6%	0.1%	28.55	0.5%	−26.8%	52.1%
Palmas, Las	57.3	40.3%	59.7%	0.1%	2.69	21.1%	34.4%	22.6%
Pontevedra	147.1	26.8%	73.2%	0.1%	7.89	−26.4%	−41.2%	61.9%
Salamanca	67.3	24.8%	75.2%	0.1%	10.17	8.0%	−24.9%	40.4%
S.C.d Tenerife	48.7	45.2%	54.8%	0.0%	2.48	112.2%	−38.0%	80.6%
Cantabria	189.8	22.6%	77.4%	0.2%	15.56	−21.3%	−15.7%	51.3%
Segovia	106.6	12.7%	87.3%	0.1%	32.25	39.1%	−25.8%	54.2%
Sevilla	375.2	22.6%	77.4%	0.4%	10.47	−9.8%	−32.2%	44.0%
Soria	55.3	13.3%	86.7%	0.1%	25.20	30.4%	−13.3%	34.5%
Tarragona	327.4	12.5%	87.5%	0.3%	14.70	−21.7%	−23.9%	65.8%
Teruel	104.8	25.9%	74.1%	0.1%	31.52	−2.4%	−20.8%	47.8%
Toledo	232.8	15.4%	84.6%	0.2%	19.64	−10.1%	−43.1%	86.2%
Valencia	629.6	22.4%	77.6%	0.6%	11.64	−7.4%	−28.5%	54.6%
Valladolid	166.6	15.0%	85.0%	0.2%	13.38	13.0%	−25.2%	27.3%
Bizkaia	247.4	14.6%	85.4%	0.2%	7.45	−37.9%	−26.1%	49.8%
Zamora	84.1	26.0%	74.0%	0.1%	24.78	39.8%	−34.0%	53.8%
Zaragoza	349.3	19.9%	80.1%	0.3%	14.26	10.1%	−18.8%	45.7%
Ceuta	0.4	0.0%	100.0%	0.0%	0.25	1045.1%	3950.7%	−36.0%
Melilla	0.2	0.0%	100.0%	0.0%	0.16	1931.9%	570.2%	76.0%
Total-Spain	**10,105.2**	**36.6%**	**63.4%**	**100.0%**	**9.36**	**−10.0%**	**−27.4%**	**51.1%**

Source: Own elaboration.

Table A4. Products covered by the C-intereg database.

Code	Product
1	Live animals
2	Cereals
3	Unprocessed food
4	Wood
5	Processed food products
6	Oil (food)
7	Tobacco
8	Drinks
9	Coal
10	Minerals (not ECSC)
11	Liquid fuels
12	Minerals (ECSC)
13	Steel products (ECSC)
14	Steel products (not ECSC)
15	Rocks, sand and salt
16	Cement and limestone
17	Glass
18	Construction materials
19	Fertilizers
20	Chemical products
21	Plastics and rubber
22	Machinery (non-electric)
23	Machinery (electric)
24	Transport equipment
25	Textile and clothing
26	Leather and footwear
27	Paper
28	Products of wood and cork
29	Furniture, other goods

Source: Own elaboration based on C-intereg (www.c-intereg.es).

References

1. European Commission. The Road from Paris: Assessing the Implications of the Paris Agreement and Accompanying the Proposal for a Council Decision on the Signing, on Behalf of the European Union, of the Paris Agreement Adopted under the United Nations Framework Convention on Climate Change. 2016. Available online: https://ec.europa.eu/clima/policies/international/negotiations/paris_en#tab-0-1 (accessed on 10 July 2018).
2. European Commission. A European Strategy for Low-Emission Mobility. Commission Staff Working Document. 2016. Available online: https://ec.europa.eu/clima/policies/transport_en#tab-0-1 (accessed on 10 July 2018).
3. European Commission. Communication: A Policy Framework for Climate and Energy in the Period from 2020 to 2030. 2014. Available online: https://ec.europa.eu/clima/policies/strategies/2030_en#tab-0-1 (accessed on 10 July 2018).
4. European Commission. White Paper: Roadmap to a Single European Transport Area—Towards a Competitive and Resource Efficient Transport System. 2011. Available online: https://ec.europa.eu/transport/themes/strategies/2011_white_paper_en (accessed on 10 July 2018).
5. Transport and Environment. Too Big to Ignore: Truck CO_2 Emissions in 2030. 2015. Available online: https://www.transportenvironment.org/publications/too-big-ignore-%E2%80%93-truck-co2-emissions-2030 (accessed on 10 July 2018).
6. Davydenko, I.; Ehrler, V.; de Ree, D.; Lewis, A.; Tavasszy, L. Towards a global CO_2 calculation standard for supply chains: Suggestions for methodological improvements. *Transp. Res. Part D Transp. Environ.* **2014**, *32*, 362–372. [CrossRef]

7. MAPAMA. Sistema Español de Inventario de Emisiones. 2017. Available online: http://www.mapama. gob.es/es/calidad-y-evaluacion-ambiental/temas/sistema-espanol-de-inventario-sei-/volumen2.aspx (accessed on 10 July 2018).

8. Ferng, J.J. Allocating the responsibility of CO_2 over-emissions from the perspectives of benefit principle and ecological deficit. *Ecol. Econ.* **2003**, *46*, 121–141. [CrossRef]

9. Cristea, A.; Hummels, D.; Puzzello, L.; Avetisya, M. Trade and the greenhouse gas emissions from international freight transport. *J. Environ. Econ. Manag.* **2013**, *65*, 153–173. [CrossRef]

10. Lenzen, M.; Pade, L.L.; Munksgaard, J. CO_2 multipliers in multi-region input–output models. *Econ. Syst. Res.* **2004**, *16*, 391–412. [CrossRef]

11. Mongelli, I.; Tassielli, G.; Notarnicola, B. Global warming agreements, international trade and energy/carbon embodiments: An input–output approach to the Italian case. *Energy Policy* **2006**, *34*, 88–100. [CrossRef]

12. Sánchez-Chóliz, J.; Duarte, R. CO_2 emissions embodied in international trade: Evidence for Spain. *Energy Policy* **2004**, *32*, 1999–2005. [CrossRef]

13. Alises, A.; Vassallo, J.A. Comparison of road freight transport trends in Europe. Coupling and decoupling factors from an Input–Output structural decomposition analysis. *Transp. Res. Part A* **2015**, *82*, 141–157. [CrossRef]

14. López, L.A.; Cadarso, M.A.; Gómez, N.; Tobarra, M.A. Food miles, carbon footprint and global value chains for Spanish agriculture: Assessing the impact of a carbon border tax. *J. Clean. Prod.* **2015**, *103*, 423–436. [CrossRef]

15. Llano, C.; De la Mata, T.; Díaz-Lanchas, J.; Gallego, N. Transport-mode competition in intra-national trade: An empirical investigation for the Spanish case. *Transp. Res. Part A* **2017**, *95*, 334–355. [CrossRef]

16. McKinnon, A.C.; Piecyk, M.I. Measurement of CO_2 emissions from road freight transport: A review of UK experience. *Energy Policy* **2009**, *37*, 3733–3742. [CrossRef]

17. Anderson, J.E.; van Wincoop, E. Gravity with gravitas: A solution to the border puzzle. *Am. Econ. Rev.* **2003**, *93*, 170–192. [CrossRef]

18. Llano, C.; Esteban, A.; Pulido, A.; Pérez, J. Opening the Interregional Trade Black Box: The C-intereg Database for the Spanish Economy (1995–2005). *Int. Reg. Sci. Rev.* **2010**, *33*, 302–337. [CrossRef]

19. Zanni, A.M.; Bristow, A.L. Emissions of CO_2 from road freight transport in London: Trends and policies for long run reductions. *Energy Policy* **2010**, *38*, 1774–1786. [CrossRef]

20. Cadarso, M.A.; Gómez, N.; López, L.A.; Tobarra, M.A. CO_2 emissions of international freight transport and offshoring: Measurement and allocation. *Ecol. Econ.* **2010**, *69*, 1682–1694. [CrossRef]

21. López-Navarro, M.Á. Environmental Factors and Intermodal Freight Transportation: Analysis of the Decision Bases in the Case of Spanish Motorways of the Sea. *Sustainability* **2014**, *6*, 1544–1566. [CrossRef]

22. Pérez-Mesa, J.C.; Céspedes, J.J.; Salinas, J.A. Feasibility study for a Motorway of the Sea (MoS) between Spain and France: Application to the transportation of perishable cargo. *Transp. Rev.* **2010**, *30*, 451–471. [CrossRef]

23. Demir, E.; Bektas, T.; Laporte, G. A review of recent research on green road freight transportation. *Eur. J. Oper. Res.* **2014**, *237*, 775–793. [CrossRef]

24. Gallego, N.; Llano, C.; De la Mata, T.; Díaz-Lanchas, J. Intranational Home Bias in the Presence of Wholesalers, Hub-Spoke Structures and Multimodal Transport Deliveries. *Spat. Econ. Anal.* **2015**, *10*, 369–399. [CrossRef]

25. Monzón, A.; Pérez, P.; Di Ciommo, F. La Eficiencia Energética y Ambiental de los Modos de Transporte en España. Consejo Superior de Cámaras. 2009. Available online: https: //www.researchgate.net/publication/265927638_LA_EFICIENCIA_ENERGETICA_Y_AMBIENTAL_ DE_LOS_MODOS_DE_TRANSPORTE_EN_ESPANA (accessed on 10 July 2018).

26. Kristensen, H.O. Cargo transport by sea and road-technological and economic environmental factors. *Mar. Technol.* **2002**, *39*, 239–249.

27. Kamakaté, F.; Schipper, L. *Trends in Truck Freight Energy Use and Carbon Emissions in Selected OECD Countries from 1973 to 2003*; Transportation Research Board, TRB, National Research Council: Washington, DC, USA, 2008.

28. Steenhof, P.; Woudsma, C.; Sparling, E. Greenhouse gas emissions and the surface transport of freight in Canada. *Transp. Res. Part D* **2006**, *11*, 369–376. [CrossRef]

29. Wee, B.V.; Janse, P.; Brink, R.V.D. Comparing energy use and environmental performance of land transport modes. *Transp. Rev.* **2005**, *25*, 3–24. [CrossRef]

30. European Conference of Ministers of Transport (ECMT). *Cutting Transport CO₂ Emissions: What Progress?* European Conference of Ministers of Transport, OECD Publications: Paris, France, 2007; p. 264.

31. Transport and Environment Database System (TRENDS). *Calculation of Indicators of Environmental Pressure Caused by Transport, Main Report, European Commission*; Office for Official Publications of the European Communities: Luxembourg, 2003.

32. Pulido San Román, A.; Pérez, J.; García, J. (Eds.) *Modelos Econométricos*; Piramide: Madrid, Spain, 2001.

33. Moral Cancedo, J.; Pérez García, J. Feeding Large Econometric Models by a Mixed Approach of Classical Decomposition of Series and Dynamic Factor Analysis: Application to Wharton-UAM Model. *Estudios de Economía Aplicada* **2015**, *3*, 487–512.

34. Head, K.; Mayer, T. Gravity Equations: Workhorse, Toolkit, and Cookbook. In *The Handbook of International Economics*; Gopinath, G., Helpman, E., Rogoff, K., Eds.; Elsevier: New York, NY, USA, 2014; Chapter 3; Volume 4, pp. 131–195.

35. LeSage, J.P.; Llano-Verduras, C. Forecasting spatially dependent origin and destination commodity flows. *Empir. Econ.* **2014**, 1–20. [CrossRef]

36. Garmendia, A.; Llano-Verduras, C.; Minondo, A.; Requena-Silventre, F. Networks and the Disappearance of the Intranational Home Bias. *Econ. Lett.* **2012**, *116*, 178–182. [CrossRef]

37. Feenstra, R. Border effect and the gravity equation: Consistent methods for estimation. *Scott. J. Political Econ.* **2002**, *49*, 1021–1035. [CrossRef]

38. Zofío, J.L.; Condeço-Melhorado, A.M.; Maroto-Sánchez, A.; Gutiérrez, J. Generalized transport costs and index numbers: A geographical analysis of economic and infrastructure fundamentals. *Transp. Res. Part A Policy Pract.* **2014**, *67*. [CrossRef]

39. Díaz-Lanchas, J.; Llano-Verduras, C.; Zofío, J. *Trade Margins, Transport Cost Thresholds, and Market Areas: Municipal Freight Flows and Urban Hierarchy*; Working Papers in Economic Theory from Universidad Autónoma de Madrid (Spain); No. 10; Department of Economic Analysis (Economic Theory and Economic History): Suitland, MD, USA, 2013.

40. Silva, J.; Tenreyro, S. The log of gravity. *Rev. Econ. Stat.* **2006**, *88*, 641–658. [CrossRef]

41. Liimatainen, H.; Hovi, I.B.; Arvidsson, N.; Nykänen, L. Driving forces of road freight CO₂ in 2030. *Int. J. Phys. Distrib. Logist. Manag.* **2015**, *45*, 260–285. [CrossRef]

42. Ministerio de Fomento. Estudio Para el Desarrollo de Autopistas Ferroviarias en la Península Ibérica. 2015. Available online: https://www.fomento.gob.es/NR/rdonlyres/58D8E964-F722-4539-A3D7-DA52FFDC9602/134015/EstudioAutopistasFerroviarias2015.pdf (accessed on 10 July 2018)

43. Ministerio de Fomento, Observatorio del Transporte Intermodal Terrestre y Marítimo. Documento Final. 2011. Available online: https://www.fomento.gob.es/NR/rdonlyres/DF10A112-74FF-482F-8953-67DE0DDF3D24/103643/OBSERVATORIO_Documento_Final.pdf (accessed on 10 July 2018).

sustainability

MDPI

Article

Decarbonization Pathways for International Maritime Transport: A Model-Based Policy Impact Assessment

Ronald A. Halim *, Lucie Kirstein, Olaf Merk and Luis M. Martinez

International Transport Forum/OECD, 2 rue André Pascal, 75775 Paris CEDEX 16, France;
Lucie.Kirstein@itf-oecd.org (L.K.); Olaf.Merk@itf-oecd.org (O.M.); Luis.Martinez@itf-oecd.org (L.M.M.)
* Correspondence: halim.ronald.a@gmail.com or ronald.halim@itf-oecd.org; Tel.: +33-617470889

Received: 1 May 2018; Accepted: 25 June 2018; Published: 29 June 2018

Abstract: International shipping has finally set a target to reduce its CO_2 emission by at least 50% by 2050. Despite this positive progress, this target is still not sufficient to reach Paris Agreement goals since CO_2 emissions from international shipping could reach 17% of global emissions by 2050 if no measures are taken. A key factor that hampers the achievement of Paris goals is the knowledge gap in terms of what level of decarbonization it is possible to achieve using all the available technologies. This paper examines the technical possibility of achieving the 1.5° goal of the Paris Agreement and the required supporting policy measures. We project the transport demand for 6 ship types (dry bulk, container, oil tanker, gas, wet product and chemical, and general cargo) based on the Organization for Economic Co-operation and Development's (OECD's) global trade projection of 25 commodities. Subsequently, we test the impact of mitigation measures on CO_2 emissions until 2035 using an international freight transport and emission model. We present four possible decarbonization pathways which combine all the technologies available today. We found that an 82–95% reduction in CO_2 emissions could be possible by 2035. Finally, we examine the barriers and the relevant policy measures to advance the decarbonization of international maritime transport.

Keywords: international shipping; maritime transport; decarbonization; Paris Agreement; freight transport model; policy measures; GHG emission; 1.5 degrees objective; carbon pricing; market-based measure

---

1. Introduction

International maritime transport has been the main mode of transport for global trade over the past century and one of the cornerstones of globalization. There have been significant improvements in the efficiency of international shipping in the past couple of decades. Ever since the industry introduced containerization and ultra-large container vessels, the unit cost of maritime transport has declined substantially due to the major improvement in economies of scale.

The Paris Climate Agreement of 2015, which, by far, has been the most successful agreement in advancing global commitments to reduce CO_2 emissions, does not include any targets for the shipping sector. The political dynamics that are shaped by differing interests in the shipping industry have left it to be the last sector to establish any CO_2 reduction goals. Recognizing the international character of the sector and the diverse regulatory challenges for different countries, governments expect the International Maritime Organization (IMO) to lead the advancement in decarbonizing the sector. Without a contribution from the shipping sector, the goals of the Paris Agreement in limiting the rise in global temperature to 1.5 °C–2 °C will be under threat. Shipping currently contributes to approximately 2% of the total CO_2 emissions, yet emissions from shipping are estimated to grow between 50 and 250% by 2050 [1], which would potentially increase shipping's emissions to up to 17% of the total greenhouse gas (GHG) emissions if no measures are taken [2].

Decarbonization of international shipping has progressed rather slowly due to fragmented and diverse ambitions and interests of stakeholders in the sector. Until recently, debates at the IMO were characterized by major disagreement as to how and whether the sector should align to the goals of the Paris Agreement. The current IMO GHG reduction roadmap indicates a decision-making process that is sluggish in implementing the necessary measures and regulations [3]. An important milestone of the roadmap is the adoption of a strategy to reduce GHG emissions, including a level of ambition and candidate short-, medium-, and long-term measures, which were announced at the 72nd IMO Marine Environment Protection Committee (MEPC) meeting in April 2018. The strategy mandates a reduction in total annual GHG emissions from shipping by at least 50% by 2050 compared to the 2008 level while pursuing efforts towards phasing them out entirely. The strategy also includes a reference to "a pathway of CO_2 emissions reduction consistent with the Paris Agreement temperature goals". The initial strategy will be revised in 2023 and reviewed again 5 years thereafter.

Without a concrete, ambitious, and enforceable target, there will be little incentive for the industry to invest in low-carbon technologies on a sufficient scale. We argue that one reason for this is the high risks and uncertainties associated with investments in the generally more costly low-carbon technologies. These policy uncertainties could, hence, also stifle innovation in low-carbon technologies and fuels. One of the key factors that has hampered progress in defining an ambitious target is the lack of thorough studies that assess the technical possibility of decarbonizing international maritime transport, especially according to the more ambitious goal of the Paris Agreement—i.e., the 1.5 °C temperature limit.

Most of the previous studies focus on above-1.5 °C scenarios, such as in [4–6], with a longer decarbonization time horizon up to 2050. A notable exception is in [7] which also includes a 1.5 °C scenario in which shipping emissions are close to zero by 2035. Some studies focus on zero-carbon shipping in a shorter term, but only for new ships. For example, a recent study by Lloyd's Register assesses the requirements for a transition to zero-emission vessels by 2030 [8].

Moreover, there are also very few studies that assess the required policy measures to support the realization of the decarbonization target. An ambitious target will generally require massive and rapid changes, often involving capital investments that might not have clear prospects for profitability. This could bring considerable economical disruptions (i.e., loss of profit) to industry stakeholders, notably ship-owners, shipbuilders, shipping service providers, and national governments. Without appropriate national and supra-national policies that can provide strong incentives and mechanisms that favor the adoption of low-carbon technologies, ambitious targets and strict regulations might face strong resistance by relevant industry stakeholders. Therefore, it is important that any targets and mitigation measures that are imposed to the industry are accompanied by incentives and supporting policies if they are to be effective and widely accepted by the stakeholders.

This paper presents a systematic assessment of the technical feasibility of decarbonizing maritime transport by 2035, and its implications for the required supporting policy measures. Specifically, we study the possible decarbonization pathways that could conform to the 1.5 °C goal, where CO_2 emissions would need to reach almost zero by 2035. This more ambitious goal is chosen since it would represent the most disruptive scale of adaptation by the industry, which also poses the biggest policy challenges. We establish a modeling framework that consists of international freight transport and emission models to study the impact of technological, operational, and alternative fuels measures on CO_2 emissions. We project the transport demand for the global trade of 25 commodities based on the Organization for Economic Co-operation and Development's (OECD's) forecast and assign it to 6 major ship types that represent the global shipping fleets (dry bulk, container, oil tanker, gas, wet product and chemical, and general cargo). Next, we test the impact of technical, operational, alternative fuels, and market-based measures on CO_2 emissions until 2035.

The contribution of this paper is twofold. First, it examines the possibility to decarbonize international maritime transport by 2035 using today's technologies. Second, it provides recommendations to the policy-makers on the combination of measures and incentives that can help to achieve the decarbonization of the shipping sector. The remaining of this paper is structured as follows. In Section 2, we review the available emission reduction measures. Section 3 presents the model-based assessment of the CO_2 reduction potential of the combination of different measures. In Section 4, we assess the barriers and market failures in decarbonizing international shipping. Section 5 highlights the implications for effective policy instruments. Finally, Section 6 concludes on the study's results, as well as their market and policy implications.

2. Review of Technical and Market-Based Measures

This section gives an overview of possible measures to achieve decarbonization of shipping by 2035. We distinguish between three types of measures: technological measures, operational measures, and alternative fuels and energy (Table 1). We use the findings from existing research to inform our modelling of possible emission reductions. The respective emission reduction potentials presented in each of the subsections are assessed individually and cannot be cumulated without considering their possible interactions.

Table 1. Overview of principal measures to reduce shipping's carbon emissions.

Type of Measures	Main Measures
Technological	Light materials, slender design, friction reduction, waste heat recovery
Operational	Lower speeds, ship size, ship–port interface
Alternative fuels/energy	Sustainable biofuels, hydrogen, ammonia, fuel cells, electric ships, wind assistance, solar energy

2.1. Technological Measures

Improving energy efficiency through technological measures is the aim of the global regulation of the energy efficiency of ships. This regulation requires ships built after 1 January 2013 to comply with a minimum energy efficiency level, the Energy Efficiency Design Index (EEDI) included in the International Convention for the Prevention of Pollution from Ships (MARPOL) Annex VI, which measures the CO_2 emitted (g/tonne mile) based on ship design and engine performance data. The EEDI level is tightened incrementally every five years with an initial CO_2 reduction level of 10% for the first phase (2015–2020), 20% for the second phase (2020–2025), and a 30% reduction mandated from 2025 to 2030.

There are various concerns related to the effectiveness of the EEDI. Since the EEDI regulation affects only newbuild ships, it takes time for the regulation to cover the global fleet. The average age of the shipping fleet is approximately 25 years, which means that the large majority of ships will be covered by EEDI only by 2040. Insofar as the EEDI acts as a target, it cannot be considered to be a very challenging target: the attained EEDIs of newbuild ships largely exceed the currently required EEDIs including Phase 3 requirements even though they are not mandatory before 2025—in particular, those of containerships and general cargo ships [9,10]. The attained scores often do not reflect the use of innovative electrical or mechanical technology, but they can be simply achieved through optimization of conventional machinery or through a change the hull design [10]. The impact of EEDI on emission reductions in shipping are estimated to be small: only a marginal difference has been found in CO_2 emissions between EEDI and non-EEDI scenarios [7]. For the EEDI regulation to have a larger impact, the mandated reductions or reference years would need to become more ambitious.

Technological measures cover technologies applied to ships that help to increase their energy efficiency beyond EEDI. Covered by a large body of literature, measures listed in Table 2 are generally considered the major technological measures to increase the energy efficiency of ships. All of these technologies are available on the market, but not all options can be applied as a retrofit. It should be noted that the reduction potentials are variable throughout different ship types, weather or engine conditions, and operational profiles. Moreover, estimations from industry sources may be exceedingly optimistic and should be taken with caution.

Table 2. Main technological measures and associated fuel savings potential.

Measures	Potential Fuel Savings
Lightweight materials	0–10%
Slender hull design	10–15%
Propulsion improvement devices	1–25%
Bulbous bow	2–7%
Air lubrication and hull surface	2–9%
Heat recovery	0–4%

Note: Emission reduction potentials are assessed individually. Ranges roughly indicate possible fuel savings depending on varying conditions such as vessel size, segment, operational profile, route, etc., hence limiting the possibilities for comparison. Numbers cannot be cumulated without considering potential interactions between the measures. Sources: [7,11–18].

2.2. Operational Measures

We cover four different operational measures: speed, ship size, ship–port interface, and onshore power (Table 3). Both slower speed and increase in ship size have contributed to a decrease in shipping emissions over the last years. The measure "ship–port interface" is related to a reduction in ship waiting time before entering a port. Ship size developments refer to ship capacity utilization. Shore power facilities are considered part of a larger set of port measures that could reduce emissions of ship operations.

Table 3. Main operational measures and whole-fleet CO_2 reduction potential.

Measures	CO_2 Emissions Reduction Potential
Speed	0–60%
Ship size and capacity utilization	0–30%
Ship–port interface	0–1%
Onshore power	0–3%

Note: Emission reduction potentials concern the cumulative reduction potential for the entire ship fleet. Numbers cannot be cumulated without considering potential interactions between the measures. Sources: [1,7,19–25].

A review of operational measures shows that slow steaming yields significant CO_2 emission reductions, e.g., a speed reduction of 10% translates into an engine power reduction of 27% [19]. Lower speeds are more effective if design speeds of ships are brought down as well [22]. Drawbacks of these measures include the potential need for additional vessels to maintain service frequency, longer lead times, and the risk of modal shift of time-sensitive shipments to rail or road transport.

The largest vessels of all ship types emit less CO_2 per tonne kilometer under conditions of full capacity utilization. CO_2 emissions could be reduced by as much as 30% at a negative abatement cost by replacing the existing fleet with larger vessels, according to [23]. The relationship between ship size and emissions is not linear, but reflects a power-law relationship with diminishing marginal emission reductions as vessel size increases. However, as the newer (and more energy-efficient) ships are often larger ships, the size effect of larger ships could be overestimated [26].

Further reductions can be achieved by smoother ship–port interfaces and onshore power supply (cold-ironing). Approximately 5% of shipping's CO_2 emissions are currently generated in ports [27]. If improved ship–port interfaces reduced ship waiting times—and their use of auxiliary engines in ports—to zero, the carbon emission reductions might amount to approximately 1% of total shipping emissions [28]. Optimized voyage planning, collaboration, and real-time data exchange can further contribute to improved berth planning. Onshore power supply (OPS) facilities in ports allow ships to turn off their engines and connect to the electricity grid to serve auxiliary power demand. However, the use of OPS requires retrofits on the ship.

2.3. Alternative Fuels and Energy

Although a range of alternative fuels and energy have lower or zero ship emissions when used for ship propulsion, upstream emissions may arise in the production process. In Table 4, we cover a range of alternative fuels and energy sources. Not all of these options have reached market maturity yet.

Table 4. Main alternative fuels and energy, and associated fuel savings potential.

Measures	Potential Fuel Savings
Advanced biofuels	25–100%
Synthetic fuels (hydrogen and ammonia)	0–100%
Liquid natural gas (LNG)	0–20%
Fuel cells	2–20%
Electricity and hybrid propulsion	10–100%
Wind assistance	1–32%

Note: Emission reduction potentials are assessed individually. Ranges roughly indicate possible fuel savings depending on varying conditions such as vessel size, segment, operational profile, route, weather conditions, etc., hence limiting the possibilities for comparison. Numbers cannot be cumulated without considering potential interactions between the measures. Considering upstream emissions of synthetic fuels and electricity, an almost 100% emission reduction can be reached only if generated from renewable energy sources. Sources: [7,11,29–34].

The emission reduction potential of biofuels and synthetic fuels depends to a great extent on their production methods. Advanced biofuels from both the second and third generations could, in theory, reduce potential adverse social and environmental effects by using degraded land or residual biomass. Yet, more knowledge on their performance and physical properties, as well as more testing and standardization, would be required for broader use by the shipping industry [32]. Synthetic fuels can be produced via electrolysis powered by wind, hydro, or solar energy to avoid lifecycle emissions arising from production [30]. Production of synthetic fuels could hence easily develop where renewable energy sources are abundant or where they can produce a large excess output [6]. Although liquid natural gas has been shown to reduce CO_2 emissions to some extent, some doubts persist regarding its overall environmental benefits vis-à-vis heavy fuel oil (HFO), considering its methane emissions [35]. Further emission reductions could be reaped by using hybrid systems involving fuel cells and batteries. The efficiency of fuel cells greatly depends on the fuel cell type and the fuel used [36]. All-electric propulsion is currently used in short-range passenger shipping (i.e., Norway) or short-range river transport (i.e., China or Netherlands). Hybrid electric systems may provide an interesting option for longer distances, yielding potential fuel savings of 10–40% and payback times as low as one [31]. In addition, wind power applications further decrease a ship's fuel demand, combined either with other wind technologies, with slow steaming and other incremental efficiency improvements, or with photovoltaic technology [33,37]. Drawbacks include their potential interference with cargo handling. An extensive review of these measures and their potentials and disadvantages is included in [28].

2.4. Market-Based Measures

Currently, no market-based measure has been applied on an international level. As one of the first market-based solutions, the European Union (EU) implemented the world's first emissions trading scheme (ETS) in 2005. Other comparable existing ETSs can be found in Australia, New Zealand, the United States (northeast states Regional Greenhouse Gas Initiative (RGGI)), California, Quebec, Japan, and a pilot in China [38]. Shanghai is so far the only pilot region that includes the aviation and port sectors in its ETS. The EU adopted a Monitoring, Reporting, and Verification (MRV) regulation for ships larger than 5000 gross tonnage calling at European Union ports and ports within the European Free Trade Area (EFTA), which entered into force in July 2015. Data collection started on 1 January 2018 on a per-voyage basis and is managed by the European Maritime Safety Agency (EMSA) [39]. In the case that there would be no global agreement at the International Maritime Organization (IMO) until 2023, the EU considered covering the shipping sector under an EU Maritime Climate Fund, to be set up under the ETS [40]. Contributions would be based on reported emissions under the MRV regulation and the ETS carbon price. The EU ETS approach shares a number of characteristics with other proposals at the IMO, namely, those of Norway, the UK, France, and Germany. A total carbon emissions ceiling would be coupled with monitoring of the real energy performance of ships. In an ETS covering the shipping and ports sector, stakeholders could gain or purchase quotas based on their emissions, and trade this quota within or outside the sector. Auction revenues generated from the sale of emission allowances would be used in a climate change fund supporting mitigation efforts. Beside emissions trading systems (ETS) or "cap and trade", direct carbon taxes are also a form of carbon pricing. A carbon tax directly determines a price for carbon by setting a tax rate on GHG emissions [41].

3. Impact Assessment of CO_2 Mitigation Measures

3.1. Framework to Assess the Impact of Mitigation Measures to Reduce Global Shipping Emissions

We estimate the future CO_2 emission from international shipping using the International Transport Forum's International freight model (IFM) and the "ASIF" (Activity, Structure, Intensity, Emission Factor) method [42].

The ITF's International freight model is designed to project international freight transport activities (in tonne kilometers) for 19 commodities for all major transport modes and routes while taking into account different transport and economic policy measures (e.g., the development of new infrastructure networks, or the alleviation of trade barriers). The model is built on the four-steps freight transportation modelling approach and it takes the OECD trade projection as an input. The IFM is designed to be able to estimate the weight of commodities traded between countries, the choice between modes and transport routes used to transport these commodities based on transport networks characteristics, and relevant socio-economic variables such as transport costs and time. The model consists of the following components:

1. Trade flow disaggregation model;
2. Value-to-weight model;
3. Mode choice model; and
4. Route choice model.

3.1.1. OECD International Trade Model

The OECD's trade projection is produced using a Computable General Equilibrium (CGE) model called the ENV-Linkages model [43]. The model is designed to estimate the dynamic evolution of international trade, in terms of both spatial patterns and commodity composition due to the changes in the global production and consumption of commodities. It is calibrated based on the macroeconomic trends of the OECD@100's baseline scenario for the period 2013–2060 at sectorial and regional levels. As such, it projects international trade flows in values (US$) for 26 regions and 25 commodities until 2060.

3.1.2. Trade Disaggregation Model

The underlying trade projections are disaggregated into 26 world regions. This level of resolution does not allow estimating transport flows with precision as it does not allow a proper discretization of the travel path used for different types of products. Therefore, we disaggregate the regional origin–destination (OD) trade flows into a larger number of production/consumption centroids. These centroids were calculated using an adapted p-median procedure for all the cities around the world classified by United Nations in 2010 relative to their population (2539 cities). The objective function for this aggregation is based on the minimization of a distance function which includes two components: GDP density and geographical distance. The selection was also constrained by allowing one centroid within a 500 km radius in a country. This resulted in 333 centroids globally, with spatially balanced results also for all continents.

$$T_{odk}^y = T_{VLk}^y \frac{GDP_o^y}{\sum_{v=1}^V GDP_k^y} \frac{GDP_d^y}{\sum_{l=1}^L GDP_l^y} \tag{1}$$

In Equation (1),

T_{odk}^y = trade values from centroid o to centroid d in year y for commodity k,

T_{VLk}^y = trade values from origin region V to destination region L,

o, d = origin and destination centroids,

k = commodity k,

y = year of analysis,

k = centroid that belongs to the origin region V,

l = centroid that belongs to the destination region L.

3.1.3. Value-to-Weight Model

We used a Poisson regression model to estimate the rate of conversion of value units (dollars) into weight units of cargo (tonnes) by mode, calibrated using datasets from Eurostat and Economic Commission for Latin America and the Caribbean (ECLAC) data on value/weight ratios for different commodities.

We use the natural logarithm of the trade value in millions of dollars as the offset variable, with panel terms by commodity, a transport cost proxy variable (logsum calculation for maritime, road, rail, and air transport costs per ton between each pair of centroids), and geographical and cultural variables: binary variables for trade agreements and land borders used above and a binary variable identifying if two countries have the same official language. Moreover, economic profile variables were included to describe the trade relation between countries with different types of production sophistication and scale of trade intensity. We validate the output of the value-to-weight model using the UN Comtrade database that provides values and weights of all commodities traded between any countries worldwide. Table A1 provides validation of the total values and weights of global trade produced by the model.

$$w_{odk}^y = T_{odk}^y \, e^{rs_{odk}^y} \tag{2}$$

$$rs_{odk}^y = a + b_1 \, e^{gdp\%_o^y} + b_2 \, e^{gdp\%_d^y} + b_3 \, e^{gdp_c\%_o^y} + b_4 \, e^{gdp_c\%_d^y}$$

$$+b_5 \ln\left(\frac{gdp_{c_o}^y}{gdp_{c_d}^y}\right) + b_6 \, contig_{od} + b_7 \, lang_{od} + b_8 \, rta_{od} \tag{3}$$

$$+lgs_k e^{-logsum(cost_{od})}$$

In Equations (2) and (3),

w_{odk}^y = weight of commodity k that is traded between origin o and destination d for year y (in tonnes),

T_{odk}^y = value of trade for commodity k between origin o and destination d for year y (in US$),

rs_{odk}^y = value-to-weight conversion factor for commodity k, between origin o and destination d for year y (in tonnes/US$),

$gdp\%_o^y$ = GDP percentile of origin in year y,

$gdp\%_d^y$ = GDP percentile of destination in year y,

$gdp_c\%_o^y$ = GDP per capita percentile of origin in year y,

$gdp_c\%_d^y$ = GDP per capita percentile of destination in year y,

$\ln\left(\frac{gdp_{c_o}^y}{gdp_{c_d}^y}\right)$ = natural logarithm of the ratio between GDP per capita of origin and GDP per capita of destination in year y,

$contig_{od}$ = land contiguity between origin o and destination d, $contig$ = (0, 1),
$lang_{od}$ = shared language between origin o and destination d, $lang$ = (0, 1),
rta_{od} = trade agreement between origin o and destination d, rta = (0,1),
$logsum(cost_{od})$ = *logsum* variable of transport costs using different modes between origin o and destination d, d
lgs_k = *logsum* coefficient/panel term for commodity k.

3.1.4. Mode Choice Model

The mode share model (in weight) for international freight flows assigns the transport mode used for trade between any origin–destination pair of centroids. The mode attributed to each trade connection represents the longest transport section. All freight will require intermodal transport both at the origin and destination. This domestic component of international freight is usually not accounted for in the literature, but is included in our model. The model is estimated using a standard multinomial logit estimator including commodity type panel terms on travel times and cost. Both Eurostat and ECLAC datasets are used as sources of observation data for the volume of commodities and its mode of transport. Transport costs and travel times are estimated using the network model and observed data whenever available. Two geographical and economic context binary variables are added, one describing if the OD pair has a trade agreement and the other for the existence of a land border between trading partners. The mode choice model is validated by ensuring the mode share of the volume of goods transported is similar to the observed mode share for international transport in 2011 by weight. Additionally, the total tonne kilometers for all 4 major modes of transport (air, road, rail, sea) are also validated against the observed data. These observed data are obtained from reports of various organizations such as the International Maritime Organization (IMO), the International Civil Aviation Organization (ICAO), and the World Bank. Tables A1 and A2 provide detailed descriptions of the validation result.

$$u^m_{odk} = asc_m + CF_k \, TC^m_{od} + TF_k \, TT^m_{od} + Ct^m \, contig_{od} + Rt \, rta_{od} \tag{4}$$

$$P_m = \frac{e^{u^m_{odk}}}{\sum^M_{m=1} e^{u^m_{odk}}} \tag{5}$$

In Equations (4) and (5),

P_m = the choice probability of mode m,
u^m_{odk} = the choice utility of mode m for commodity k between origin o and destination d,
asc_m = alternative specific constant for mode m,
CF_k = transport cost coefficient for commodity k,
TC^m_{od} = transport cost for mode m between origin o and destination d,
TF_k = travel time coefficient for commodity k,
Ct^m = contiguity coefficient for mode m,
$contig_{od}$ = contiguity variable between origin o and destination d, $contig$ = (0, 1),
Rt = trade agreement coefficient,
rta_{od} = trade agreement variable between origin o and destination d, rta = (0, 1).

3.1.5. Route Choice Model

We used a path size logit model in combination with a path generation method to assign the volume of freight transport across all possible international shipping routes between all origins and destinations. The model does this using a shortest path algorithm and choice set creation algorithm to identify the subsegments of the complete shortest route for each port-to-port segment of a shipping line. The model accounts both for maritime connections between two countries and for overland connections between the centroids. The route and port choice algorithms use a path size logit model which takes overlaps between the alternative routes into account and distinguishes the transport costs associated with these alternatives properly. The basis of this model can be found in [44]. The model is calibrated by minimizing the difference between observed and modelled port throughputs for more than 400 major ports in the world. A detailed description on the model can be found in [45] or in [46]. The formal definition of the cost model is delineated below:

$$C_r = \sum_{p \in r} A_p + \sum_{l \in r} c_l + \alpha \left(\sum_{p \in r} T_p + \sum_{l \in r} t_l \right) \tag{6}$$

In Equation (6):

C_r = unit cost of route r from origin centroid to destination centroid (US\$/Twenty-equivalent unit, TEU),

p = ports used by the route,

l = links used by the route,

A_p = unit cost of transhipment at port p (US\$/TEU),

c_l = unit cost of transportation over link l (US\$/TEU),

T_p = time spent during transhipment at port p (days/TEU),

t_l = time spent during transportation over link l (days/TEU),

α = value of transport time (US\$/day).

The model accounts both for maritime connections between two countries and for overland connections between these countries. The route and port choice algorithms use a path size logit model which takes overlaps between the alternative routes into account and distinguishes the transport costs associated with these alternatives properly. The basis of this model can be found in [44]. The following is the formal definition of the route choice model. The route probabilities are given by

$$P_r = \frac{e^{-\mu \ (C_r + \ln S_r)}}{\sum_{h=1}^{H} e^{-\mu \ (C_h + \ln S_h)}} \tag{7}$$

while the path size overlap variable S is defined as

$$S_r = \sum_{a \in LK_r} \frac{Z_a}{Z_r} \frac{1}{N_{ah}} \tag{8}$$

In Equations (7) and (8):

P_r = the choice probability of route r,

C_r = generalized costs of route r,

C_h = generalized costs of route h within the choice set,

CS = the choice set with multiple routes,

h = path indicator/index, $h \in CS$,

μ = logit scale parameter,

a = link in route r,

S_r = degree of path overlap,

Lk_r = set of links in route r,

Z_a = length of link a,

Z_r = length of route r,

N_{ah} = number of times link a is found in alternative routes.

3.1.6. CO$_2$ Mitigation Impact Assessment Model

The ASIF framework is used to assess the impact of the maximum possible technical, operational, and alternative fuels measures on the total CO$_2$ emissions of international shipping (Figure 1).

The output of IFM provides the tonne kilometers for different commodities ("Activity") and the projections for future transport demand scenarios which we assign to possible ship types to estimate the activities for each ship type. We consider 6 ship types in our model: dry bulk, container, oil tanker, gas, wet product and chemical, and general cargo. We estimate the vehicle kilometers for each ship type using the ship's load factor data and the projection for the changes in ship size until 2035 ("Structure"). Furthermore, we compute the fuel consumptions of all ship types using engine efficiency improvement pathways together with the distribution of fuel types across different ship types ("Intensity"). The resulting fuel consumption for each ship type is then used to estimate the total CO_2 emissions using carbon factor and energy content data for different fuel types ("Emission factor").

$$E_{qf} = A_q S_{qf} I_q F_f \tag{9}$$

$$S_{qf} = Sh_{qf} LF_q \tag{10}$$

$$TCO_2 = \sum_{q=1}^{Q} \sum_{f=1}^{F} E_{qf}. \tag{11}$$

In Equations (9)–(11),

TCO_2 = total CO_2 emissions from international shipping,
E_{qf} = total CO_2 emissions from ship type q using fuel type f (in tonnes CO_2),
A_q = total annual activity for ship q (in tonne kilometers),
S_{qf} = total vehicle kilometers for ship type q which uses fuel type f (in vkm),
SH_{qf} = share of ship type q which uses fuel type f (in %),
LF_q = load factor of ship type q (in tonnes/vehicle),
I_q = engine intensity of ship type q (in MJ/vkm),
F_f = emission factor of fuel type f (in tonnes CO_2/MJ).

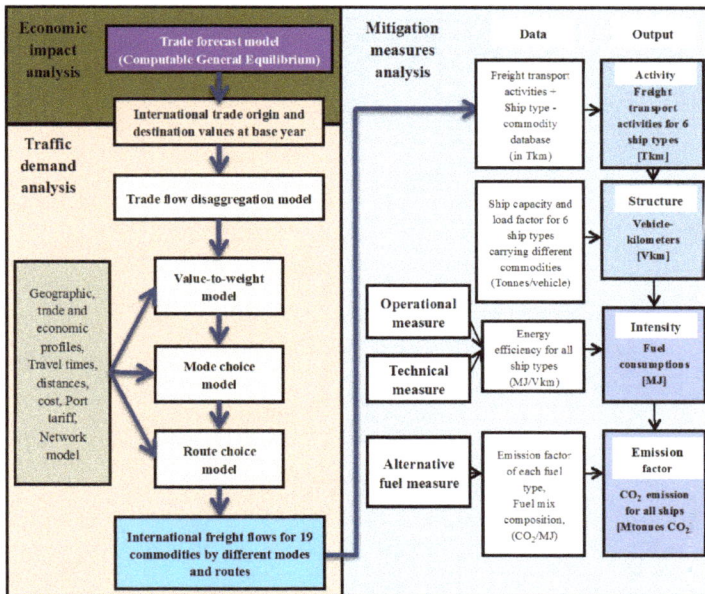

Figure 1. Modeling framework used to estimate international shipping CO_2 emissions.

3.2. Data

We use data from various datasets to estimate each component of the model and produce the baseline CO_2 emissions for each ship type until 2035 using the ASIF framework.

The weight-to-value model and the mode choice model are estimated using observed international trade flows from Eurostat and ECLAC datasets. The Eurostat dataset registers trade flows between Europe and the rest of the world that are obtained from the customs of each EU country, and the ECLAC dataset records trade data between Latin American countries and countries worldwide. These datasets combined provide more than 17,427 observed trade flows in values (US$) and weights (tonnes) and their modes of transport. We use the UN Comtrade data that record trade flow data for all commodities, grouped in more than 97 product types (chapters) both in value and weight, to validate the output of the model. It is necessary to aggregate the trade flows both in values and weights at the country-to-country level for all commodities in the Harmonized Commodity Description and Coding Systems (HS) to enable a comparison with the output of our model. The GDP and population data, including their projection until 2050, were obtained from the environment directorate of the OECD.

Given the baseline transport demand projection for 6 ship types, the ships' vehicle kilometers are estimated using load factor data for each ship type. The load factor data for each ship type are estimated by multiplying the average freight capacity and average utilization rate of each ship type. We obtained average freight capacity data from the UNCTAD Review of Maritime Transport 2013 [47] and ships' utilization rate data from the 3rd IMO GHG study [1]. We estimated the evolution in ships' load factor by taking into account the future evolution in ship size (Figure A1). This projection of ship size is based on the observed historical pattern of ship sizes from 1996 to 2015. The pathways for ships' engine efficiency for each ship type for the baseline scenario were obtained from a UMAS study [7,48]. Furthermore, the emission factor data for different fuel types were obtained from the International Energy Agency's Mobility Model [49] (Table A3). By multiplying the engine efficiency (in MJ/vkm) with the emission factor for different fuel types (in CO_2/MJ), we obtain the carbon intensity of each ship type (Figure 2).

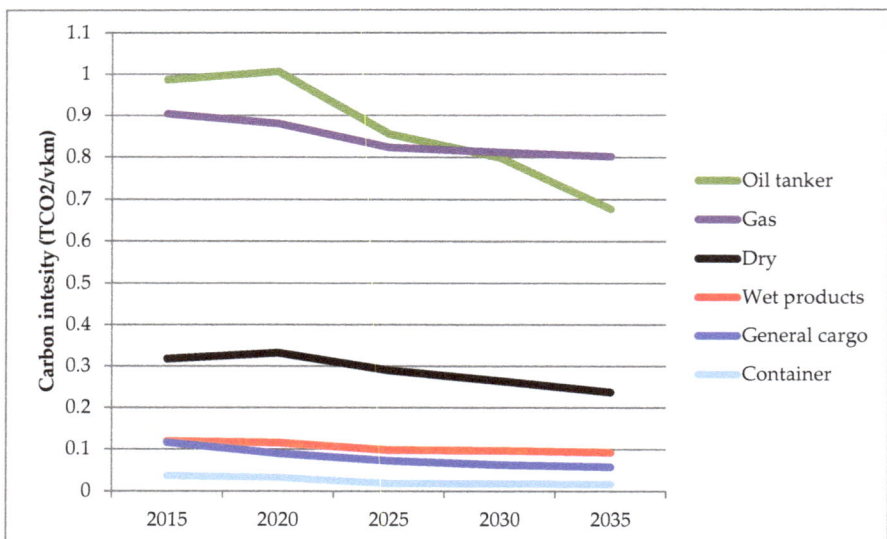

Figure 2. Evolution of ships' carbon intensity in the baseline scenario.

4. Results

4.1. Baseline Emission Scenario: The Impact of Less Fossil Fuel Trade and the Rise of Trade Regionalization

Carbon emissions from global shipping are projected to reach approximately 1090 million tonnes by 2035 in a baseline scenario without additional policy measures. This would represent a 23% growth of emissions by 2035 compared with 2015. The baseline scenario incorporates the impact of existing international regulations, including that on the energy efficiency of ships. A geographical representation of shipping emissions and their evolution shows that a large share of carbon emissions in the baseline scenario is generated along main East–West trade lanes (Figure 3). In our study, we incorporate two possible developments in the baseline scenario: a strong reduction of trade in fossil fuels and further regionalization of trade. We will show that this results in a downward adjustment of shipping emissions when these developments are taken into account.

One of the impacts of the Paris Agreement is the rise in commitments by countries and various subnational governments to reduce the use of fossil fuel commodities such as coal and oil. This development is reflected in certain scenarios for global energy demand. For instance, the sustainable development scenario in the World Energy Outlook 2017 of the International Energy Agency (IEA) projects that global energy demand from coal will decline up to 41% by 2040, and oil up to 22% [50]. The declining use of fossil fuels would have a significant impact on maritime trade due to the quantities of coal, oil, and gas shipped over long distances. The decrease of worldwide coal production of about 2.9% in 2015 translated to a decline in seaborne coal trade of 4.3%, which represents around 50 million tonnes of cargo by sea transport [51]. We assume that a reduction in global coal and oil trade will take place gradually from 2015 onwards and could lead to 50% and 33% reductions of coal and oil trade volume, respectively, by 2035. This reduction factor is similar to one of the sustainable pathways in IMO scenario RCP 2.6, which projects a decline of about 48% in transport demand for coal trade and 28% for liquid bulk trade, including oil.

Furthermore, global outsourcing has driven much of the trade growth of the last decades, but current developments might indicate a more regionalized trade system in the future. Emerging economies have gained a larger share in global trade and increasingly trade with each other. One of the major trends in trade policy is the continuous increase in preferential trade agreements at a regional level. In Asia, intra-regional trade has increased in relative and absolute terms [52]. The share of Chinese exports directed to emerging and developing Asian countries has grown considerably in the last decade. Such shifts in trade patterns could significantly alter the global demand for seaborne transport. In addition, maritime cost increases related to the 2020 sulphur cap might have effects on regionalization of trade. This cap will reduce the allowed sulphur content in ship fuel from 3.50% to 0.50% and could translate into increases in import prices. These changes might be substantial enough to lead to changes in trade flows. Depending on price elasticities—most of which are unknown—one could assume that these cost increases lead to a shortening of certain supply chains, considering that the increase in maritime transport costs makes nearby sourcing more attractive. Taking into account the potential impact of the rise in intra-regional trade and in transport cost due to the sulphur cap, we assume a 20% rise in intra-regional trade flows by 2035 replacing intercontinental trade flows, and thus resulting in a reduction of tonne kilometers as compared to the baseline scenario. Figure 4 presents the adjusted transport demand projection for 2035 if a reduction in fossil fuel trade and trade regionalization are taken into account.

More intra-regional trade, combined with a further reduction of trade in fossil fuels, could reduce baseline CO_2 emissions to around 850 million tonnes by 2035. This adjusted level of emissions will be used in the remaining parts of the paper as benchmark for the CO_2 emissions that would need to be reduced in order to reach full decarbonization of shipping by 2035 (Figure 5).

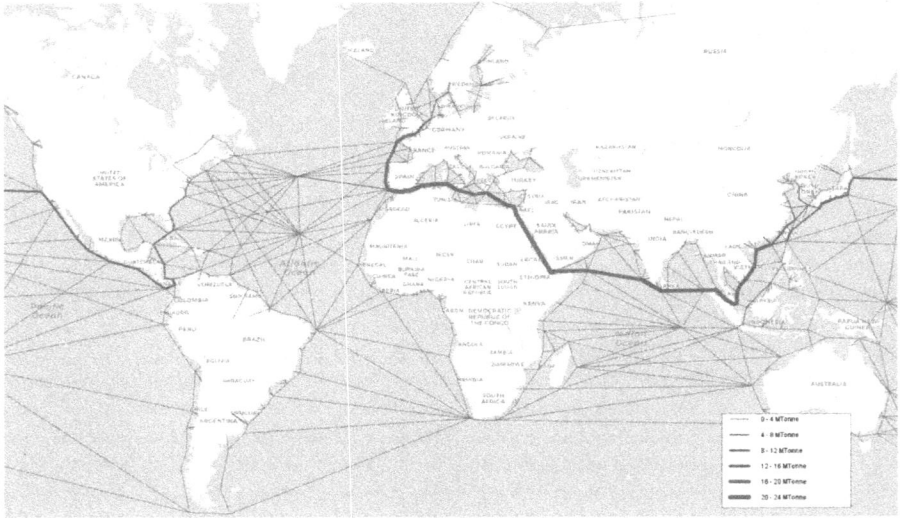

(a)

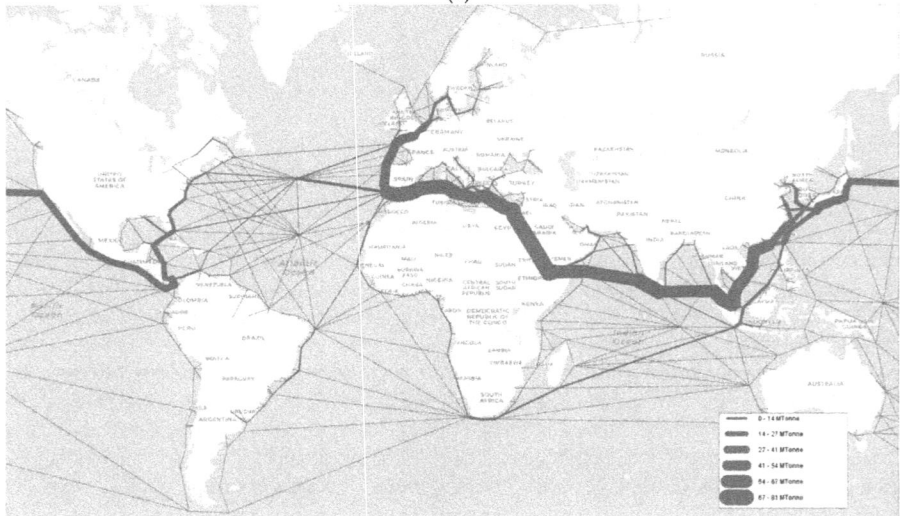

(b)

Figure 3. Visualization of CO_2 emission across global shipping routes in 2015 (**a**) and 2035 (**b**).

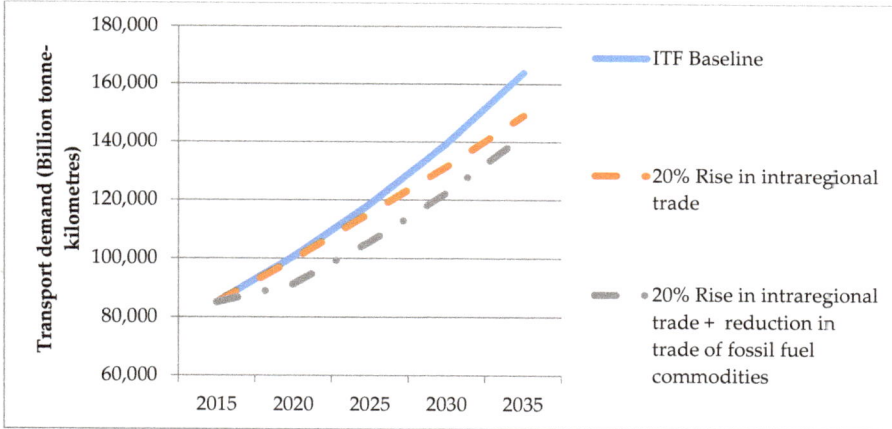

Figure 4. Transport demand scenarios estimated using the ITF international freight model.

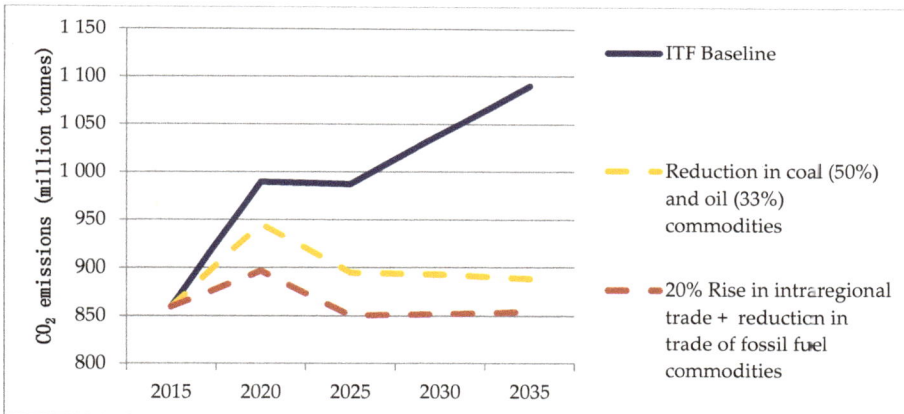

Figure 5. Impact of reduction in fossil fuel trade by 2035.

4.2. Combination of CO_2 Mitigation Measures

Table 5 provides an overview of different technologies, operational measures, and alternative fuels that are included in our modeling framework. We use the measures that are assessed in the UMAS study [7] as the base input for our modelling framework and add several relevant additional measures, such as ship size and port–ship interface. The technological measures listed in the table do not represent an exhaustive set of measures. The implementation of one measure might be incompatible with the application of other measures. We therefore used the detailed assessment of these measures, as well as a compatibility matrix, as provided in the appendix of the study by UMAS [7]. Beside technological and operational measures and fuels, we also assume an implementation of a market-based measure in the form of carbon pricing that is applied to both global and sectorial markets such as shipping.

Table 5. Main types of measures to reduce shipping's carbon emissions.

Type of Measures	Main Measures
Technological (based on Smith et al., 2016)	Contra rotating propeller, air lubrication, main engine turbo compounding propeller, aux turbo compounding series, Organic Rankine Cycle waste heat recovery, Flettner rotors, kites, engine derating, speed control of pumps and fans, block coefficient improvement
Operational	Lower speeds, ship size increase, ship–port interface
Alternative fuels/energy	Liquefied natural gas (LNG), advanced biofuels, hydrogen, ammonia, electric ships, wind assistance

While the individual measures can deliver a significant reduction in CO_2 emissions, it is unlikely that one single measure on its own would be the most efficient and cost-effective way to achieve decarbonization of shipping by 2035: a combination of measures would be needed, which generate different decarbonization pathways. This section sets out these possible pathways to decarbonize international shipping by 2035.

The implementation of one measure might be incompatible with other measures. This is especially the case for the possible technologies that can be combined on a single ship, for instance combining different wind and solar technologies. Furthermore, certain technical measures will also not allow certain operational measures to be implemented. More detailed information on the possible combination of technologies that can be installed on a ship is provided in a compatibility matrix in [7].

The projections for the possible decarbonization pathways are based on the results of the Whole Ship Model used in [7]. As the starting point for our modelling exercise, we use their data on reduction levels of ships' carbon intensity associated with the application of operational, technological, and alternative fuel measures. Furthermore, we studied the impact of additional operational and alternative fuels measures such as those listed in Table 5 above on the possible further reduction of ships' carbon intensity. We focus on the reduction in carbon intensity levels (EEOI) that can be achieved by combining all possible technical, operational, and alternative fuel measures without explicitly considering the cost-effectiveness of the measures. Furthermore, we do not include the dynamic feedback that might exist between measures due to their interactions. This allows us to assess the maximal possible carbon emission reductions by 2035.

Different pathways to reach carbon emission reductions can be constructed by combining the operational and technical measures with the use of alternative fuels at different times and degrees. On the operational side, we consider speed reduction as a key measure to reduce the carbon intensity of ships. We consider two possible alternatives for implementing speed reduction: moderate and maximum speed reduction. "Moderate" speed reduction implies reductions of 6% (for container ships) and 9% (for tankers and bulk carriers) of the standard operational speed for different ship types, which was assumed to be 12.8 knots for bulk carriers and tankers and 18.4 knots for container ships, in line with the study by Smith et al. (2016).

In the case of "maximum" speed reduction, we consider strong speed reductions of 26% (for container ships), 30% (for tankers), and 65% (for bulk carriers) of the standard operational speed. Even though it is technically possible to attain such low operating speeds, navigators will prioritize safety, stability, and maneuverability of the ship, especially when operating in difficult weather conditions. Another operational measure that has been integrated in this modelling framework is optimized ship berth planning. This relatively low-cost measure is aimed at reducing the waiting time of ships at port before berthing. According to our estimation, this measure could deliver around a 1% reduction of the total CO_2 emissions. We assume that the operational measures, especially speed reductions, could be implemented from 2020 onwards to yield maximum potential by 2030, which would require decision-making by 2018.

In terms of technical measures, we apply maximum ship design specifications that can lead to the highest reduction in a ship's carbon intensity, taking into account the speed reduction measures described above. This maximum specification entails the implementation of a series of technologies encompassing ship engine design and hydrodynamic improvements that can increase a ship's energy efficiency as described in Smith et al. (2016). This pathway includes a range of technological measures such as wind assistance and block coefficient improvements to reduce resistance which can help to deliver additional CO_2 reductions (up to 30%). When speed reduction is implemented, the energy efficiency savings gained from measures such as an improved block coefficient and wind assistance will diminish. We take this interaction into account and assume that the increase in energy efficiency will take place gradually (in a linear fashion) between 2020 and 2035.

Furthermore, we apply two additional measures: the uptake of electric ships and Onshore Power Supply (OPS). We include a scenario in which the pace of innovation in battery technology will sharply reduce battery costs (according to various projections such as Bloomberg New Energy Finance [53]), which could drive electrification of around 10% by 2035. We assume that most of these electric ships will be used to serve international short-distance shipments between countries. In such a scenario, the penetration rate of electric ships is assumed to see a gradual increase from 1% in 2025 to 10% in 2035. The second additional measure is Onshore Power Supply (OPS) which can help to reduce the carbon emissions from ships at berth during the loading and unloading process. OPS is already fairly widely used and is likely to be expanded due to favorable regulation, e.g., in the European Union where it will become mandatory by 2025 for European "core ports". As OPS could also be used as a charging facility for electric ships, we assume that uptake of electric ships and OPS will coincide. The implementation of these measures would be facilitated by a market-based mechanism introduced in 2025.

Three different levels of fuel carbon factor reduction are considered as possible pathways: 50%, 75%, and 80%. While the first two are taken from Smith et al. (2016), the third reduction level is estimated by assuming the uptake of alternative fuels such as ammonia. The level of carbon factor reduction presented here indicates the average reduction in carbon content of the fuel (gram of carbon dioxide per megajoule of energy) compared with the baseline that can be achieved by the use of alternative fuels. Here, high carbon factor reduction is used to indicate high uptake of alternative fuels such as advanced biofuels, hydrogen, and ammonia. In the case of 80% carbon factor reduction, it is assumed that hydrogen and ammonia will form around 70% of the fuel mix in ship propulsion. This, along with an assumed increase in the uptake of biofuels (22%) and LNG (5%), could significantly diminish the use of oil-based fossil fuels to around 3% by 2035 (Figure 6). While the gradual uptake of these fuels starts from 2015, zero-carbon alternative fuels such as hydrogen and ammonia are expected to see a stronger uptake after 2025, when we assume the start of a market-based measure such as carbon pricing. In this scenario, we assume the adoption of a carbon price based on Lloyd's Register study [8] where carbon emissions are priced at around 500 US\$/tonne by 2035 in order to make zero-carbon fuels competitive. For simplicity's sake, we assume that the increase in price is linear over the 2025–2035 period.

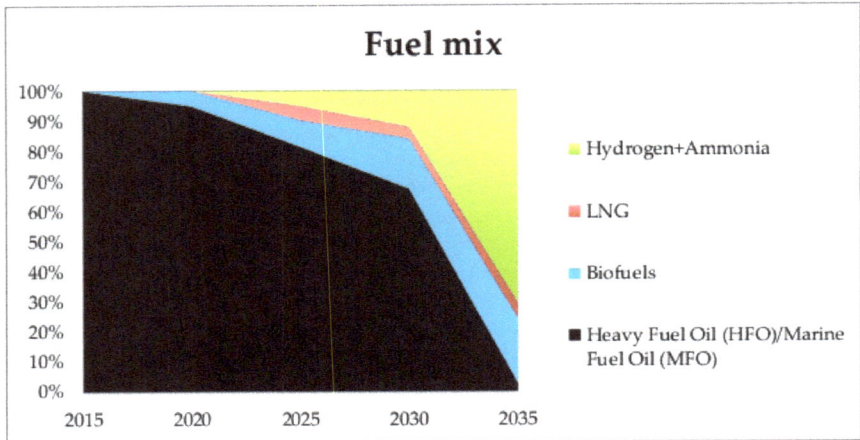

Figure 6. Fuel mix evolution between 2015 and 2035 for 80% carbon factor reduction.

Another measure that is included in this pathway is the increase in ship size that will lead to higher ship capacity. Unlike other measures which might require additional incentives and stimuli, the changes in ship size already form part of the shipping industry's strategy to seize economies of scale. We assume that the trend of ship size increases over 1996–2015 (per different ship types) and can be extrapolated towards 2035.

4.3. Four Possible Pathways to Decarbonize Maritime Transport

We consider four different pathways based on possible combinations of the measures considered in this study (Table 6). All pathways assume maximum application of the possible technical measures. The main differences between the pathways are related to speed reductions (moderate or maximum) and the application of zero-carbon fuels and electric ships, ranging from very high to more moderate assumptions.

Table 6. Four potential decarbonization pathways and their components.

Pathway	Operational Measures	Technical Measures	Carbon Factor Reduction Due to Alternative Fuels	Electric Ship Penetration
"Maximum intervention"	Maximum	Maximum	80%	10%
"Zero-carbon technology"	Moderate	Maximum	80%	10%
"Ultra-slow operation"	Maximum	Maximum	50%	-
"Low-carbon technology"	Moderate	Maximum	75%	-

The "maximum intervention" pathway represents the most ambitious reduction trajectory to reach zero emissions, where maximum speed reduction will be implemented starting from 2020 and reach its maximum reduction level by 2030, while the other measures such as energy efficiency improvements and zero-carbon fuels are implemented gradually (Figure 7). If we ignore the possible negative impact on international trade (such as increased transport time), drastic speed reduction could reduce CO_2 emissions by 43% by 2030. However, the effect of speed reduction alone will not be sufficient to reach zero carbon emissions by 2035. The estimated growth in international trade will offset the reduction impact of this measure starting by 2030. On the other hand, the application of technical measures will help to maintain a downward trend in the emissions between 2030 and 2035. The additional reduction that can be delivered by energy-saving technologies will be relatively low when the ship is operating at ultra-low speed. The increase in ship size together with the application of zero-carbon fuels, especially from 2025 onward, will help to reduce CO_2 emissions further by 95% from the adjusted demand level, which leads to remaining emissions of 44 million tonnes by 2035. However, to achieve this level of decarbonization by 2035, zero-carbon fuels such as hydrogen and ammonia would have to see a rapid uptake and should constitute the majority of the fuel mix by 2035 (more than 70%). Additionally, we assume that electric ships could constitute around 10% of the global ship fleet by 2035, which contributes to the reduction in total CO_2 emissions.

The "zero-carbon technology" pathway is as ambitious as the previous scenario with regards to zero-carbon technologies but assumes only a moderate speed reduction that helps to reduce emissions in the short run (Figure 8). Reducing speed will lower emissions by 4% in the short term and the implementation of technical measures can help reduce emissions in the medium to long term by 46%. Similar to the "maximum intervention" pathway, the use of electric ships and zero-carbon fuels will be the key measure to reach a 92% emission reduction by 2035. The combination of these measures will help to bring CO_2 emission levels down to 56 million tonnes, which is equivalent to a 93% emissions reduction from the adjusted demand level. Both the "maximum intervention" pathway and the "zero-carbon technology" pathway would allow a strong reduction in CO_2 emissions by 2035.

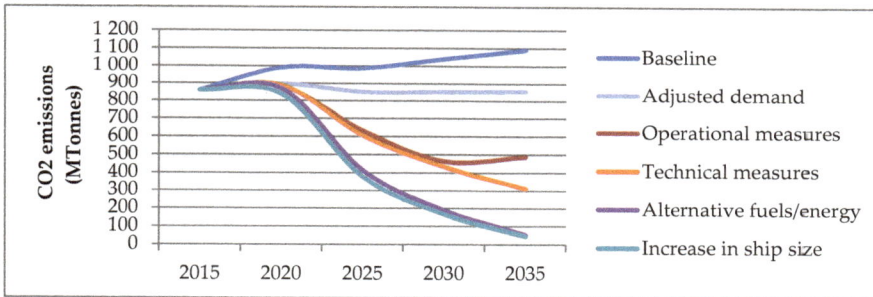

Figure 7. "Maximum intervention" pathway.

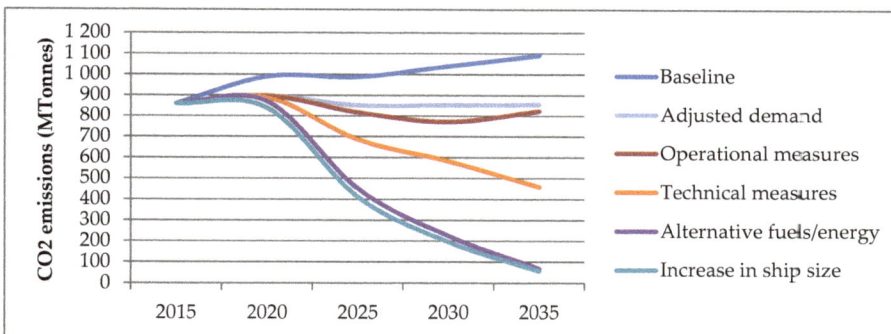

Figure 8. "Zero-carbon technology" pathway.

The "ultra-slow operation" pathway represents a scenario that relies heavily on speed reduction and sets a less ambitious target for energy-saving technologies and zero-carbon fuel adoption (Figure 9). The overall pattern of this pathway is similar to the "maximum intervention" pathway where most of the reduction comes from a drastic speed reduction, followed by the gradual implementation of other measures. In this pathway, we assume that electric ships will fail to penetrate the global ship fleet and might serve only domestic purposes. The uptake of zero-carbon fuels in this scenario is also foreseen to be less strong than in the other pathways, which could reflect insufficient investments in infrastructure and commitments to ensure sufficient availability of biofuels, hydrogen, and ammonia to replace the conventional fuels. This pathway would lead to an 82% emissions reduction from the adjusted demand level to reach around 156 million tonnes in 2035.

The "low-carbon technology" pathway represents a scenario that aims to balance moderate speed reductions with the use of zero-carbon fuels. This scenario reflects a strong uptake in zero-carbon fuels in the medium to long term, but with a less optimistic view on the penetration rate of electric ships and the uptake of ammonia. With the application of moderate speed reductions, the overall trajectory of this pathway resembles that of the "zero-carbon technologies" pathway, with a less rapid emissions decline to 2035 (Figure 10). This pathway will result in an 86% reduction of CO_2 emissions from the adjusted demand level, reaching 123 million tonnes by 2035.

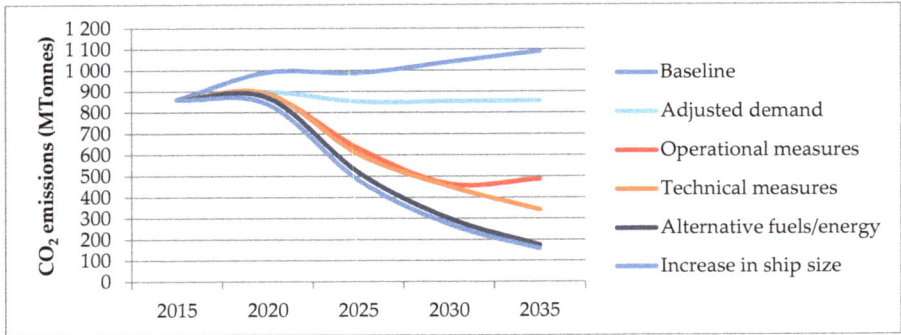

Figure 9. "Ultra-slow operation" pathway.

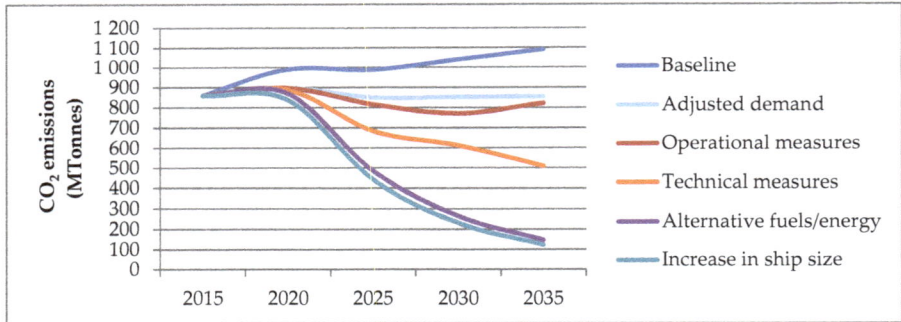

Figure 10. "Low-carbon technology" pathway.

We have presented four different pathways that could lead to the decarbonization of maritime transport with CO_2 emissions approaching zero by 2035, with remaining shipping emissions ranging from 44 to 156 million tonnes by 2035 (Figure 11). These pathways demonstrate that targeted interventions using a combination of possible measures can reduce CO_2 emissions from international shipping between 82% ("ultra-slow operation") and 95% ("maximum intervention") from the adjusted demand level. Table 7 presents the total CO_2 emission reductions for the four pathways by 2035. At the aggregate level, it is observable that two similar initial trajectories can be distinguished based on the application of speed reduction measures in the short term. The "maximum intervention" and "ultra-slow operation" pathways represent an extreme reduction in speed, while the "zero-carbon" and "low-carbon technology" pathways represent a more moderate speed reduction. Furthermore, the level of decarbonization in these pathways by 2035 depends on the extent to which zero-carbon fuels and technologies are applied. As demonstrated by the nearly-zero-carbon pathways ("maximum intervention" and "zero-carbon technology"), the use of zero-carbon fuels and technology is indispensable to achieving full decarbonization.

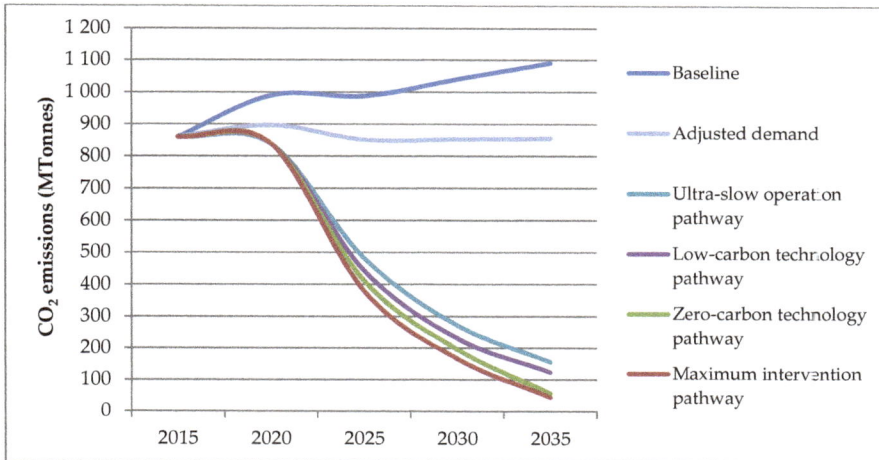

Figure 11. Four different decarbonization pathways for shipping.

Table 7. Total CO_2 emission reductions by 2035 for the four decarbonization pathways.

Pathways	CO$_2$ Reduction (in Million Tonnes)	Reduction Percentage (%)
Maximum intervention	810	95
Zero-carbon technology	798	93
Ultra-slow operation	698	82
Low-carbon technology	731	86

4.4. Impact of Increase in Maritime Transport Cost on Modal Share and CO$_2$ Emissions

The implementation of CO_2 mitigation measures as detailed in Table 6 is very likely going to increase transport costs for international shipping due to higher fuel costs, and increase capital expenditures to retrofit the ships and to install other low-carbon technologies. Furthermore, when slow steaming measures are applied extensively, this can cause transport costs to increase further due to longer shipping times, which escalate time-related costs. In such a case, shippers might consider other modes that offer lower travel times and transport costs to be more attractive, especially for highly time-sensitive goods such as fashion, electronics, car parts, and perishable goods, such as food.

This section provides an analysis of the potential impact of the increase in transport costs for international shipping on maritime transport demand. While the increase in transport costs might impact the modal share of international transport, a drastic increase of such a cost for the longer term could also induce changes in global trade patterns (e.g., increase in intra-regional trade, or reduction in total global trade volume). Our analysis does not cover the impact of increase in transport costs on global trade, as it requires a comprehensive study that should take into account the dynamic feedback between trade and transport model. For simplicity's sake, we assume that the values of international trade will remain the same as in the baseline scenario.

The exact increase in transport cost will be difficult to estimate since it depends on uncertain factors such as investments and commitments in establishing the infrastructure needed to ensure an adequate supply of green technologies and zero-carbon fuels, and the implementation of market-based measures such as carbon pricing. In order to analyze the impact of an increase in transport cost on the demand for maritime transport, we test a scenario where there is a 100% increase in unit transport cost for sea transport by 2030. Specifically, we assume that sea transport cost will increase from 0.0016 $/tonne km to 0.0032 $/tonne km and we apply a 25–65% speed reduction on the sea transport mode based on the maximum intervention scenario.

We focus our analysis on trades between China and Europe and global trade. China–Europe trade represents one of the major global trade flows which can potentially see a shift in its mode share when sea transport becomes a lot slower due to slow steaming. Furthermore, since the launch of China's Belt and Road initiative, rail cargo transport between China and Europe has gathered political momentum to undergo a major capacity expansion. One of the most rapidly developing rail corridors between China and Europe is the Trans-Siberian railways via Kazakhstan's rail system. Compared with sea transport, this railway connection can reduce travel time up to 42%. To reflect this potential development, we incorporate a 50% reduction in transport cost and time for the rail mode between China and Europe. Table 8 presents the modal share of China–Europe transport by 2030 under baseline and increased cost scenarios. The result shows that a 100% increase in sea transport cost causes a slight reduction in the mode share of maritime transport (1.4%), which represents 8.7 MTonnes of freight volume. The majority of this volume is estimated to shift to rail transport (7.8 MTonnes), which is projected to see an increase of 1.23% in its share. Although the reduction in share of maritime transport is relatively small, the shift to rail mode represents a roughly 15% increase in the total volume of rail transport.

Table 8. Impact of increased sea transport cost on modal share of China–Europe transport.

	Baseline 2030 (MTonnes)	Share (%)	100% Increase in Maritime Transport Cost (MTonnes)	Share (%)	Difference in Share (%)	Differences in Weights (MTonnes)
Air	2.59	0.41	2.82	0.44	0.04	0.22
Rail	51.25	8.07	59.05	9.30	1.23	7.80
Road	13.23	2.08	13.90	2.19	0.11	0.67
Sea	567.80	89.44	559.10	88.07	−1.37	−8.70

On a global scale, the impact of higher sea transport cost on the modal share is less significant compared with China–Europe transport (Table 9). The share of sea transport could decline around 0.16%; this represents approximately 34 Mtonnes of freight volume. The majority of the shifts are from sea mode to both road mode (13 MTonnes) and rail mode (18 MTonnes). One of the reasons for this is that sea transport remains the cheapest transport mode that serves major trade lanes such as those between Europe and Asia and between the U.S. and Europe.

Table 9. Impact of increased sea transport cost on modal share of international freight transport.

	Baseline 2030 (MTonnes)	Share (%)	100% Increase in Maritime Transport Cost (MTonnes)	Share (%)	Difference in Share (%)	Differences in Weights (MTonnes)
Air	70	0.33	72	0.34	0.01	2.60
Rail	598	2.84	611	2.90	0.06	13.17
Road	2539	12.06	2557	12.15	0.09	17.92
Sea	17,813	84.65	17,780	84.49	−0.16	−33.70
Waterways	22	0.11	22	0.11	0.00	0.01

Next, we estimate the impact of a shift in modal share on the total transport demand for maritime transport. In the scenario where transport cost will increase gradually (19% annually) from 2020 until 2035, a slight reduction (0.10–0.14%) in maritime transport activity is foreseen. Table 10 presents the estimated changes in transport activities (Tkm) due to an increase in transport cost and its associated CO_2 emissions. By 2035, reduced sea transport demand could lead to a reduction of approximately 1.2 MTonnes of CO_2 emissions from international shipping, which is equivalent to approximately 2.3% of the projected CO_2 emissions under the maximum intervention scenario.

Table 10. Impact of increased sea transport cost on maritime transport activities and CO_2 emissions.

	2020	2025	2030	2035
Baseline (Billion Tkm)	91,187.14	105,690.92	122,416.98	142,131.47
Increased maritime transport (Billion Tkm)	91,095.96	105,574.66	122,257.83	141,932.49
Difference (Billion Tkm)	91.19	116.26	159.14	198.98
CO_2 emission (Ktonnes)	897	936	1108	1196

In conclusion, the impact of a major increase in sea transport cost on CO_2 emissions for international shipping is going to be likely marginal. However, the small shift from sea mode to other modes such as rail, road, and air could increase the transport demand for these modes considerably (as exemplified by rail mode for China–Europe trade). In the case where there are no sufficient CO_2 mitigation measures implemented for the other modes, the total CO_2 emissions from international transport might increase since maritime transport is generally less carbon-intensive than rail and road transport. This result underlines the importance of minimizing the shift from maritime transport to other modes that are more carbon-intensive such as air, road, and rail, especially if there are not adequate CO_2 mitigation measures being implemented for these modes.

5. Barriers and Market Failures in Decarbonizing International Shipping

Prevailing conditions and incentive systems in the shipping sector prevent firms from making optimal environmental choices. In this section, we highlight the main barriers and market failures that lead to a delay in adoption of technologies and fuels with higher environmental performance.

5.1. Sunk Costs and Path Dependence in the Shipping Sector

The average life of a ship is approximately 25 years, which means that a significant share of current ships will still be in operation by 2035. Even if all ship owners would from now on order zero-emission vessels, there would still be a considerable share of ships that would not be zero-carbon. Decarbonization of the sector will depend to an important extent on the level of fleet renewal that is possible, which depends on the extent of scrappage of old vessels and the capacity to retrofit existing vessels. This causes important sunk costs. The potential for fleet renewal is larger if maritime trade is expanding and could also be subject to policy interventions to speed up the process and mitigate excessive economic harm that sudden changes could cause.

Significant use of CO_2 mitigation measures also assumes sufficient adaptation of infrastructure and production capabilities to future demand for alternative fuels and energy that might not take place immediately, considering the path dependence in the shipping sector. Choices made on the basis of temporary conditions can persist long after these conditions change, especially when the capital has a high life span. This durability of invested capital makes major changes particularly impractical and costly. Examples of far-reaching adaptations that might be needed include the wider energy infrastructure and production capabilities related to advanced biofuels, hydrogen, ammonia, and other zero-carbon fuels. In addition, ships would require the relevant facilities for bunkering and energy provision, e.g., charging systems for electric ships. There might be two distinctive concerns in this respect: first, concerns about the maximum supply potential within a given time period (e.g., of advanced biofuels); and second, how to reach sufficient scale for measures to become commercially viable, e.g., with regards to synthetic fuels, wind technology, and ship batteries.

5.2. Carbon Emissions as Negative Externality: The Climate as Unpriced Public Good

A negative externality arises when an individual or a firm takes an action but does not bear the costs imposed on a third party. In the case of the shipping industry, pollution imposes health, environmental, and economic costs on the whole of society without bearing the cost of it. Since costs are never borne entirely by the emitters and they are not obliged to compensate those who lose out because of climate change, they face little or no economic incentive to reduce emissions. Human-induced climate change and associated GHG emissions have been described as the greatest market failure in history [54]. A particular challenge with this "market failure" is the uncertainty about the exact size, timing, and location of the effects on environment, society, and the economy. Furthermore, climate change is a global phenomenon in both its causes and consequences and is therefore politically extremely challenging to address.

Climate change risks are not internalized in the price of maritime transport, especially since ship fuel is not taxed, in contrast, for example, to fuels for the road sector. Taxing fuels would be a way to internalize part of the externalities of carbon emissions. This lack of taxation is also hampering the uptake of alternative options for ship propulsion: heavy fuel oil for ships is not taxed but generates sizable negative externalities, whereas some of the alternative energy sources (e.g., electricity) with much less of these externalities are actually taxed. This complicates the transition from HFO to alternative technologies and fuels.

5.3. Split Incentives

Split incentives represent a type of principal–agent problem and occur when participants in an economic transaction do not have the same priorities and incentives. A classic example is the split incentive between charterer and ship owner in the time charter market, where the ship owner provides a vessel, but the fuel costs are borne by the charterer as part of the operational costs [55]. The difference between the actual level of energy efficiency and the higher level that would be cost-effective from the firm's point of view is often referred to as the efficiency gap (IEA, 2007). Usually, minimizing the capital cost of the vessel, the ship owner does not have an incentive to choose the most energy efficient technology, which often leads to suboptimal investment choices in environmental terms.

Whether time charter is used depends on the shipping segment. They are most prevalent in the container and dry bulk sectors where about 70% and 60% of ships, respectively, are run under time charter agreements whose duration mostly does not exceed one year, which is too short to amortize green investments [55]. However, charterers may also decide to reward owners for their investments in clean technologies or engage in longer charter contracts to allow for a sufficient payback time for these technologies. For example, the purchase of LNG bunkering vessel Coralius from Sirius Shipping was facilitated by Skangas (part of the Finnish state-owned Gasum Group) who agreed to a 15-year charter agreement. Preem and ST1 were willing to engage in long-term charters, making it possible for Terntankers to order LNG-powered vessels. Similarly, some shippers accepted paying higher charter rates to compensate for higher costs related to more environmentally friendly vessels [56].

5.4. Imperfect Information and Information Asymmetry

Imperfect, insufficient, or false information can cause firms to make suboptimal investments in energy efficiency. This type of market failure is particularly relevant in this case, as it contributes to preventing the uptake of greener technologies. Previous research has shown that the quality of knowledge and the level of technological know-how acquired through R&D activities are vital for the diffusion of technologies [57]. There is, however, a shortage of detailed and audited performance data of new technical measures with low market maturity, which acts as a barrier to their uptake [15]. The lack of reliable information on performance in actual operating conditions then leads to a typical chicken-and-egg problem in which no firm is ready to adopt a technology—or no financier ready to finance a zero-carbon ship—because there is a lack of strong proof of its efficiency and commercial viability.

This deficiency can be explained by the wide array of factors that influence fuel consumption of ships, namely, weather conditions, draught, machinery conditions, or operational aspects, which lead to highly variable performance data even for a single ship. In some cases, ship owners may have an incentive to make overly optimistic efficiency claims towards the charterer. Finally, the quality of measurement might also vary, although the industry is gradually shifting towards more frequent and reliable data collection methods, including continuous monitoring systems, which could potentially also discourage misrepresenting performance data [15]. The shortage of fuel efficiency data under real operating conditions also highlights the considerable market failure of a suboptimal level of resources allocated to technological and scientific knowledge. Subsidies to R&D or partnerships between government, research, and industry would be a possible solution to produce more evidence supporting or discarding a certain technology, or to develop alternative, nonincremental options.

5.5. Access to Finance

Currently, there is a risk premium to implementing innovative technologies and ship owners therefore face high barriers to upgrading to a more energy-efficient fleet. Ship owners need to effectively convince financiers that the additional costs of greener technology will be recovered. A study by UMAS and Carbon War Room [4] looked at implications of climate mitigation policies for ship owners and financiers under different market conditions and identified several actions that would need to be taken to both understand and manage these risks. This could include, for example, integrating risks associated with evolving climate regulations into financing decisions and identifying opportunities that environmentally responsible investments represent for financiers, such as the substantial expected demand for capital for vessel modifications. Although the general awareness amongst shipping financiers of climate-related stranded asset risks is rising, only very few are actively managing those risks [58]. In the longer term, however, the conditions for loans for zero-carbon ships could become more favorable than for ships powered on fossil fuels, considering their risk to become stranded carbon assets once stricter environmental regulation is enforced.

6. Implications in Designing Effective Policy Instruments

Although the vision of a trade-off between environment and competitiveness has been blurred over the recent years, government intervention remains indispensable for the broader adoption of low-carbon technologies and fuels in shipping. Greater policy certainty due to the target that has been set at the recent IMO 72nd MEPC meeting will encourage investment and can stimulate innovation [59]. In this context, the target would send an important signal to industry and research that investing in decarbonization will be profitable. It is likely that there will be future regulations on an international level aimed at achieving an at least 50% reduction of CO_2 emissions by 2050 compared with the 2008 level. It is also noteworthy that the second goal of the third initial strategy is aimed at phasing out CO_2 emission as soon as possible, consistent with the Paris Agreement goal. This development has helped to create the much-needed certainty for the industry and opened ways for stronger CO_2 mitigation policies to be adopted as part of the strategy that will be announced in 2023.

Decisions in the coming years at the IMO could include the adoption of a carbon pricing scheme for global shipping, thus leaving it to the market to allocate resources optimally. A carbon price has the potential to reduce the price gap between conventional and more sustainable fuel options. Using the example of wind assistance, calculations by Lloyd's Register show that technologies with a 10% fuel savings potential would become commercially viable only at higher fuel prices from 1000 USD/metric ton [60]. For instance, with the adoption of a carbon price, wind assistance could become an interesting option, especially if prices of alternative fuels are high. Receipts from a carbon-pricing scheme could fund further research and development in green shipping or ship retrofitting programs. Many countries still lack the economic and institutional resources to face radical decarbonization and implement and enforce new regulations effectively. Therefore, such a fund could also assist in transposing regulations and mitigating adverse impacts of decarbonization on trade in least-developed countries and small island developing states, for example, through compensation or technical assistance. Finally, the nature and function of the target are also important characteristics that will shape how decarbonization will progress considering uncertain future developments. Some research has highlighted that a floating target would have the necessary flexibility to encompass uncertainties in future maritime trade volumes, thereby mitigating quota volatility [3].

Although a large number of measures that could increase energy efficiency are available at negative net costs, available options often require high upfront investments [61]. A possible way to overcome high upfront investments and long payback times of technologies is the "savings as a service" model in which technology is rented and paid for entirely out of fuel savings [62]. Furthermore, with the recent adoption of the IMO target, companies that were delaying their investments in low-carbon technologies will have to adapt their business strategies to ensure sustainability and profitability in the longer term. They can start joining other companies which have proactively engaged in the development of greener technologies and identified a clear business case for doing so [56]. For instance, cargo owners have increasingly demanded higher transparency on environmental performance from carriers. A type of emerging technology push has also been observed in the LNG and biofuels sector that manifests itself in joint initiatives by governments and industry in order to develop new revenue streams [28]. On the financial side, initiatives such as the Task Force on Climate-related Financial Disclosures (TCFD) have provided additional transparency and company information on climate-related issues that investors and asset managers use for their investment decisions in order to avoid risks of stranded carbon assets. Split incentives between charterers and ship owners could be mitigated by longer charter contracts.

The barriers that prevent the market from spontaneously moving towards decarbonization might also require additional incentives and fiscal instruments. National or regional incentive schemes could complement carbon pricing at a global level. Applying stricter targets at a national or regional level first (patchwork approach) has been argued to play a catalytic role for progress on a global level and should not be despised as "illegitimate unilateralism" [3]. Governments could provide financial incentives for green shipping, e.g., via public procurement and temporary exemptions of electricity taxes for electric ships, or reduce trade tariffs for energy-efficient technologies [63]. In turn, electricity (either directly used for ship propulsion or for synthetic fuel production) could be bound to renewable portfolio standards found, for instance, in Germany and Chile [41]. Government action might also entail collaboration with financial institutions such as domestic or international development banks to create targeted financial instruments for green shipping, similar to existing schemes of the European Investment Bank. To improve access to finance for companies willing to adopt low-carbon technology, governments also have the possibility to create favorable conditions for financial instruments such as "Blue Bonds" that aim to channel private finance towards "green" shipping. If carefully designed, supplementary policies can be helpful in addressing market barriers and the burden of a potential carbon price. Given the high upfront costs to adapt to increasingly stringent environmental objectives, transitional assistance for some industries may be appropriate. This can include, for instance, support for technological research and development and implementation, as well as delivering the necessary low-carbon infrastructure (i.e., for sustainable biofuels and synthetic fuels). Furthermore, actions taken should be decided in close consultation with industry stakeholders and ports in order to avoid undesired and unforeseen economic effects. This should, however, take into account eventual risks of regulatory capture.

Transition towards a low-carbon shipping sector also implies developing adequate regulations and standards for technological measures and fuels, which could be the result of joint efforts by governments, as well as the naval and fuel industries. The need for carbon pricing would be less imminent if strong standards would be developed. For example, low-carbon fuel standards could be developed for the shipping sector, similar to the fuel standards that have been developed for road transport in many countries [28]. For example, this could also encompass wider adoption of more reliable fuel consumption and CO_2 emissions monitoring systems, which—if harmonized on an international level—can better inform performance evaluations of certain technologies and subsequent investment decisions [55]. However, while sensor technologies used on ships generate a vast array of data, these operational and technical insights are not often publicly available for investors, analysts, or researchers. This could be solved by developing a protocol to pool data on a platform without compromising business secrets [3], which nonetheless would require a great amount of collective action. A similar effort has already been made in the manufacturing sector [64].

Many new technologies and alternative fuels still require research and development, particularly to develop their commercial viability. Moreover, operational procedures would need to become streamlined and harmonized to ensure safety and interoperability. For instance, there is no standardized design and fueling procedure for hydrogen-powered ships and its bunkering infrastructure and remaining safety design issues with regards to the volatility of the fuel need to be resolved [14,36]. The low energy density of hydrogen requires very sizeable fuel tanks, which increase the capital cost and may reduce cargo space in commercial shipping. In addition, if relevant volumes of synthetic fuels, such as hydrogen and ammonia, are used by 2035, it would be essential to ensure that production processes are based on renewable electricity generation. Otherwise, no improvement in CO_2 emissions compared to conventional HFO could be guaranteed.

7. Conclusions

This paper examines the technical possibility of achieving the ambitious 1.5° goal of the Paris Agreement and the required supporting policy measures. We found that maximum deployment of technologies that are currently available could make it possible to reach almost full decarbonization by 2035. We formulated four possible decarbonization pathways for shipping, which foresee remaining carbon emissions ranging from 44 to 156 million tonnes by 2035 with CO_2 emission reduction ranging from 82 to 95% of the projected 2035 level. A major part of the required reductions could be realized via alternative fuels and renewable energy. Technological measures are available to increase the energy efficiency of ships and could yield a substantial part of emission reductions. Finally, operational measures could also achieve an important share of the required emission reductions.

Government intervention can help to accelerate the commercial viability and technical feasibility of certain measures. Various policies or regulations could support the shift to zero-carbon operations, including more stringent energy efficiency targets, a speed limit, and a low-carbon fuel standard. These policies could be introduced globally and by IMO member states. Governments and ports could provide necessary infrastructure, e.g., for shore power facilities, electric charging systems, and bunkering facilities for alternative fuels. Governments could also encourage green shipping domestically, stimulate research and development on zero-carbon technologies, and design programmes to increase commercial viability of these technologies. Financial institutions could develop green finance programmes to stimulate sustainable shipping. Shippers could be further encouraged to assess the carbon footprint of their supply chain and target zero-carbon shipping options.

Financial incentives are essential to reducing the price gap between conventional and more sustainable fuel options. These incentives could include adopting a carbon price for global shipping, leaving it to the market to allocate resources to maximum effect. Receipts from such a scheme could also be used (in part) for further decarbonization of the sector, e.g., to facilitate research and development in green shipping, facilitate ship retrofit programmes, and compensate for potential adverse trade impacts in least-developed countries and small island developing states. Carbon pricing at a global level could be complemented with incentive schemes at the national or regional level. Governments could also provide financial incentives for green shipping, e.g., greening the procurement of maritime transport falling under public service agreements, temporary exemptions of electricity taxes for electric ships, etc. Ports could also provide financial incentives for green shipping via differentiation of their port fee tariffs based on environmental criteria. Governments might partner with financial institutions or encourage domestic development banks to develop targeted financial instruments for green shipping.

Author Contributions: R.A.H. led the drafting and quantitative analysis of the article. He conceptualized the analytical framework used to assess the impact of CO_2 mitigation measures, and implemented and validated the model. Furthermore, he also analyzed different scenarios presented in this article. L.M.M. developed the earlier version of the ITF International Freight Model and supervised the methodology used in this article. L.K., R.A.H.,

O.M. conducted a literature review and collected data used for policy analysis. The original draft was prepared and written by R.A.H., L.K., and O.M. L.M., L.K. and O.M. reviewed and edited the article. Finally, O.M. acquired the funding of the project, and provided supervision for the project.

Funding: This research was made possible by the European Climate Foundation through a voluntary contribution to the International Transport Forum.

Acknowledgments: Valuable comments on a draft version of the article were provided by Stephen Perkins, Jari Kauppila (ITF), Tristan Smith (UCL Energy Institute), and Renske Schuitmaker (International Energy Agency). Francisco Furtado contributed to the validation of the model.

Conflicts of Interest: The authors declare no conflict of interest. The funding sponsors had no role in the design of the study; in the collection, analyses, or interpretation of data; and in the decision to publish the results.

Appendix A. Model Validation

Table A1. Validation of mode share by weights.

Weight	Model Output for 2011 (Million Tones)	Available Statistics (Million Tones)	Source
Air	31.93	31.8	ICAO report 2010
Sea	8579	8784	UNCTAD Review of Maritime Transport 2011
Road	1352	-	
Rail	289	-	

Sources: [65,66].

Table A2. Validation of transport demand by modes.

International Transport Demand	Model Output 2011 (Billion Tonne km)	Reference (Billion Tonne km)	Source	Mode Share (%)
Maritime	75 551	75 022	UNCTAD review of Maritime Transport 2016	90
Air	155	146	ICAO, WB	0.3
Road	6 642	-		7.8
Rail	1 875	-		2.2

Sources: [67,68].

Appendix B. Supplementary Data

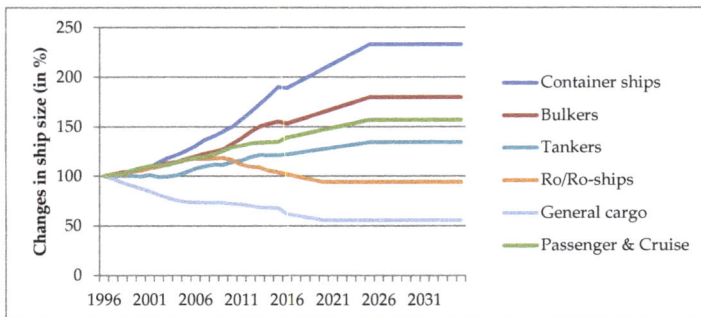

Figure A1. Historical (1996–2015) and estimated (2016–2035) changes in ship size.

Table A3. Emission factors of different fuel types.

Fuel Type	Emission Factors (kgCO$_2$/MJ)
HFO	0.081
MDO/MFO	0.072
LNG	0.810
Hydrogen	0
Biodiesel	0.522
Ammonia	0

Sources: IEA.

References

1. Smith, T.; Jalkanen, J.; Anderson, B.; Corbett, J.; Faber, J.; Hanayama, S.; Pandey, A. Third Imo Greenhouse Gas Study 2014. 2014, p. 327. Available online: http://www.imo.org/en/OurWork/Environment/PollutionPrevention/AirPollution/Documents/Third%20Greenhouse%20Gas%20Study/GHG3%20Executive%20Summary%20and%20Report.pdf. [PubMed]
2. Cames, M.; Graichen, J.; Siemons, A.; Cook, V. Emission reduction targets for international aviation and shipping. In *Directorate General for Internal Policies*; European Parliament—Policy Department A; Economic and Scientific Policy: Bruxelles, Belgium, 2015.
3. Wan, Z.; el Makhloufi, A.; Chen, Y.; Tang, J. Decarbonizing the international shipping industry: Solutions and policy recommendations. *Mar. Pollut. Bull.* **2018**, *126*, 428–435. [CrossRef] [PubMed]
4. Raucci, C.; Prakash, V.; Rojon, I.; Smith, T.; Rehmatulla, N.; Mitchell, J. *Navigating Decarbonisation: An Approach to Evaluate Shipping's Risks and Opportunities Associated with Climate Change Mitigation Policy*; UMAS: London, UK, 2017.
5. GL, D. Low Carbon Shipping Towards 2050. Available online: www.dnvgl.com/publications/low-carbon-shipping-towards-2050-93579 (accessed on 20 February 2018).
6. IEA. *Energy Technology Perspectives 2017*; IEA: Paris, Farnce, 2017.
7. Smith, T.; Raucci, C.; Hosseinloo, S.H.; Rojon, I.; Calleya, I.; De La Fuente, S.; Wu, P.; Palmer, K. *CO$_2$ Emissions from International Shipping: Possible Reduction Targets and Their Associated Pathways*; UMAS: Lordon, UK, 2016.
8. LR/UMAS. *Zero-Emission Vessels 2030. How Do We Get There?* UMAS/Lloyd's Register: London, UK, 2018.
9. Environment, T. *Statistical Analysis of the Energy Efficiency Performance (EEDI) of New Ships*; Transport and Environment: Brussel, Belgium, 2017.
10. Hoen, M.; Faber, J. *Estimated Index Values of Ships 2009–2016: Analysis of the Design Efficiency of Ships that Have Entered the Fleet Since 2009*; CE Delft: Delft, The Netherlands, 2017.
11. Bouman, E.A.; Lindstad, E.; Rialland, A.I.; Strømman, A.H. State-of-the-art technologies, measures, and potential for reducing GHG emissions from shipping—A review. *Transp. Res. Part D Transp. Environ.* **2017**, *52*, 408–421. [CrossRef]
12. Gilbert, P.; Bows-Larkin, A.; Mander, S.; Walsh, C. Technologies for the high seas: Meeting the climate challenge. *Carbon Manag.* **2015**, *5*, 447–461. [CrossRef]
13. IMarEST. *Marginal Abatement Costs and Cost Effectiveness of Energy-Efficiency Measures*; IMO Document MEPC 62/INF.7: London, UK, 2011.
14. Lindstad, H.; Eskeland, G.S. Low carbon maritime transport: How speed, size and slenderness amounts to substantial capital energy substitution. *Transp. Res. Part D Transp. Environ.* **2015**, *41*, 244–256. [CrossRef]
15. Rehmatulla, N.; Calleya, J.; Smith, T. The implementation of technical energy efficiency and CO$_2$ emission reduction measures in shipping. *Ocean Eng.* **2017**, *139*, 184–197. [CrossRef]
16. Carlton, J.; Aldwinkle, J.; Anderson, J. *Future Ship Powering Options: Exploring Alternative Methods of Ship Propulsion*; Royal Academy of Engineering: London, UK, 2013.
17. Tillig, F.; Mao, W.; Ringsberg, J. *Systems Modelling for Energy-Efficient Shipping*; Chalmers University of Technology: Gothenburg, Sweden, 2015.
18. Van Kluijven, P.C.; Kwakernaak, L.; Zoetmulder, F. *Contra-Rotating Propellers*; Rotterdam Mainport University of Applied Sciences RMU: Rotterdam, The Netherlands, 2013.
19. Faber, J.; Huigen, T.; Nelissen, D. *Regulating Speed: A Short-Term Measure to Reduce Maritime GHG Emissions*; CE Delft: Delft, The Netherlands, 2017.

20. Golias, M.M.; Saharidis, G.K.; Boile, M.; Theofanis, S.; Ierapetritou, M.G. The berth allocation problem: Optimizing vessel arrival time. *Marit. Econ. Logist.* **2009**, *11*, 358–377. [CrossRef]
21. Kiani, M.; Bonsall, S.; Wang, J.; Wall, A. A break-even model for evaluating the cost of container ships waiting times and berth unproductive times in automated quayside operations. *WMU J. Marit. Aff.* **2006**, *5*, 153–179. [CrossRef]
22. Lindstad, H.; Asbjørnslett, B.E.; Strømman, A.H. Reductions in greenhouse gas emissions and cost by shipping at lower speeds. *Energy Policy* **2011**, *39*, 3456–3464. [CrossRef]
23. Lindstad, H.; Asbjørnslett, B.E.; Strømman, A.H. The Importance of economies of scale for reductions in greenhouse gas emissions from shipping. *Energy Policy* **2012**, *46*, 386–398. [CrossRef]
24. Lindstad, H. *Strategies and Measures for Reducing Maritime CO_2 Emissions*; NTNU: Trondheim, Norway, 2013.
25. Psaraftis, H.N.; Kontovas, C.A. Ship speed optimization: Concepts, models and combined speed-routing scenarios. *Transp. Res. Part C Emerg. Technol.* **2014**, *44*, 52–69. [CrossRef]
26. ITF. *The Impact of Mega-Ships*; ITF/OECD: Paris, France, 2015.
27. Merk, O. *Shipping Emissions in Ports*; ITF/OECD: Paris, France, 2014.
28. ITF. *Decarbonising Maritime Transport: Pathways to Zero-Carbon Shipping by 2035*; ITF/OECD: Paris, France, 2018.
29. Anderson, M.; Salo, K.; Fridell, E. Particle-and gaseous emissions from an LNG powered ship. *Environ. Sci. Technol.* **2015**, *49*, 12568–12575. [CrossRef] [PubMed]
30. Bicer, Y.; Dincer, I. Clean fuel options with hydrogen for sea transportation: A life cycle approach. *Int. J. Hydrog. Energy* **2018**, *43*, 1179–1193. [CrossRef]
31. GL, D. *DNV GL Handbook for Maritime and Offshore Battery Systems*; DNV GL: Oslo, Norway, 2016.
32. Hsieh, C.; Felby, C. *Biofuels for the Marine Shipping Sector, An Overview and Analysis of Sector Infrastructure, Fuel Technologies and Regulations*; IEA Bioenergy: Paris, France, 2017.
33. Traut, M.; Gilbert, P.; Walsh, C.; Bows, A.; Filippone, A.; Stansby, P.; Wood, R. Propulsive power contribution of a kite and a Flettner rotor on selected shipping routes. *Appl. Energy* **2014**, *113*, 362–372. [CrossRef]
34. Veerbeek, R.; Ligterink, N.; Meulenbrugge, J.; Koornneef, G.; Kroon, P.; de Wilde, H.; Kampman, B.; Croezen, H.; Aarnink, S. *Natural Gas in Transport: An Assessment of Different Routes*; CE Delft: Delft, The Netherlands, 2013.
35. Smith, T. Why LNG as the Ship Fuel of the Future is a Massive Red Herring. Available online: http://splash247.com/lng-ship-fuel-future-massive-red-herring/ (accessed on 28 March 2017).
36. GL, D. *Study on the Use of Fuel Cells in Shipping*; Study Commissioned by the European Maritime Safety Agency: Lisbon, Portugal, 2017.
37. Teeter, J.L.; Cleary, S.A. *Decentralized Oceans: Sail—Solar Shipping for Sustainable Development in Sids*; Natural Resources Forum; Wiley Online Library: New York, NY, USA, 2014; pp. 182–192.
38. Swartz, J. China's national emissions trading system. In *ICTSD Series on Climate Change Architecture*; ICTSD: Geneva, Switzerland, 2016; pp. 20–23.
39. GL, D. EU MRV Regulation. Available online: www.dnvgl.com/maritime/eu-mrv-regulation/index.html (accessed on 28 March 2018).
40. Environment, T. *Decarbonising Shipping Sector in EU: ETS Maritime Climate Fund and 2030 Targets*; Transport & Environment: Brussel, Belgium, 2017.
41. CPLP. What is Carbon Pricing? Available online: https://www.carbonpricingleadership.org/what/ (accessed on 28 March 2018).
42. Schipper, L.; Marie-Lilliu, C.; Gorham, R. *Flexing the Link between Transport and Greenhouse Gas Emissions—A Path for the World Bank*; International Energy Agency: Paris, France, 2000.
43. Chateau, J.; Dellink, R.; Lanzi, E. *An Overview of the OECD ENV-Linkages Model*; OECD: Paris, France, 2014.
44. Ben-Akiva, M.; Bierlaire, M. Discrete Choice Methods and their Applications to Short Term Travel Decisions. In *Handbook of Transportation Science*; Hall, R., Ed.; Springer: New York, NY, USA, 1999; Volume 23, pp. 5–33.
45. Halim, R.; Kwakkel, J.; Tavasszy, L.A. A scenario discovery study of the impact of uncertainties in the global container transport system on European ports. *Futures* **2016**, *81*, 148–160. [CrossRef]
46. Tavasszy, L.A.; Minderhoud, M.; Perrin, J.-F.; Notteboom, T. A Strategic Network Choice Model for Global Container Flows: Specification, Estimation and Application. *J. Transp. Geogr.* **2011**, *19*, 1163–1172. [CrossRef]
47. UNCTAD. *Review of Maritime Transport 2013*; UNCTAD: Geneva, Switzerland, 2013.
48. LR/UMAS. *Global Marine Technology Trends 2030*; Lloyd's Register: London, UK, 2017.
49. International Energy Agency. *Momo ETP 2014*; IEA Energy Technology Policy Division: Paris, France, 2014.

50. IEA. *World Energy Outlook 2017*; International Energy Agency: Paris, France, 2017.
51. VDKi. First estimation of World trade, consumption and import of hard coal in 2015. In *Press Release No. 02/2016*; VDKi: Hamburg, Germany, 2016.
52. IMF. Direction of Trade Statistics (DOTS). Available online: www.data.imf.org/?sk=9D6028D4-F14A-464C-A2F2-59B2CD424B85 (accessed on 22 January 2018).
53. Finance, B.N.E. Lithium-Ion Battery Costs and Market: Squeezed Margins Seek Technology Improvements and New Business Models. Available online: https://data.bloomberglp.com/bnef/sites/14/2017/07/BNEF-Lithium-ion-battery-costs-and-market.pdf (accessed on 5 March 2018).
54. Stern, N. *Stern Review on the Economics of Climate Change*; HM Treasury: London, UK, 2006; Volume 30.
55. Rehmatulla, N.; Parker, S.; Smith, T.; Stulgis, V. Wind technologies: Opportunities and barriers to a low carbon shipping industry. *Mar. Policy* **2017**, *75*, 217–226. [CrossRef]
56. ITF. *Reducing Shipping GHG Emissions: Lessons from Port-Based Incentives*; ITF/OECD: Paris, France, 2018.
57. Costantini, V.; Crespi, F.; Martini, C.; Pennacchio, L. Demand-pull and technology-push public support for eco-innovation: The case of the biofuels sector. *Res. Policy* **2015**, *44*, 577–595. [CrossRef]
58. Mitchell; Rehmatulla, N. Dead in the Water. In Proceedings of the Shipping in Changing Climates (SCC) Conference, London, UK, 24 November 2015.
59. Porter, M.E.; Van der Linde, C. Toward a new conception of the environment-competitiveness relationship. *J. Econ. Perspect.* **1995**, *9*, 97–118. [CrossRef]
60. LR. *Wind-Powered Shipping: A Review of the Commercial, Regulatory and Technical Factors Affecting Uptake of Wind-Assisted Propulsion*; Lloyd's Register Marine: London, UK, 2015.
61. Buhaug, Ø.; Corbett, J.; Endresen, Ø.; Eyring, V.; Faber, J.; Hanayama, S.; Lee, D.; Lee, D.; Lindstad, H.; Markowska, A. *Second IMO GHG Study 2009*; IMO: London, UK, 2009.
62. Alliance, S.G.S. Shipping Systems Fit for the Future. Available online: https://www.smartgreenshippingalliance.com/ (accessed on 29 March 2018).
63. Rehmatulla, N.; Smith, T.; Tibbles, L. The relationship between EU's public procurement policies and energy efficiency of ferries in the EU. *Mar. Policy* **2017**, *75*, 278–289. [CrossRef]
64. Kusiak, A. Smart manufacturing. *Int. J. Prod. Res.* **2018**, *56*, 508–517. [CrossRef]
65. UNCTAD. *Review of Maritime Transport 2011*; UNCTAD: Geneva, Switzerland, 2011.
66. ICAO. *Annual Report of the Council*; ICAO: Montreal, QC, Canada, 2011.
67. UNCTAD. *Review of Maritime Transport 2016*; UNCTAD: Geneva, Switzerland, 2016.
68. Bank, T.W. Air Transport, Freight. Available online: https://data.worldbank.org/indicator/IS.AIR.GOOD.MT.K1 (accessed on 2 February 2018).

sustainability

MDPI

Article

Energy Efficiency in Logistics: An Interactive Approach to Capacity Utilisation

Jessica Wehner

Department of Technology Management and Economics, Chalmers University of Technology,
41296 Gothenburg, Sweden; jessica.wehner@chalmers.se; Tel.: +46-(0)31-772-1719

Received: 30 April 2018; Accepted: 23 May 2018; Published: 25 May 2018

Abstract: Logistics operations are energy-consuming and impact the environment negatively. Improving energy efficiency in logistics is crucial for environmental sustainability and can be achieved by increasing the utilisation of capacity. This paper takes an interactive approach to capacity utilisation, to contribute to sustainable freight transport and logistics, by identifying its causes and mitigations. From literature, a conceptual framework was developed to highlight different system levels in the logistics system, in which the energy efficiency improvement potential can be found and that are summarised in the categories activities, actors, and areas. Through semi-structured interviews with representatives of nine companies, empirical data was collected to validate the framework of the causes of the unutilised capacity and proposed mitigations. The results suggest that activities, such as inflexibilities and limited information sharing as well as actors' over-delivery of logistics services, incorrect price setting, and sales campaigns can cause unutilised capacity, and that problem areas include i.a. poor integration of reversed logistics and the last mile. The paper contributes by categorising causes of unutilised capacity and linking them to mitigations in a framework, providing a critical view towards fill rates, highlighting the need for a standardised approach to measure environmental impact that enables comparison between companies and underlining that costs are not an appropriate indicator for measuring environmental impact.

Keywords: energy efficiency; capacity utilisation; logistics; road freight transport; sustainability; system level; systems perspective

1. Introduction

Actors in logistics and freight transport face increased pressure to reduce the climate impact of their operations and to become more environmentally sustainable. According to the European Commission [1], road transport alone accounts for 70 percent of all greenhouse gas (GHG) emissions from transport. To tackle the problem, EU member countries committed to reducing GHG emissions in the transport sector by at least 60 percent by 2050 compared to 1990. In 2013, transport alone consumed 63 percent of the world's oil [2]. A significant objective to increase environmental sustainability is the reduction of energy consumption in the freight transport sector [2]. A key challenge for managers is to respond to the increase in transport volume [3], which stems from consumers' desire for more products as well as longer transport distances because of global supply chains and international production [4–6]. One way to foster sustainable development in freight transport is to focus on increasing the energy efficiency [7]. Here, energy-efficient freight transport needs to be approached in its wider system that is, the logistics system. To radically decrease the energy consumption from transport, technological advances alone will not be enough; the task also requires changes in behaviour and structure of the logistics system [8,9], as well as inclusion of the end-consumer in a wider and extended system [10].

Logistics systems can become more energy-efficient through behavioural changes among end-consumers, shippers, and logistics service providers (LSPs). One area of potential improvements is to consider unutilised capacity [11,12]. The current literature has a strong focus on reducing the energy consumption in freight transport by increasing the load factor [12,13], that is, offering a narrow view of transport capacity.

This paper extends the view on capacity utilisation in freight transport by considering the larger system in which it operates. The need to view freight transport in its wider system when enhancing sustainability has been recognised in the current body of knowledge, such as in the sustainability framework by Turki, et al. [14], who have approached transport in connection to manufacturing, remanufacturing, and warehousing. Accordingly, this research not only considers the capacity utilisation in freight transport, but also in the adjacent logistics activities. By applying a systems perspective, the interactive nature of the different components of capacity in the logistics systems becomes more apparent, which implies that capacity utilisation is related to the different levels of the logistics system. Accordingly, this paper builds upon the notion that energy efficiency in every logistics activity can be increased if the capacity is used to its full potential. This may, for example, be the case through supply-chain-related initiatives, such as collaboration and supplier education, as a means to address energy efficiency [15]. Capacity as an interactive concept within the overall logistics system, and especially between the actors in the supply chain, has not been studied in detail. Viewing capacity from a systems perspective is necessary in order to identify all of the improvement potential. This research tries to shed light on the following research questions, namely: (1) Where in the logistics system is unutilised capacity available that will improve energy efficiency? (2) How can this unutilised capacity be mitigated? Against this background, the purpose of this paper is to take an interactive approach to capacity utilisation so as to contribute to sustainable freight transport and logistics, by identifying the causes and mitigations of unutilised capacity.

By taking an interactive approach to capacity utilisation, this paper contributes insights to the literature on sustainable logistics and road freight transport, most practically by developing a framework of capacity utilisation to increase the energy efficiency in logistics. At the same time, by highlighting where the unused capacity in logistics systems can be found and proposing mitigations, the paper offers important implications for the transport industry, by broadening the understanding of capacity utilisation.

In what follows, literature is reviewed and presented in a frame of reference in the next section. In addition, a conceptual framework is proposed, which stems from the literature. The following section explains the method, including the sampling, data collection and analysis, and research quality. The findings from the interviews are presented in 'Results'. After that, a discussion of the findings is presented, and the framework is developed. Lastly, the paper concludes with managerial implications and theoretical contributions and proposes directions for future research.

2. Frame of Reference

To illustrate the role of capacity utilisation in logistics and road freight transport, a conceptual framework was developed. The framework's building blocks were derived from a review of the literature on freight transport, logistics, and supply chain management (SCM), with a focus on the logistics-energy domain and environmental sustainability. A combination of keywords (logistics, energy logistics, sustainable logistics, supply chain, supply chain management, freight transport, energy efficiency, energy, and sustainability) were used, and depending on the type of journal, the keywords were searched for alone or in combination. For this, the top-ten ranked journals of logistics and SCM in the Nordic countries [16], one journal focusing on sustainability, and two journals focusing on the energy domain were chosen, namely: the European Journal of Purchasing and Supply Management, International Journal of Logistics Management, International Journal of Logistics: Research and Applications, International Journal of Physical Distribution and Logistics Management, International Journal of Retail & Distribution Management, Journal of Business Logistics, Journal of Supply Chain Management, Supply Chain Management: An International Journal, Logistics Management and Supply Chain Management Review, and Sustainability, Energy and Journal of Cleaner Production. Further papers were added through 'snowballing', that is, searching more openly based on references and keywords from the articles that were identified in the structured part of the review. Abstracts and papers were read to identify those that were related to the research. In a further step, this literature was read and analysed regarding how it addressed 'capacity utilisation' in connection with 'energy efficiency/consumption'.

2.1. Energy Efficiency in Logistics and Road Freight Transport

Energy is a source of power that is needed to operate logistics activities [17]. Although road freight transport can be powered with different energy sources (e.g., fossil fuels, bio fuels, electricity from nuclear, or alternative energy), fossil fuels are still the most common form of energy source and can be traced back to the era when fossil fuels were considered inexpensive and plentiful [18].

Halldórsson and Kovács [19] point out that energy efficiency plays an important role in logistics and SCM. A positive research trend on the topic has been highlighted by Centobelli, et al. [7]. The European Parliament and the Council of the European Union [20] has defined energy efficiency as "the ratio of output of performance, service, goods or energy, to input of energy" and energy efficiency improvement as "an increase in energy efficiency as a result of technological, behavioral and/or economic changes". By extension, achieving a high level of energy efficiency means reducing the total energy consumption given a particular level of output [21]. The logistics literature mainly addresses energy efficiency in three broad terms, namely: the interplay of activities that influence energy efficiency (what), the inclusion of all actors (who), and the consideration of system boundaries (where) to measure energy efficiency.

Firstly, the literature focuses on the interplay of different activities that increase the energy efficiency in logistics. Aronsson and Huge-Brodin [8] have identified consolidation, standardisation, information flow, and virtual warehousing as drivers of efficiency and environmental performance. Later, Wolf and Seuring [22] identified collaboration (i.e., integration, cooperation, and information sharing) as the most important component for the successful management of supply chains. More specifically, Pfohl and Zöllner [23] described efficiency in logistics as the result of factors, such as environmental relations, product lines, production, technology, and the size of organisations, while Piecyk and McKinnon [24] described energy efficiency as being influenced by the weight of goods, empty running, and average vehicle energy consumption, among other factors. In contrast, Kalenoja, et al. [25] identified energy consumption, delivery times, transport speed, flexibility, reliability, and vehicle load as influential components in energy efficiency. According to Plambeck [15], efficiency can be increased by the harmonisation and coordination of the different operations in supply chains. More recently, Bottani, et al. [26] outlined an integrated approach for achieving efficiency that involves the pooled management of packaging, procurement, warehousing, and transport activities.

Secondly, when discussing the actors' involvement in energy efficiency in logistics and road freight transport, the literature focused foremost on the logistics service providers (LSPs) and shippers [15,22,23] and the harmonisation, coordination, and collaboration between them (e.g., know-how transfer to suppliers and long-term commitment) [15] and the supplier selection [27]. Energy efficiency also depends on information exchange between the actors; Yuan, et al. [28] developed a model to study the effect of the carbon emission information asymmetry between different actors on the supply chain performance from carbon trading. However, the impact of the end-consumer on energy efficiency in logistics and road freight transport, especially in connection to the last-mile deliveries and consumer transport, has been attracting more interest in literature over the years [10,29,30]. Since different actors have different impacts on energy efficiency, it is important to examine who takes what action.

Thirdly, clear system boundaries have to be defined, and indicators for measurement must be chosen [25] so as to determine where in the system the energy efficiency is measured and the activities are taken. Kalenoja, et al. [25] suggested different settings of system boundaries; for example, a narrower definition of system boundaries that includes only inbound and outbound logistics or a broader view including several suppliers and reverse logistics. The literature further suggests an expansion of system boundaries through the inclusion of, for example, top-management as enabler for environmental-friendly purchasing of transport services [31], reverse logistics [32], transport during the last mile [30] and, in particular, consumer transport [10,29,33,34], and other suppliers in the network [15,22]. Browne, et al. [10], who assessed the energy consumption of different product supply chains, found that the energy consumption that was used to transport goods during the last mile to the point of consumption was greater than that of all of the upstream transport activities combined. Rizet, et al. [35] and Brown and Guiffrida [34] identified the potential of saving energy in the last mile. The reason for such inefficiency in private transport during the last mile was unutilised capacity in private vehicles. Aronsson and Huge-Brodin [8] emphasised the complexity of logistics systems and therefore a macro-perspective to view the supply chain.

Guided by the aim and the research questions of the study to analyse where the unutilised capacity can be found in the logistics system and how it can improve energy efficiency, several themes have emerged from the literature review. The literature was set in relation to the capacity utilisation. An overview of themes that emerged from literature is provided in a review of the literature regarding energy efficiency (Table 1).

Table 1. Review of literature regarding energy efficiency.

Emergent Themes	Relevant Literature	Dimensions of Energy Efficiency	Relation to Capacity Utilisation
Measuring energy efficiency and goal-setting	Centobelli, et al. [7], Kalenoja, et al. [25], Liimatainen and Pöllänen [36], Liimatainen, et al. [37], McKinnon and Ge [38], Wu and Dunn [32]	Measuring energy efficiency, setting goals	Defining system boundaries, taking a broad approach
Measuring energy consumption	Browne, et al. [10], Browne, et al. [29], Piecyk and McKinnon [24]	Measuring energy consumption, reducing fuel and energy consumption, positioning energy as an essential cost driver	Vehicle load factor, empty running and transport distance, weight of goods influence capacity utilisation, interplay of different components
Collabo-ration between actors	Björklund [31], Bottani, et al. [26], He and Zhang [27], Plambeck [15], Wolf and Seuring [22], Yuan, et al. [28]	Discussing collaboration and information sharing. connecting collaboration and energy efficiency	Collaboration enables the use of unutilised capacity
End-consumer	Brown and Guiffrida [34], Browne, et al. [10]	Identifying the end-consumer's role, raising end-consumers' awareness of their implications on energy consumption	End-consumers' behaviour creates unutilised capacity. need to include consumer transport when viewing supply chain
Logistics system	Aronsson and Huge-Brodin [8], Kalenoja, et al. [25], McKinnon [39], Pfohl and Zöllner [23]	Establishing responsibilities for the environment and transported products, just-in-time deliveries, returns	Taking macro-perspective of supply chain, slowing down the supply chain
Last mile	Brown and Guiffrida [34], Kin, et al. [30], Rizet, et al. [35]	Discussing e-commerce Handling last-mile delivery	Transport in the last mile is not used to its full capacity

The literature was reviewed to address the energy efficiency in the logistics system starting in road freight transport and considering the adjacent logistics activities. The selected literature was then analysed regarding the capacity utilisation, which has been discussed in the following section.

2.2. Capacity Utilisation in Logistics and Road Freight Transport

Often, the capacity is associated with the loading capacity (i.e., the load factor), which is the physical ability of a vehicle to carry a freight for a certain length of time [40]. In that view, trucks are key units of capacity in road freight transport. However, capacity captures more than what the simple description outlines. Hayes, et al. [41] described capacity as a complex interaction of different components, including physical space, equipment, operating rates, human resources, system capabilities, company policies, and the rate and dependability of the suppliers. According to the literature, the concept of capacity developed from an understanding of the physical space [9,10] and only later became an indicator of the energy efficiency [25] and a factor in the logistics systems [34].

Wu and Pagell [42] found that different initiatives, including reducing the number of trips, activating an efficient information system, and pursuing collaboration, could increase the capacity utilisation and reduce the impact of energy use on the environment. Chapman [9] highlights the combining loads of different operators as a way to increase the capacity utilisation. Furthermore, Liimatainen and Pöllänen [36] discuss the components that are closely connected to capacity utilisation, including the average load on laden trips, empty running, and average vehicle energy consumption, which helps to improve energy efficiency in logistics. Rizet, et al. [35] pinpoint the time that is taken for activities, costs, service levels, and polices as factors influencing the capacity utilisation, and with that, energy consumption. Brown and Guiffrida [34] highlight that the particular capacity in private vehicles is not used to its full capacity. Furthermore, just-in-time delivery, with its small loads in rapid time, is responsible for unutilised capacity [9,39,43]. Vehicle loading, empty running, vehicle time utilisation, and deviation from schedule, are proposed by McKinnon and Ge [38] as key performance indicators (KPIs) in order to evaluate capacity utilisation.

The literature addresses the capacity utilisation as a narrow concept and only views it on one system level at a time. This research extends this view and proposes a system perspective on the capacity utilisation.

2.3. Systems Perspective on Capacity

A systems perspective on capacity entails diverse components and acknowledges their complexity and interactivity, which, in turn, can help illuminate the interactive nature of capacity utilisation. Supply chains, as well as environments, consist of various, complex subsystems [33] whose management can be difficult. This is based on the assumption that the systems are open [44]. The supply chain actors have to cooperate and share information as well as risk. Systems thinking enables individual actors to plan their logistics operations (e.g., transport, inventory, and purchasing) more efficiently and effectively. To acknowledge humans' roles and purposes in supply chains, this approach belongs to the school of soft systems thinking [44].

Using the systems perspective can also clarify the interconnection of energy efficiency and capacity utilisation in the logistics system, starting with road freight transport. In short, capacity utilisation leads directly to energy efficiency and fosters sustainable development. To interlink different components that set the conditions for capacity utilisation and understand its systemic nature, the research addresses the causes and mitigations at various levels of the logistics system.

The literature was reviewed to address the energy efficiency in the logistics system and was then analysed regarding the capacity utilisation as a means to energy efficiency. The concept of the capacity was chosen to highlight the improvement potential, following the structure of the activities (what), actors (who), and areas in the logistics system, where improvement is possible (Figure 1). This conceptual framework has served as the basis for constructing the interview guide and collecting data. The factors that set the conditions for capacity utilisation within the categories of activities, actors, and areas in the logistics system were derived from the empirical data.

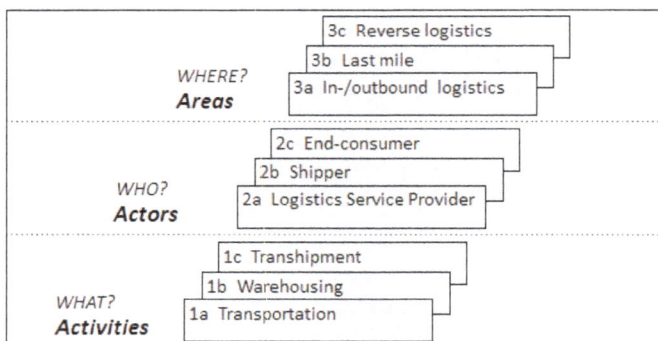

Figure 1. System levels and categories in the logistics system.

3. Method

To validate and develop the conceptual framework, qualitative data in the form of semi-structured interviews were collected. The data generation aimed to investigate the causes of unutilised capacity on different system levels, by taking an interactive approach to capacity utilisation. From the empirical data, the causes of the unused capacity in logistics and mitigations were derived, which were linked to each other.

3.1. Sampling

To access different perspectives on the logistics system, a multi-actor approach was used, with a sample of LSPs, their customers (i.e., shippers), and a consultant, who worked to improve energy efficiency by implementing lean management, an approach that seeks to reduce waste and improve services. By interviewing different LSPs and their direct customers, different product supply chains with cooperating actors could be reconstructed, which in turn facilitated an investigation into the interactive nature of capacity, with energy efficiency positioned at the intersections between the LSPs and their customers.

In total, 17 semi-structured interviews, with representatives of nine different companies, were conducted (Table 2). All of the interviewees, except for the consultant on lean management, had worked as logistics or supply chain managers with high-level managerial responsibility in their respective organisations. Holding such a position was a selection criterion, thereby ensuring a profound knowledge of the company's logistics activities. The consultant was added to the sample so as to gather an additional perspective and to provide ideas on utilising capacity in the logistics system.

The first company was chosen out of convenience [45]. Other companies were added to encompass the desired range for a multi-actor approach, until theoretical saturation was reached (i e., when the interviews had provided no further new information) [46]. Table 2 provides a brief overview of the size and scope of participation of all of the sample companies.

Table 2. Sample of companies interviewed.

Company	Description	Size *	Number of Interviews Conducted
A	Manufacture of machine elements and logistics service provider	Large	1
B	Manufacturer of packaging, processing, and provider of distribution solutions	Large	1
C	Manufacturer of paper and tissues	Large	1
D	Garment retailer with physical stores and e-commerce presence	Medium	3 (+ visit to distribution centre)
E	E-grocery retailer and deliverer	Small	1 (+ visit to distribution centre)
F	World-leading logistics service provider	Large	4
G	Nordic logistics service provider	Large	2
H	Nordic logistics service provider	Medium	1
I	Lean energy consultancy	Small	3

* Small: <1000 employees; medium: 1000–9999; large: >10,000 employees.

The interviews focussed on the transport into and within urban areas, which was regarded as the energy-intensive part of the supply chain, for example [10], and the focus was set on the road freight transport and its adjacent logistics operations.

3.2. Data Collection and Analysis

Each of the interviews lasted 60 to 90 min, were semi-structured and involved open-ended questions that encouraged the interviewees to elaborate upon their experiences regarding capacity utilisation. The interview guide was led by the aim of the study, the previous review of the literature, and the conceptual framework. During the interviews, comprehensive notes were taken; only one interview was audio recorded and transcribed, because all of the other interviewees indicated that they would feel more comfortable if the interview was not audio recorded.

The data that were collected during the interviews were analysed with the qualitative data analysis software NVivo, which involved coding the interview transcription and notes into nodes. As a start, the themes that emerged from the literature review (Table 1) were used as nodes, and other nodes were added during the analysis process, when necessary. The data were repeatedly analysed and sorted under the nodes. Then, the nodes were further reduced and could be summarised in the three categories from the conceptual framework, namely, the actors, activities, and areas in the logistics system. The categories explained who (i.e., actors) and what (i.e., activities) created the unutilised capacity and where (i.e., areas in the logistics system) it was created. This reduction of nodes helped to provide a better overview of the problematic areas [47]. Moreover, the framework could be validated by starting out with the themes from the literature and developing them further, which again resulted in the system levels from the conceptual framework.

The data were analysed, and both the causes and mitigations were derived following the idea behind the framework by Lee, et al. [48] on causes and counteractions. Their framework was used, given its similarities between the specific problems that were investigated. The causes of the unutilised capacity derived from the interview data and were grouped into the three categories, which simplified the subsequent process of identifying the mitigations. Alongside the causes, excerpts were extracted from the data as examples. The presented data clearly showed which result was extracted from which interview. These excerpts were reduced during an iterative process that involved repeatedly reviewing the data. In another step, the causes, which were derived from the data, were linked to the corresponding mitigations. During the data analysis, the literature was used as a reference so as to generate a deeper understanding of the specific issues that were investigated, as well as to determine whether the suggested mitigations had already been proposed in previous research.

3.3. Research Quality

To ensure a high quality of research, the design of the study was constructed carefully. The literature review not only provided an overview of the topic and insights into the systems thinking, but it also informed the analytical framework, based on Lee, et al. [48].

Since the basis for the interview guide emerged from the literature, a high relevance of the interview questions, relative to the studied topic, was ensured. Since the empirical data was to map the different perceptions of the actors in logistics, the qualitative criteria—trustworthiness and its four dimensions of credibility, transferability, dependability, and confirmability [49]—were used to evaluate the quality of research. Firstly, to ensure credibility, the research findings were validated with the study's participants by submitting the findings to the participants, so as to ensure correction of the understood world. Secondly, transferability, or the general applicability of the findings, was ensured by generating a sufficient richness of detail, by repeating the interviews with four of the nine participants. Thirdly, dependability was ensured by keeping records of all of the phases of the research process and documenting all of the method-related decisions. Furthermore, the author's peers viewed and discussed those materials with the author. Fourthly, and finally, confirmability was ensured by confirming that the findings were free of bias by comparing the data that were gathered with the data in the related literature.

4. Results

The data from interviews were structured around three categories, namely, the activities, actors, and areas in the logistics system, so as to identify the causes of the unutilised capacity and corresponding mitigations.

4.1. Causes of Unutilised Capacity

To identify and conceptualise the causes of the unutilised capacity in the logistics system that created inefficiencies, the data were structured around the categories and system levels.

4.1.1. Activities

The activities refer to what causes unutilised capacity in the logistics system and thus increases the energy consumption. In general, unutilised capacity stems from how parcels and other shipments are handled during transport, in the warehouse, and during transhipment (see Table 3).

Table 3. Activities that cause unutilised capacity.

Levels	Causes
Transport	Product characteristics and fit in vehicle Labour regulations Redundant transport of air and shipping hanging garments lead to low fill rates Delivery peaks during mornings and afternoons (i.e., rush hours) Last-minute changes in routing due to express deliveries High volumes of parcels are needed to fill the system and are taken from more energy-efficient systems Imbalances in volume flow and empty running
Warehousing	Human error during order picking Automation and standardisation leads to inflexibility Dysfunctional information technology
Transhipment	Difficulty sharing distribution capacity among shippers Limited internal and external information sharing Rules set by stronger actors and divergent interests Prohibited collaboration of larger logistics service providers (LSPs) (i.e., anti-competition law)

Transport: The way in which the trucks are loaded is crucial to energy efficiency in transport, and is mostly limited by the goods and their characteristics, as it was explicitly stated by the representatives of companies D and F. Another restriction was that the goods may not be stacked above shoulder height because of the labour regulations (evidence came from company D, F, G, and H). Other causes of unutilised capacity that were discussed during the interviews were the empty running, hanging garments, and idle loading units at the wrong places (such as by companies D, F, and G). The most popular pickup times are during the mornings and the delivery times are during the afternoons, which overlaps with rush hours, and the drivers are already exposed to constant pressure to meet deadlines. Companies E, F, and G mentioned that the constant pressure of time and traffic congestion could decrease the capacity utilisation. Last-minute changes in routing because of, for instance, additional express deliveries, were explicitly mentioned by companies F and G as the cause of inefficiencies. Additionally, the ability to load shipments depended on the departure and arrival times. If times had to be strictly kept, then a unit might not have been loaded on a truck in time. In that case, a shipment might have to wait an entire day at a terminal until it was distributed, thereby creating unutilised capacity in the truck, which would have to leave without it. Company F pointed out that a high volume of parcels and groupage which are needed to fill the logistics system, because the margins for single shipments are so low that only large volumes make the logistics business profitable, lead to further problems. Filling one system with a sufficient volume of shipments often meant that the shipments were transferred from other systems. For instance, filling a truck might require taking products from a more energy-efficient system (e.g., rail), which would leave behind unutilised capacity. Moreover, over-ordering products increases the return trips. Imbalances in the volume flow and empty running were additional causes of unutilised capacity (from companies B, F, G, and H).

Warehousing: During the picking process in the warehouse, human errors could occur, as stated by companies D and E. Large volumes that suddenly need handling, incorrectly implemented automation, and dysfunctional information technology (IT) could also generate inflexibility and unutilised capacities in warehouses, as could the instances when standardisation was impossible (explicitly mentioned by companies B and D). However, the interviewees from companies F, G, and H indicated exactly the contrary—that, in some cases, the standardised processes were responsible for unutilised capacity.

Transhipment: The interviewees from companies F and H revealed that the consolidation of goods was often criticised by the shippers, who did not want to share the distribution capacity that is, after all, a factor of their competitiveness. Furthermore, information was often not sufficiently shared in either internal communications or communications among the different actors, as stressed by the interviewees from companies A and B. In particular, the information flow often lacked the details that were necessary for all of the actors and departments to meet their various interests, the result of which was unutilised capacity. The relationships among the different actors rarely occurred on an equal footing. As the interviews with companies F and G showed, collaboration often involved rules that were dictated by stronger, larger actors, whereas the weaker actors had no choice but to follow the lead. For example, a shipper might demand certain routes, loading specifications, and delivery times, which prompted unutilised capacities for the transport provider. The interviewees from companies F and D also revealed that the so-called 'big players' in logistics cannot collaborate because of the laws against cartelisation, which prevented the LSPs from forging a unified strategy with an environmental-friendly agenda and reducing energy consumption.

4.1.2. Actors

The system levels under the category of Actors encompassed the causes that were directly because of the LSPs, shippers, or end-consumers, if not a combination of those parties. These causes are referred to as 'who' is responsible for the unutilised capacity in logistics and, in particular, road freight transport (Table 4).

Table 4. Actors that cause unutilised capacity.

Levels	Causes
LSP	Over-delivery of services Incorrect price setting and pricing model unaligned with real costs (e.g., round prices, 'free' home deliveries) Priorities to fulfil customer demands lead to compromises and adaption of own logistics processes Offer a broad range of services that is uncompetitive with niche actors responsibilities for fill rates delegated to the transport provider Inflexibility with mixing certain shipments
Shipper	Narrow delivery and pickup timeframes for LSPs Requirements, inflexibility, and lack of compromises Demands to receive goods early and post late Over-ordering capacity
End-consumer	Lack of awareness of consequences of own behavior Lack of information on transport's GHG footprint Sales campaigns with free shipping and sending along retour papers Increasing demand for express deliveries and increasing returns of goods High expectations for narrow timeframes for home deliveries

Logistic service provider: LSPs continuously expand their service offerings and often provide an over-delivery of services, as stressed by the companies F and G. For example, the LSPs delivered the products in express deliveries when the express deliveries were unnecessary, offered services at unprofitable prices, and pursued volumes that they needed to fill the system, all to maintain or strengthen their market share. However, an interviewee from company F emphasised that, since the prices do not often reflect the real costs, the services are often offered in excess of the customer demands, meaning that unnecessary capacities are created. In short, the LSPs' top priority was often to fulfil the customers' demands regardless of the energy consumption and its consequences (evidence from companies F, G, and H). For example, whereas small shippers contracted standardised services, larger ones wanted to influence the logistics process by, for instance, specifying certain routes. In response, LSPs had to adapt their own logistics processes and resources, which often produced inefficient solutions. Companies F and G mentioned that large LSPs tend to offer a broad range of services, but cannot compete in terms of the price and efficiency with the smaller actors that offer niche services, but by trying to keep their dominant market share unutilised capacity is generated. When an LSP outsourced transport to a transport provider, it also delegated all of the responsibility for the fill rates and fuel consumption to the transport provider (evidence came from companies A,

B, F, and G). Eschewing responsibility, being inflexible in making changes in transported volumes, and being unable or unwilling to mix groupage and parcels in the same vehicles (as mentioned by companies B and F) created further unutilised capacity in the system.

Shipper: Among the actors, the shippers cause unutilised capacity because of the imposing requirements (as mentioned by companies F, G, and D), for example, for delivery and pickup timeframes that LSPs and transport providers could not meet during rush hours or because of congestion. If delivery timeframes were not met, then the LSPs often had to return to the point of delivery after agreeing to a new delivery time. Another cause, which was mentioned by the same three companies, was inflexible delivery timeframes among the different shippers at the same location. For example, although the different shops in a mall all received goods from the same LSP, because some shops wanted to receive deliveries one hour before opening and others during business hours, the LSP had to make several trips to the same address. Moreover, unutilised capacity was created, as mentioned by four of the companies (D, E, F, and G), since the customers preferred to receive goods as early as possible and to post the deliveries as late as possible. This made it impossible to deliver and collect the parcels at once. Furthermore, shippers, as commented on by companies B and F, often over-ordered to ensure enough capacity in the case of high demand; however, it often remained unutilised.

End-consumer: The behaviour of the end-consumers drives the energy consumption in transport, although the end-consumers are often unaware or dismissive of their impact on fuel consumption, as mentioned by companies F and G. Furthermore, the same interviewees explained that, although products might contain information about their organic origins, they do not contain information about their GHG footprint that is related to transport. The end-consumers are often unaware of the consequences of their product choices, relative to GHG emissions. Furthermore, the sales campaigns that offered 'free' home deliveries, shipments without declared surcharges for delivery, and packages including retour papers, encouraged end-consumers to order and return more goods (evidence from companies D, E, and F). The interviewees from companies D and F explained that, because of the increased demand for express deliveries and the increased expectations about the exact delivery times, as well as a high failure rate of unattended home deliveries, the end-consumers were responsible for a great deal of energy consumption in the last mile of the supply chain. Almost all nine of the interviewees highlighted that end-consumers almost always preferred the fastest, most-convenient logistics solution.

4.1.3. Areas in Logistics System

The areas in the logistics system describe where the unused capacity originated, particularly in the context of in- and out-bound logistics, last-mile distribution, and reverse logistics (see Table 5).

Table 5. Areas in logistics system from where unutilised capacity originates.

Levels	Causes
In- and out-bound logistics	Increased demand for short lead times (i.e., just-in-time) High energy consumption because of many small shipments (i.e., no economies of scale)
Last mile	Increased number of small shipments instead of full pallets Standardised boxes often larger than necessary High failure rate of home deliveries
Reverse logistics	Reverse logistics poorly integrated in flow to end-consumers Unprofitable returns

Inbound and outbound logistics: A major cause of unutilised capacity in the supply chain, as stated by companies A, B, C, D, and F, is the steady demand for short lead times and just-in-time deliveries. The interviewees mentioned that the end-consumers increasingly requested express deliveries, which affects the whole supply chain. Furthermore, companies D and G mentioned that shippers therefore had to order smaller batches and could not fill truck loads, which precluded any economy of scale. The just-in-time deliveries are fuel intensive, given the high number of small shipments that are involved, and the vehicles often cannot be loaded to their full capacity. Such just-in-time deliveries originated from the existence of smaller warehouses and the inability or unwillingness of shippers tie up capital in products.

Last mile: With e-commerce, many small parcels were sent to end-consumers' homes instead of in full pallets to the retailers (evidence from C, D, F, and G). The steady growth in the abundance of such parcels has increased demands for transport and, in turn, the total fuel consumption and traffic congestion in the last mile. In e-commerce, the standardised boxes that are used for packaging were often unnecessarily large and thus generated unutilised capacity, as explained by the interviewee from company G. Another cause occurred when big trucks delivered goods to the point of consumption. Even with high fill rates, the chain imbalances caused the vehicles to be not fully utilised once they dropped off shipments during milk rounds. Narrow delivery timeframes also increased the unutilised capacities and increased the failure rates of the home deliveries even further, as pointed out by companies D, E, and G.

Reverse logistics: The return of unwanted or damaged products, and the recycling of products at the end of their lifetimes, required additional transport and handling of goods. Often, reverse flows were poorly integrated with flows towards the end-consumers, as explained by companies D and H. Furthermore, the interviewees from companies D and F pointed out that with e-commerce, in particular, the returns represented unprofitable business, for they were often free of charge for the end-consumers and were exploited by the end-consumers who merely wanted to test products. Even when the returns posed a small fee for the end-consumers, they rarely covered the real cost for the retailers. In short, return policies that favoured the end-consumers tended to create redundant transport and unutilised capacity.

4.1.4. Additional Cause

Cost was seen as an overlapping cause across all of the three categories. During the interviews, two insights were highlighted (first and foremost stressed by companies D, F, G, and H), as follows: that energy was not the crucial cost driver when it came to transport, and that the LSPs often did not correctly calculate prices, which is an unprofitable practice that ultimately harms them and encouraged the customers to over-order and use transport in excess. Interviewees generally indicated that, given the low cost of fossil fuels, energy was not the crucial factor when trying to keep the total costs low. Instead, the costs are driven by time, administration fees, salaries, the handling of goods, and the range of product assortments. As a result, detours and low fill rates are widely tolerated. At the same time, company F pointed out that LSPs offer round prices, which means that average prices are applied across Sweden, although the remote areas are more expensive to reach. Normally, the deliveries to urban areas are balanced against the prices for remote areas. However, if an LSP was chosen only for deliveries to remote locations and not to urban areas, while the deliveries in urban areas were performed by niche actors that offered lower prices than the LSP, then the LSP would have lost business and profit. The pricing model of LSPs, therefore, did not reflect the real costs. Additionally, the end-consumers did not see the costs of transport, which were often hidden in the product prices and appeared to them as being free of charge, despite the reality. As a result, the end-consumers over-ordered products and transport services and thus created unutilised capacity.

4.2. Mitigations of Unutilised Capacity

The mitigations of unutilised capacity were derived from the empirical data. Linking the causes of unutilised capacity with mitigations can increase the energy efficiency and reduce the GHG emissions from road freight transport and logistics, as detailed in the following subsections.

4.2.1. Activities

Table 6 summarises the mitigations concerning the activities in the logistics system, meaning what could be done to utilise capacity.

Table 6. Suggested mitigations within the category of activities.

Levels	Suggested Mitigations
Transport	Avoid peak deliveries (e.g., incentivise delivery during off-peak times) Ensure efficient routing Track real-time need for transport Consolidate and combine heavy products but little volume with voluminous but light products Receive fewer but fuller trucks Utilise the whole height of a truck (e.g., double-stack pallets)
Warehousing	Standardise foldable and stackable boxes Label and pack products arriving at distribution centres in advance Devise alternatives to hanging garments Reduce picking errors Change product designs and sizes to better fit pallets
Transhipment	Order necessary volumes only Use platform and information technology to support internal and external information flows Concentrate all logistics-related knowledge in one division instead of spreading it over several divisions Use an online marketplace to sell or buy free capacity Encourage collaboration (e.g., petition the political system)

Transport: The interviewees from companies F and G suggested delivering products during periods with less traffic congestion, such as during off-peak delivery times, a behaviour which could be encouraged through incentives. Additionally, companies E, F, and G mentioned that routing could easily be calculated with the right software, and real-time tracking could show the available capacity in vehicles. Also, consolidation, which was explicitly mentioned by companies B, F, and G, should be encouraged by retailers, for consolidation, in addition to saving energy, can reduce the number of trucks arriving at terminals and, thus, congestion. By combining heavy and light goods on trucks, better fill rates and better prices could be secured for the LSPs. The interviewee from company F gave the example that, in a best case, 80 percent of the weight, but only 20 percent of the volume should be placed on the bottom, whereas 20 percent of the weight but 80 percent of volume should be placed on top. The product designs and sizes could be altered to better fit on pallets. As such, pallets should be high enough—for example, double-stacked pallets directly fitted into one truck—and thus use the entire height (a suggestion from companies B, C, D, F, and G).

Warehousing: The interviewees mentioned several mitigations for handling and loading. Firstly, it was recommended to work with standardised, foldable, stackable boxes for delivering products from the distribution centres to stores, where the boxes could be folded and stacked so as to minimise the use of space until they were returned (suggested by companies F and D). Secondly, companies D and G mentioned that to expedite handling, products should be labelled with prices and packed in store-ready batches before their delivery to the distribution centres. Thirdly, alternatives for hanging garments that prevent wrinkling and use less space, should be considered (advice from company D). Fourthly, company E suggested that human error when picking orders could be reduced or even eliminated with systems such as the picking-by-bag system, which is controlled via an IT component, and further training. The system adds product sizes and weights and can thus tell the pickers which products fit together in a given number of bags. Above all, the mission of maximising the capacity utilisation should span the entire product flow, from developing products that fil the standardised boxes exactly to ensuring a good pallet fit (suggested by companies B, D, and H).

Transhipment: It is important that the shippers would not over-order unnecessary volumes, as mentioned by companies D and F. Regarding the improved information flow, interviewees from companies B, E, and F suggested using platforms and IT, which could help both the internal and external information flow. Additionally, company B recommended improving the internal information flow by concentrating logistics-related knowledge at the shippers in one division instead of spreading it across several divisions. That way, one division is responsible for sourcing the transport work and contract management, which helps concentrate all of the knowledge and discover the synergies. Collaboration is another way to utilise the untapped capacity. For example, the interviewees from companies F and H suggested that the unutilised capacity in trucks could be sold to other companies in an online marketplace. Companies D and F also suggested that the political system should encourage collaboration, which it has mainly hindered to date. That way, companies could create common strategies with environmental-friendly agendas.

4.2.2. Actors

Table 7 summarises the results regarding the mitigations that the various actors could implement to take advantage of unutilised capacity.

Table 7. Suggested mitigations within the category of actors.

Levels	Suggested Mitigations
LSP	Outsource transport from retailers to LSPs Use the same transport operator for several shippers (i.e., economies of scale) Handle bookings electronically Use electric cars for distribution in urban areas
Shipper	Report all emissions and follow up Expand flexibility in delivery timeframes Set clear requirements early on (i.e., in tendering process)
End-consumer	Educate end-consumers on the consequences of their behaviour Communicate CO_2 footprint of transport to end-consumers Make transport costs visible to end-consumers

Logistics service provider: The transport was outsourced from the retailers to the LSPs, who could deliver products more energy-efficiently than the retailers could with their own solutions (suggested by companies D, F, G, and H). Economies of scale could be achieved, which were mentioned by company F and G, when one transport operator served several shippers, especially in a given region. Company G recommended that IT systems should be implemented to facilitate the booking process. Another suggestion to decrease the emissions directly involved using electric cars for parcel and mail distribution in urban areas (suggested by companies D, E, F, and G).

Shipper: Since the GHG emissions are of great concern to the market, keeping them low is a critical objective. Companies F, D, and G proposed that, in order to raise awareness of and gauge the extent to which the emissions targets are met, all of the emissions from transport during the year should be reported to the shippers, and followed up, and the decreased emissions should be rewarded. As the interviews with companies F and G revealed, the shippers should allow a greater flexibility in terms of the time slots and delivery timeframes, and be more willing to discuss and develop environmental-friendly solutions together. Companies B and D said that in order to ensure the fulfilment of the emissions targets, clear criteria for the LSP should be set in the tendering process, concerning fuel consumption, quality, product safety, and employee safety, so the LSP has clear performance expectations.

End-consumer: The interviewees from companies E, D, F, and G stated that the end-consumers must be made aware of how their behaviour affected the environment; they suggested educating end-consumers about their behaviour's impact and making information of the CO_2 footprint from transport available to the end-consumers. Furthermore, they also proposed educating the end-consumers on the hidden cost of transport, by making that cost visible. For example, although home deliveries were often free for the end-consumers, they needed to recognise that last-mile transport has its price.

4.2.3. Areas in Logistics System

As summarised in Table 8, this category presents where in the logistics system certain mitigations could be implemented in order to utilise unutilised capacity.

Table 8. Suggested mitigations within the category of areas in the logistics system.

Levels	Suggested Mitigations
In- and outbound logistics	Decrease demand for short lead times and high speeds Only deliver just-in-time when truly necessary
Last mile	Use better-fitting packaging to avoid air transport Disconnect deliverers and end-consumers Use tracking systems for end-consumers Extend timeframes depending on proximity to distribution centres Avoid one hub in the supply chain
Reverse logistics	Prices should better reflect costs Re-shelve returned products from e-commerce instead of returning them to manufacturers

In- and out-bound logistics: Capacity utilisation can be improved by decreasing the demands on the short lead times and on the high speeds in the supply chain, that is, reducing the just-in-time deliveries (suggested by companies B, D, F, and G). Speed is often triggered by an increased demand for express deliveries from the end-consumers. They need to become aware that express deliveries are costly and created unutilised capacity, and they should pay a higher price for them.

Last mile: The interviewees from companies D and G noted that standard-sized packaging, although typically appreciated, often includes too much air and should be replaced in certain cases with customised packaging that wraps items as tightly as possible. Additionally, an interviewee from company F suggested disconnecting delivery from the end-consumers by not having deliverers and end-consumers meet at points of reception, but instead delivering goods to pickup points. By relocating the points of reception from the front door to the pickup points, large commercial vehicles do not have to drive the extra leg, and energy consumption could be reduced. Additionally, tracking systems and apps, as mentioned by companies E and G, could make delivery arrivals visible to the end-consumers so as to minimise unsuccessful home deliveries. Company E suggested that the delivery timeframes should depend on the proximity to the distribution centre and increase with distance. For example, one-hour delivery timeframes could be allowed only in city centres and close to distribution centres, two-hour timeframes farther away from city centres, and three-hour deliveries could be allowed if the consumer lives even beyond a certain point. Furthermore, redundant transport could be decreased by avoiding one hub in the supply chain, as proposed by company F.

Reverse logistics: In order to avoid redundant transport from returns in e-commerce, interviewees from companies D, F, and G proposed that end-consumers should have to pay for the service so as to discourage their overuse. Companies G and H recommended that returned products from e-commerce should not be returned to the manufacturers; instead, the LSPs should relabel and re-shelve them for direct sale.

4.2.4. Additional Mitigation

Another important mitigation that was mentioned by all of the interviewees, which stretched across all of the categories, was the adaption of the cost structure of the LSPs. Since fuel prices do not reflect the damage that is caused to the environment, the prices for transport should be revaluated. In addition, because LSPs often offer unprofitable prices, a new cost structure that better reflects the real costs should be devised.

5. Discussion

In the following section, the conceptual framework is further developed and summarises the causes and mitigations of unutilised capacity. Additionally, four key contributions are discussed.

5.1. Framework of Causes and Mitigations of Unutilised Capacity

The results from the interview data have been summarised and added to the origin conceptual framework (Figure 2). Herein, capacity utilisation is viewed in a larger logistics system, on different system levels, and an interactive approach to capacity utilisation was taken.

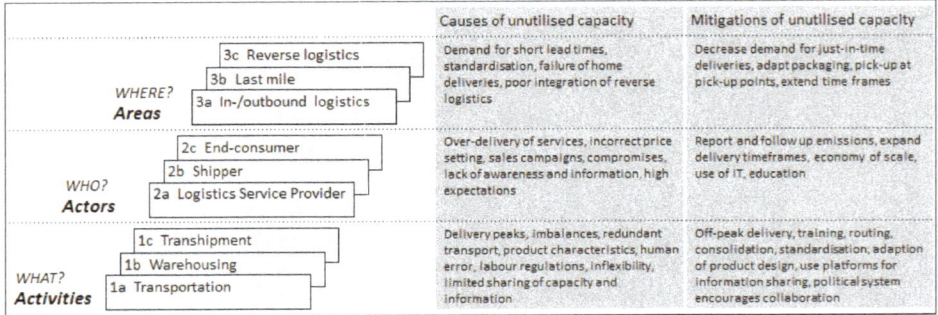

Figure 2. Framework of the causes and mitigations of unutilised capacity.

Activities that caused unutilised capacity are delivery during peak hours, imbalances in the flow, redundant transport, product characteristics, human error in the warehouse, labour regulations, inflexibility, and limited sharing of capacity and information. The mitigations include i.a. off-peak delivery, training, efficient routing, standardisation (although exceptions were mentioned), and a change in the political environment.

The causes that were because of actors are, for example, over-delivery of services, incorrect price setting, sales campaigns, compromises, lack of awareness and information, and too-high expectations. However, regular follow-up on emissions, expanded delivery timeframes, economies of scale, use of IT, and education can counter these causes.

Additionally, unutilised capacity arose, for example, through the demand for short lead times, standardisation, high failure rates of home deliveries, and poor integration of reverse logistics. This can be mitigated by decreasing the demand for short lead times, adaption of packaging, disconnecting deliverers and end-consumers, and extension of time frames.

The capacity is conceptualised as an interactive concept in an open system. This was exemplified by the interviewee from company D, who pointed out that the company's freight transport operation illustrated the interplay of the components. Briefly, products were designed to fill boxes completely, and the boxes were designed for a good fit on the pallets and were arranged so that the double-stacked pallets filled exactly one truck's height, with three pallets next to each other across the width of the truck. The utilisation of the capacity has to be followed up throughout each system level and needs to begin as early as possible.

As the interviews clarified, unutilised capacity was especially apparent at the intersections and overlapped between different components and actors. As such, the responsibility for energy consumption was often passed on to other actors; consequently, the actual problems of energy overconsumption and emissions have remained unsolved. Therefore, a broad approach that addresses the different causes and changes at every level of the logistics system [8,36], as well as between the actors, was necessary. Many causes, such as inflexibilities, limited information sharing, and compromises, which lead to inefficiencies for at least one actor, resulted from the relationship between the actors. Sallnäs [50] described those dependencies between the LSPs and the shippers, and stated that if both parties had high environmental ambitions for their relationship, then the coordination of their environmental practises became more likely. This was confirmed through the interviews, however, the responsibilities were too-often eschewed and forwarded to another actor.

The framework has presented the causes and mitigations for unutilised capacity at different system levels through a holistic picture. The borders between the different causes or mitigations at each system level were often difficult to draw, because the system levels could meld together. The framework is limited in illustrating that the causes can affect each other and the mitigations can also influence the causes on a different system level.

5.2. Key Contributions

Returning to the purpose of taking an interactive approach to examine capacity utilisation to contribute to sustainable freight transport and logistics, this paper developed a conceptual framework from the literature and expanded it with empirical data so as to identify the causes and mitigations. Four key contributions have been suggested here.

Firstly, by categorising the causes of the unutilised capacity in three categories—the actors, activities, and areas in logistics systems—and on nine system levels, this paper has provided a simple but comprehensive framework to identify the causes of unutilised capacity. The identification of the causes was crucial to mitigate them and work towards sustainable freight transport and logistics. As shown in the frame of reference, the categories were identified from the literature, but the body of knowledge had not linked them together. This paper moved beyond a simple description of energy-efficient approaches in logistics. In contrast, its findings conceptualised ways to approach the problem of high energy consumption and GHG emissions in freight transport and logistics, by addressing one of its major symptoms, namely, unutilised capacity. The causes of unutilised capacity are countered with the improvement efforts that were derived from the empirical data.

Secondly, this paper provided a critical view towards the fill rates. High fill rates are often sought in the pursuit of increased capacity utilisation and energy efficiency, but if this means that products are taken from a more energy-efficient system (e.g., rail) or that the fill rates result in higher return rates, they should be avoided.

The third contribution highlighted the need for a standardised approach so as to measure the environmental impact from freight transport and logistics, and to make the data comparable between companies. The interviews revealed that the companies lacked sufficient knowledge about calculating the exact impact of their operations, largely because of the difficulty of measuring energy efficiency, which is of importance to improve the logistics operations [25,38]. This was because of the individual difficulties of collecting appropriate data, working with various indicators, and defining the system boundaries. For one, the collected numbers were often assumed and standard values were applied. Moreover, difficulties arose because of the variety of fuel types and numerous possibilities that were used to calculate the fill rates. This paper suggests applying system-wide measurements. Individual measurement components are redundant; they do not consider the impact raised by the entire system.

Fourthly, this paper underscored that the costs are not an appropriate indicator for measuring environmental impact, although they were often used by companies to track the energy consumption and thus energy efficiency, often with the belief that the reduced costs for freight transport are accompanied by lower emissions [8]. Several difficulties occurred when the costs were used as an indicator for energy efficiency; for example, LSPs often set round prices for their logistics services, and the end-consumers remained unaware of the actual costs of transport. However, although the LSP's objective is to operate with the lowest costs possible, saving energy does not always suit that objective. Typically human resources and time, not energy, were crucial cost drivers in road freight transport, given the low market price for oil worldwide. Several interviewees suggested that higher taxes on fuel would help to reduce energy consumption. Indeed, carbon taxes have been suggested in the literature as a disincentive to generating emissions [5].

6. Conclusions

In recent years, energy efficiency has gained attention in the literature on logistics and supply chain management [7,19,51]. Other than identifying the different drivers for the energy-efficient management of the supply chain, including collaboration [22], consolidation and standardisation [8], the weight of goods, and empty running [24], this paper contributes to the understanding of how energy efficiency in logistics, starting with in road freight transport, can be achieved in a broad system, by identifying and countering the causes of unutilised capacity. By identifying the three categories, namely, actors, activities, and areas in the logistics system, and highlighting the various system levels, the origin of unutilised capacity could be identified.

This paper has two significant managerial implications. Firstly, it offers the logistics managers, from the shippers and LSPs, an overview of the problem of low energy efficiency in logistics, which elucidates how responsibilities cannot simply be forwarded to other actors, but that a holistic approach is needed. Secondly, it conceptualises capacity and presents it in a simple framework, highlighting the system levels. By providing those three categories, the causes can be conceptualised, which can help the logistics managers to identify where improvements are possible and to go beyond the obvious fill rates of vehicles to increase capacity utilisation. In addition, the paper shows how minor changes in the logistics system can affect the system's overall energy consumption.

The theoretical contributions are twofold. Firstly, by approaching capacity as an interactive concept, the paper elucidates the importance of each component in an interlinked system. It suggests that the components can be horizontally aligned—for example, the fit of the products in a box, the fit of boxes on pallets, and the fit of pallets in trucks—and vertically aligned, as in collaboration among actors. Secondly, by investigating both the LSPs and shippers, the paper has been able to view the problem from multiple perspectives. At the intersection of the LSPs and their customers lies great improvement potential for energy efficiency, when the actors collaborate in a long-term commitment on equal footing and work together towards environmental sustainability [50]. Following suit, this paper contributes to the body of knowledge by taking an interactive approach to capacity utilisation and presenting multiple interpretations of energy efficiency.

Limitations arose during the interviews regarding the use of the term energy, which the interviewees emphasised could have been interpreted differently. Whereas this paper addresses energy efficiency in freight transport, the operational term that was used during the interviews should have been the input (i.e., fuel) or the output (i.e., CO_2 emissions). However, the broadness of the term energy encouraged the interviewees to talk more freely during the interviews. A further limitation was the exclusion of the end-consumers from the data collection. Although a multi-actor approach was implemented, only shippers and LSPs were interviewed, owing to the difficulty of collecting data from private consumers.

The role of the end-consumer regarding energy efficiency in logistics is underdeveloped and calls for future research, such as the investigation of different distribution options in the last mile and the impacts on the supply chain that are triggered by the end-consumers' behaviour. Furthermore, the need for a common approach to measure the environmental impact that enables the comparison between companies, calls for future research.

Acknowledgments: This work was supported by the Swedish Energy Agency and the Logistik och Transportstiftelsen (LTS). The support is gratefully acknowledged.

Conflicts of Interest: The author declares no conflict of interest. The founding sponsors had no role in the design of the study; in the collection, analyses, or interpretation of data, in the writing of the manuscript, and in the decision to publish the results.

References

1. European Commission. *A European Strategy for Low-Emission Mobility*; Communication, Ed.; European Commission: Brussels, Belgium, 2016.
2. Organization for Economic Cooperation and Development; International Energy Agency. *Key World Energy Statistics*; International Energy Agency: Paris, France, 2015; pp. 1–77.
3. Organization for Economic Cooperation and Development; International Energy Agency. *Transport, Energy and CO2: Moving Towards Sustainability*; International Energy Agency: Paris, France, 2009; pp. 1–414.

4. Creazza, A.; Dallari, F.; Melacini, M. Evaluating logistics network configurations for a global supply chain. *Supply Chain Manag. Int. J.* **2010**, *15*, 154–164. [CrossRef]
5. Gurtu, A.; Searcy, C.; Jaber, M.Y. Emissions from international transport in global supply chains. *Manag. Res. Rev.* **2017**, *40*, 53–74. [CrossRef]
6. Zhang, S.; Wang, J.; Zheng, W. Decomposition analysis of energy-related CO_2 emissions and decoupling status in china's logistics industry. *Sustainability* **2018**, *10*, 1340. [CrossRef]
7. Centobelli, P.; Cerchione, R.; Esposito, E. Environmental sustainability and energy-efficient supply chain management: A review of research trends and proposed guidelines. *Energies* **2018**, *11*, 275. [CrossRef]
8. Aronsson, H.; Huge-Brodin, M. The environmental impact of changing logistics structures. *Int. J. Log. Manag.* **2006**, *17*, 394–415. [CrossRef]
9. Chapman, L. Transport and climate change: A review. *J. Transp. Geogr.* **2007**, *15*, 354–367. [CrossRef]
10. Browne, M.; Allen, J.; Rizet, C. Assessing transport energy consumption in two product supply chains. *Int. J. Log. Res. Appl.* **2006**, *9*, 237–252. [CrossRef]
11. McKinnon, A.; Ge, Y. The potential for reducing empty running by trucks: A retrospective analysis. *Int. J. Phys. Distrib. Log. Manag.* **2006**, *36*, 391–410. [CrossRef]
12. Rizet, C.; Cruz, C.; Mbacke, M. Reducing freight transport CO_2 emissions by increasing the load factor. *Proced. Soc. Behav. Sci.* **2012**, *48*, 184–195. [CrossRef]
13. Rogerson, S.; Santén, V. Shippers' opportunities to increase load factor: Managing imbalances between required and available capacity. *Int. J. Log. Res. Appl.* **2017**, *20*, 1–23. [CrossRef]
14. Turki, S.; Didukh, S.; Sauvey, C.; Rezg, N. Optimization and analysis of a manufacturing–remanufacturing–transport–warehousing system within a closed-loop supply chain. *Sustainability* **2017**, *9*, 561. [CrossRef]
15. Plambeck, E.L. Reducing greenhouse gas emissions through operations and supply chain management. *Energy Econ.* **2012**, *34*, 64–74. [CrossRef]
16. Kovács, G.; Spens, K.M.; Vellenga, D.B. Academic publishing in the nordic countries—A survey of logistics and supply chain related journal rankings. *Int. J. Log. Res. Appl.* **2008**, *11*, 313–329. [CrossRef]
17. Halldórsson, Á.; Svanberg, M. Energy resources: Trajectories for supply chain management *Supply Chain Manag. Int. J.* **2013**, *18*, 66–73. [CrossRef]
18. Rogers, Z.; Kelly, T.G.; Rogers, D.S.; Carter, C.R. Alternative fuels: Are they achievable? *Int. J. Log. Res. Appl.* **2007**, *10*, 269–282. [CrossRef]
19. Halldórsson, Á.; Kovács, G. The sustainable agenda and energy efficiency: Logistics solutions and supply chains in time of climate change. *Int. J. Phys. Distrib. Log. Manag.* **2010**, *40*, 5–13. [CrossRef]
20. European Parliament and Council of the European Union. *Directive 2012/27/EU of the European Parliament and of the Council of 25 October 2012 on Energy Efficiency, Amending Directives 2009/125/EC and 2010/30/EU and Repealing Directives 2004/8/EC and 2006/32/EC*; European Parliament and Council of the European Union: Brussels, Belgium, 2012; Volume 4, p. 56.
21. Cullen, J.M.; Allwood, J.M.; Borgstein, E.H. Reducing energy demand: What are the practical limits? *Environ. Sci. Technol.* **2011**, *45*, 1711–1718. [CrossRef] [PubMed]
22. Wolf, C.; Seuring, S. Environmental impacts as buying criteria for third party logistical services. *Int. J. Phys. Distrib. Log. Manag.* **2010**, *40*, 84–102. [CrossRef]
23. Pfohl, H.C.; Zöllner, W. Organization for logistics: The contingency approach. *Int. J. Phys. Distrib. Log. Manag.* **1997**, *27*, 306–320. [CrossRef]
24. Piecyk, M.I.; McKinnon, A.C. Forecasting the carbon footprint of road freight transport in 2020. *Int. J. Prod.Econ.* **2010**, *128*, 31–42. [CrossRef]
25. Kalenoja, H.; Kallionpaa, E.; Rantala, J. Indicators of energy efficiency of supply chains. *Int J. Log. Res.Appl.* **2011**, *14*, 77–95. [CrossRef]
26. Bottani, E.; Rizzi, A.; Vignali, G. Improving logistics efficiency of industrial districts: A framework and case study in the food sector. *Int. J. Log. Res. Appl.* **2014**, *18*, 1–22. [CrossRef]
27. He, X.; Zhang, J. Supplier selection study under the respective of low-carbon supply chain: A hybrid evaluation model based on FA-DEA-AHP. *Sustainability* **2018**, *10*, 564. [CrossRef]
28. Yuan, B.; Gu, B.; Guo, J.; Xia, L.; Xu, C. The optimal decisions for a sustainable supply chain with carbon information asymmetry under cap-and-trade. *Sustainability* **2018**, *10*, 1002. [CrossRef]
29. Browne, M.; Rizet, C.; Anderson, S.; Allen, J.; Keïta, B. Life cycle assessment in the supply chain: A review and case study. *Transp. Rev.* **2005**, *25*, 761–782. [CrossRef]

30. Kin, B.; Ambra, T.; Verlinde, S.; Macharis, C. Tackling fragmented last mile deliveries to nanostores by utilizing spare transportation capacity—A simulation study. *Sustainability* **2018**, *10*, 653. [CrossRef]
31. Björklund, M. Influence from the business environment on environmental purchasing—Drivers and hinders of purchasing green transportation services. *J. Purch. Supply Manag.* **2011**, *17*, 11–22. [CrossRef]
32. Wu, H.J.; Dunn, S.C. Environmentally responsible logistics systems. *Int. J. Phys. Distrib. Log. Manag.* **1995**, *25*, 20–38. [CrossRef]
33. Abbasi, M.; Nilsson, F. Themes and challenges in making supply chains environmentally sustainable. *Supply Chain Manag. Int. J.* **2012**, *17*, 517–530. [CrossRef]
34. Brown, J.R.; Guiffrida, A.L. Carbon emissions comparison of last mile delivery versus customer pickup. *Int. J. Log.Res. Appl.* **2014**, *17*, 503–521. [CrossRef]
35. Rizet, C.; Browne, M.; Cornelis, E.; Leonardi, J. Assessing carbon footprint and energy efficiency in competing supply chains: Review—Case studies and benchmarking. *Transp. Res. Part D Transp. Environ.* **2012**, *17*, 293–300. [CrossRef]
36. Liimatainen, H.; Pöllänen, M. Trends of energy efficiency in finnish road freight transport 1995–2009 and forecast to 2016. *Energy Policy* **2010**, *38*, 7676–7686. [CrossRef]
37. Liimatainen, H.; Hovi, I.B.; Arvidsson, N.; Nykanen, L. Driving forces of road freight co2 in 2030. *Int. J. Phys. Distrib. Log. Manag.* **2015**, *45*, 260–285. [CrossRef]
38. McKinnon, A.; Ge, Y. Use of a synchronised vehicle audit to determine opportunities for improving transport efficiency in a supply chain. *Int. J. Log. Res. Appl.* **2004**, *7*, 219–238. [CrossRef]
39. McKinnon, A. Freight transport deceleration: Its possible contribution to the decarbonisation of logistics. *Transp. Rev.* **2016**, *36*, 419–436. [CrossRef]
40. Konings, R.; Priemus, H.; Nijkamp, P. *The Future of Intermodal Freight Transport: Operations, Design and Policy*; Edward Elgar Publishing Limited: Cheltenham, UK, 2008; Volume 9.
41. Hayes, R.; Pisano, G.; Upton, D.; Wheelwright, S. *Operations, Strategy, and Technology: Pursuing the Competitive Edge*; John Wiley & Sons: Hoboken, NJ, USA, 2005; p. 344.
42. Wu, Z.; Pagell, M. Balancing priorities: Decision-making in sustainable supply chain management. *J. Oper. Manag.* **2011**, *29*, 577–590. [CrossRef]
43. Perboli, G.; Musso, S.; Rosano, M.; Tadei, R.; Godel, M. Synchro-modality and slow steaming: New business perspectives in freight transportation. *Sustainability* **2017**, *9*, 1843. [CrossRef]
44. Lindskog, M. Systems theory: Myth or mainstream? *Log. Res.* **2012**, *4*, 63–81. [CrossRef]
45. Flick, U. *An Introduction to Qualitative Research*, 5rd ed.; SAGE: London, UK, 2014.
46. Bryman, A.; Bell, E. *Business Research Method*, 3rd ed.; Oxford University Press: Oxford, UK, 2011.
47. Ellram, L.; Tate, W.L. Redefining supply management's contribution in services sourcing. *J. Purch. Supply Manag.* **2015**, *21*, 64–78. [CrossRef]
48. Lee, H.L.; Padmanabhan, V.; Whang, S. The bullwhip effect in supply chains. *Sloan Manag. Rev.* **1997**, *38*, 93–102. [CrossRef]
49. Halldórsson, Á.; Aastrup, J. Quality criteria for qualitative inquiries in logistics. *Eur. J. Oper. Res.* **2003**, *144*, 321–332. [CrossRef]
50. Sallnäs, U. Coordination to manage dependencies between logistics service providers and shippers. *Int. J. Phys. Distrib. Log. Manag.* **2016**, *46*, 316–340. [CrossRef]
51. Marchi, B.; Zanoni, S. Supply chain management for improved energy efficiency: Review and opportunities. *Energies* **2017**, *10*, 1618. [CrossRef]

sustainability

MDPI

Article

Possible Impact of Long and Heavy Vehicles in the United Kingdom—A Commodity Level Approach

Heikki Liimatainen [1,*], Phil Greening [2], Pratyush Dadhich [2] and Anna Keyes [2]

[1] Transport Research Centre Verne, Tampere University of Technology, Tampere 33720, Finland
[2] Centre for Sustainable Road Freight, Heriot-Watt University, Edinburgh EH14 4AS, UK;
    p.greening@hw.ac.uk (P.G.); p.dadhich@hw.ac.uk (P.D.); a.keyes@hw.ac.uk (A.K.)
* Correspondence: heikki.liimatainen@tut.fi; Tel.: +358-40-849-0320

Received: 14 June 2018; Accepted: 1 August 2018; Published: 4 August 2018

Abstract: The potential effects of implementing longer and heavier vehicles (LHVs) in road freight transport have been studied in various countries, nationally and internationally, in Europe. These studies have focused on the implementation of LHVs on certain types of commodities and the experience from countries like Finland and Sweden, which have a long tradition of using LHVs, and in which LHVs used for all types of commodities have not been widely utilised. This study aimed to assess the impacts of long and heavy vehicles on various commodities in the United Kingdom based on the Finnish experiences in order to estimate the possible savings in road freight transport vehicle kilometres, costs, and CO_2 emissions in the United Kingdom if LHVs would be introduced and used similarly to in Finland in the transport of various commodities. The study shows that the savings of introducing longer and heavier vehicles in the United Kingdom would be 1.5–2.6 billion vehicle kms, £0.7–1.5 billion in transport costs, and 0.35–0.72 Mt in CO_2 emissions. These findings are well in line with previous findings in other countries. The results confirm that considerable savings in traffic volume and emissions can be achieved and the savings are very likely to outweigh possible effects of modal shift from rail to road.

Keywords: longer heavier vehicles; road freight transport; CO_2 emissions; transport costs

1. Introduction

Road freight transport contributes significantly to global greenhouse gas emissions and its importance is likely to increase in the future as passenger vehicles may be electrified more easily and the energy sector increasingly utilises renewable energy sources in order to mitigate climate change. Hence, greenhouse gas emissions, particularly carbon dioxide (CO_2) emissions, from road freight should also be reduced. Various possible measures to achieve emission reductions in road freight have been identified and analysed both nationally (e.g., [1]) and internationally (e.g., [2]). Measures can be broadly catergorised using the ASIF framework to avoiding journeys (A), modal shift (S), lowering transport energy intensity (I), and reducing carbon intensity of fuels (F) [2]. One of the most effective ways to avoid journeys and reduce energy intensity per unit of payload transported of road freight, resulting in reduced CO_2 emissions, is to increase the size of road freight vehicles. Provided their payload capacity is fully utilized, larger vehicles are always significantly more fuel efficient per tonne of payload than the smaller vehicles they replace [3].

1.1. Long and Heavy Vehicles (LHVs)

Lorries are used in various tractor-trailer combinations around the world from over 30-m long Australian B-triple vehicles with 90t gross vehicle weight (GVW) to European semitrailer combinations with 16.5 m length and 40t GVW (Figure 1). Also, the height of the vehicles varies from 4 m to 4.8 m of

British double-deck semitrailers. Semitrailer vehicles are commonly, as well as in this study, referred to as heavy goods vehicles (HGVs), while the 25.25 m long vehicle combinations are commonly referred to as long and heavy vehicles (LHVs). LHVs may have various maximum gross vehicle weights and trailer combinations following the European modular system (EMS). In this study, LHVs mean the Scandinavian rigid truck–trailer combination with 60t GVW and HGVs mean the British semitrailer with 44t GVW and generally 16.5-m length, although there are currently also 18.75 m semitrailers on British roads due to the on-going 'longer semitrailer trial' (https://www.gov.uk/government/collections/longer-semi-trailer-trial).

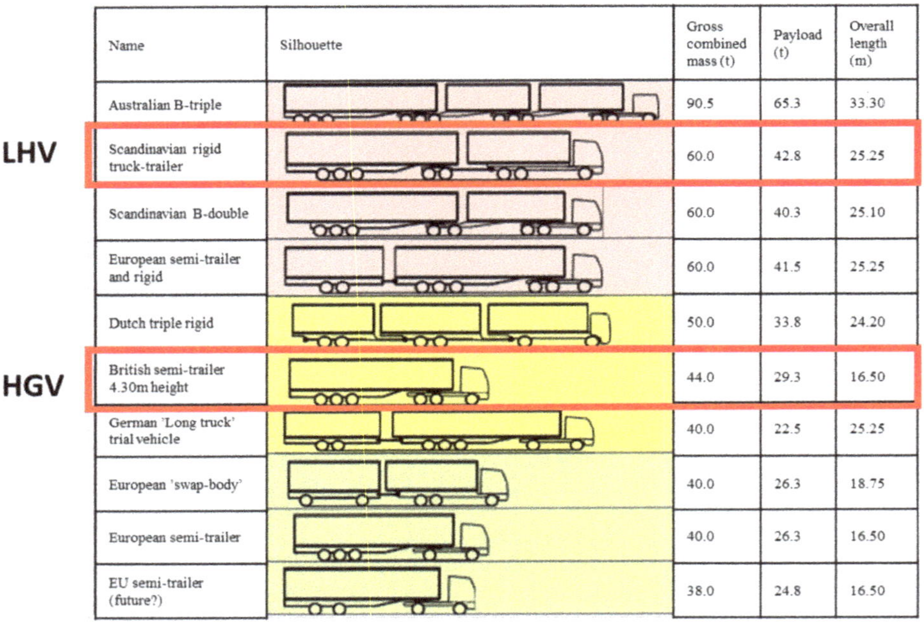

	Name	Silhouette	Gross combined mass (t)	Payload (t)	Overall length (m)
	Australian B-triple		90.5	65.3	33.30
LHV	Scandinavian rigid truck-trailer		60.0	42.8	25.25
	Scandinavian B-double		60.0	40.3	25.10
	European semi-trailer and rigid		60.0	41.5	25.25
	Dutch triple rigid		50.0	33.8	24.20
HGV	British semi-trailer 4.30m height		44.0	29.3	16.50
	German 'Long truck' trial vehicle		40.0	22.5	25.25
	European 'swap-body'		40.0	26.3	18.75
	European semi-trailer		40.0	26.3	16.50
	EU semi-trailer (future?)		38.0	24.8	16.50

Figure 1. Some major vehicle combinations (background table from the literature [4], highlights and heavy goods vehicles (HGV) and longer and heavier vehicles (LHV) texts added).

High-capacity, multiply-articulated long and heavy vehicles (LHVs) are routinely used in Scandinavian countries, the Netherlands, Germany, as well as Australia, Canada, South Africa, and the USA because of their superior productivity and low CO_2 emissions per tonne-km. Such vehicles typically have 20–30% lower fuel consumption and CO_2 emissions per unit of freight transport than their conventional tractor–semitrailer counterparts [4].

Finland has vast experience of the use of LHVs. Vehicles of 25.25 m in length with a maximum gross vehicle weight (GVW) of 60 tonnes have been in use since 1993 and currently, 78% of tonne-kilometres are carried using LHVs [5]. These vehicles have typically been a combination of a three-axle rigid truck and a four-axle full trailer, but other combinations within the European modular system (EMS) have also been used. In October 2013, the maximum GVW was increased to 76 tonnes for nine-axle vehicles and 68 tonnes with eight-axle vehicles. This has caused a significant shift from seven-axle vehicles to eight- and nine-axle vehicles, and led to around 3.5% savings in truck vehicles kms and €100 million savings in transport costs in 2016 [6]. Given the long history of LHVs in Finland, it can provide valuable information on the actual utilization of LHVs in the freight transport sector, if those would also be allowed in the United Kingdom.

1.2. Possible Impacts of LHVs Based on Literature

Long and heavy vehicles have been a subject of strenuous political debate in Europe during the 21st century. Particularly during 2008 to 2010, several policy reports were published addressing the issue on both a national and European level [7–9]. "A Review of Megatrucks" [3] highlighted the key findings from eight of these studies and stated that "there is widespread agreement that LHVs would reduce operating costs of road freight and greenhouse gas emissions per tonne-km of goods transported". Most research also agrees that vehicle mileage, transport costs, and emissions of road freight transport will be reduced on company level and also on national aggregate level if LHVs are introduced, or that these would increase if the LHVs currently in operation in countries such as Finland, Sweden, Canada, and Australia would be replaced with standard heavy goods vehicles [3,10–13].

Some desk study reports argue that LHVs would result in major modal shift from rail to road, which would outweigh the efficiency gains within road freight transport and lead to an increase in CO_2 emissions [8,9]. Opposing evidence to these desk studies exist and Steer et al. [3] conclude that "empirical evidence is difficult to find with regards to many of the primary concerns regarding LHVs ... where empirical evidence is available, it tends to show ... lower modal shift". McKinnon [14] presented evidence supporting that of Steer et al. [3] from the United Kingdom when the maximum weight of HGVs was raised to 44 tonnes. It was estimated that increasing maximum weight of HGVs to 44 tonnes would reduce rail freight tonne-kms by 10%. However, the market share of rail freight remained fairly stable at 11% [14].

More recent research from Spain did not consider modal shift because "the market share of domestic freight rail transportation in Spain is so low that any transfer would be negligible" [15]. In a German survey, "77% of the respondents did not foresee a shift of the existing freight traffic from rail to road", while 55% expected that the road freight transport market would see higher growth than expected if LHVs would be adopted [11]. A study in Belgium concluded that "the impact of LHVs on the geographic market area of intermodal terminals can be substantial if road transport prices would decrease by up to 15 or 25%", but "it would be necessary to study the goods flows that actually qualify for a reverse modal shift to LHVs, not only based on price, but also on other logistics requirements" [10]. Overall, the evidence on modal shift remains inconclusive. Hence, a sensitivity analysis taking into account possible modal shift is included in this study.

In addition to possible modal shift, worries about LHVs' effect on infrastructure and safety have been raised. Steer et al. [3] conclude that LHVs may induce additional capital and maintenance costs for infrastructure, but these can only be assessed nationally. Ortega et al. [15] estimated the required investments in Spain from 150 to 1000 million euros, depending on the extent of road network on which LHVs would be allowed. Ericson et al. [16] estimated that road wear costs would decrease by €14–20 million if Sweden would give up LHVs and use the HGVs instead. Generally, the road wear of pavement decreases when moving from HGVs to LHVs [4], but there might be negative effects on the substructure of the road [17] and increased investment needed in road bridges [18].

Regarding the safety effects of LHVs, Glaeser and Ritzinger [4] show that LHVs have worse performance in terms of ratio of amplification of lateral acceleration of the tractor unit compared with trailer and total swept width is larger, which indicate that LHVs are more difficult to manoeuvre and thus may have greater risk per vehicle. However, Steer et al. [3] conclude that there is no evidence of increased safety risk and reduction in vehicle-kms may even outweigh increased risk per vehicle. Leach et al. [13] also see no significant impact in the United Kingdom and Ortega et al. [15] say that sensitivity costs of accidents are negligible in Spain if LHVs are introduced. During the ongoing trials with longer semitrailers (trailer length 15.65 m instead of 13.6 m) in the United Kingdom, the trial vehicles have been involved in 70% fewer personal injury collisions than average articulated HGV [19], but this could be because better than average drivers have likely been selected to the trial and trial vehicles may not have been used similarly to average HGVs.

The effects of LHVs have been studied on company level [11,12]; on sectoral level, typically focusing on intermodal transport sector [10,13,20]; and on national level [14,15], reports reviewed by the authors of [3], while international studies have been limited to technical comparison of various types of LHVs [4] and an overall study across Europe [7]. Ortega et al. [15] highlight that there is a lack of sensitivity analysis, which would identify the influence of the kind and amount of freight that would use LHVs and the percentage of empty running. Meers et al. [10] also conclude that accounting for product characteristics and the corresponding transport quality requirements would enable estimations in greater detail.

In order to fill these research gaps, the purpose of this study is to assess the impacts of long and heavy vehicles on various commodities in the United Kindom based on the Finnish experiences. The two countries are quite different in terms of the importance of various sectors on economy and freight transport needs, hence it is necessary to evaluate the use of LHVs on the greatest level of detail available, that is, on commodity level. Each commodity can be seen to have similar logistics practices and types of goods carried are similar between countries. This enables conclusions on the suitability and uptake rate of LHVs to be drawn in the United Kingdom based on Finnish experiences. Specifically, the research question to be answered in this study is the following: *What are the possible savings in road freight transport vehicle kilometres, costs, and CO_2 emissions in the United Kingdom, if LHVs were introduced and used similarly to Finland in the transport of various commodities?*

2. Materials and Methods

In order to answer the research question, continuous road freight transport surveys in the United Kingdom and Finland were used. Finland was chosen as the country of reference because it is one of the few countries that use LHVs, in particular 60 t and 25.25 m LHVs, and it has similar data available as in the United Kingdom. Another alternative could have been Sweden, but the researchers did not have access to the Swedish dataset. In the United Kingdom, the Continuing Survey of Road Goods Transport, Great Britain (CSRGT GB) is a survey that reports the operations of approximately 7000 trucks. In Finland, the Goods Transport by Road Survey (GTRS) includes approximately 2500 trucks annually. Both surveys are conducted in a similar way following the European guidelines [21].

In order to estimate the maximum benefits of using 60 t and 25.25 m vehicles in the United Kingdom, data on the tonne-kms by commodity and type and weight of vehicle were gathered from UK Department for Transport (DfT) [22]. Vehicle kms by commodity and type and weight of vehicle are not publicly available, so a request for such data from 2016 was made and fulfilled by the DfT. An assumption was made that the 60 t and 25.25 m vehicles would only affect the haulage currently carried out with over 33 t articulated vehicles. Average load on laden trips of over 33 t artics for each commodity in the United Kingdom was then calculated by dividing the tonne-kms by vehicle kms.

For Finland, the data used in this study consisted of the raw data from the GTRS from 2012. Data from 2016 was available, but because Finland allowed GVW 76 t vehicles in October 2013, it was decided that the the 2012 data would be used as reference with the U.K. data to estimate the potential of 60 t and 25.25 m LHVs. If 2016 data would have been used, the average loads by commodity would have been unrealistically high for some commodities, because allowing 76 t vehicles has resulted in significant increase in average loads [6]. The raw data included each trip and vehicle reported in the survey, so data could be analysed flexibly. The Finnish raw data was processed to produce tonne-kms and vehicle kms with the same commodity and vehicles type and weight classifications as the U.K. data. Hence, the average load on laden trips of over 33 t artics for each commodity in Finland in 2012 was calculated by dividing the tonne-kms by vehicle kms. Maximum potential of LHVs in the United Kingdom was then calculated by dividing the U.K. tonne-kms by the Finnish average load by commodity. This resulted in alternative vehicle kms and the potential vehicle kms saved by LHVs were then calculated by subtracting the new vehicle kms from original U.K. vehicle kms.

LHVs effect on empty running could not be similarly calculated, because there obviously is not a change in average load on empty runs. Hence, the share of empty running of total mileage with over 33 t artics in the United Kingdom (24%) was assumed to remain the same if LHVs would be implemented and new empty mileage was calculated based on the relative change in the laden mileage of all commodities.

The effects on transport costs were then calculated using the average per kilometre vehicle operating costs by Road Haulage Association (RHA) [23] as a baseline for current 44 t and 16.5 m HGVs and increasing those using the cost differences estimated by Vierth et al [24] for various types of commodities. Vierth et al. [24] present the total transport costs in SEK/10km and the shares of three cost components (fuel, personnel, and other) for five types of transport (part load, forest, long-haul distribution, tanker and bulk, and construction). The differences in total cost and cost components are due to differences in the distance travelled relative to working time (km/h), annual mileage per vehicle, and annual working hours per vehicle. The transport costs used in this study (Table 1) are calculated using the RHA [23] figures for 44 t tractor–semitrailer as a baseline for the long-haul type of transport.

Table 1. Transport costs and fuel consumption. RHA—Road Haulage Association; HGVs—heavy goods vehicles; LHVs—long and heavy vehicles.

	Type of Transport	Part Load	Long-Haul	Tanker and Bulk	Construction	Forestry
RHA [23]	Driver costs (£/year)			32,400		
	Fixed costs (£/year)			48,020		
Vierth et al. [24]	Distance travelled relative to working time (km/h)	51.1	44.6	33.3	30	45.6
	Annual working time (h/year)	2700	4032	3600	2352	3850
HGVs (16.5 m, 44 t)	Driver cost (£/km)	0.23	0.18	0.27	0.46	0.18
	Fixed costs (£/km)	0.35	0.27	0.40	0.68	0.27
	Other vehicle costs (£/km)	0.12	0.12	0.12	0.12	0.12
	Fuel cost (£/km)	0.40	0.40	0.46	0.49	0.53
	Total (£/km)	1.10	0.97	1.25	1.75	1.11
	Fuel consumption (l/100 km)	26.6–43.0 (empty-full load, [25])				
LHVs (25.25 m, 60 t)	Driver cost (£/km)	0.23	0.18	0.27	0.46	0.18
	Fixed costs (£/km)	0.44	0.35	0.46	0.71	0.36
	Other vehicle costs (£/km)	0.12	0.12	0.12	0.12	0.12
	Fuel cost (£/km)	0.51	0.49	0.51	0.56	0.58
	Total (£/km)	1.30	1.14	1.37	1.85	1.25
	Fuel consumption (l/100 km)	33.7–51.1 (empty-full load, [25])				

The effect on fuel consumption and CO_2 emissions are calculated based on the fuel consumption for empty and full load HGVs and LHVs presented in Table 1 based on the unit emissions database by VTT Technological Research Centre of Finland [25]. VTT's data actually contains HGV fuel consumption for 40 t GVW, but the full load consumption for 44 t semitrailer was extrapolated assuming linear increase in consumption between 40t and 44t. The VTT's database was chosen because it has long tradition in providing fuel consumption data for both HGVs and LHVs based on both their own measurements and long term collaboration with other laboratories under the European Research on Mobile Emission Sources (ERMES) group and the emissions reported in the Handbook Emission Factors for Road Transport (HBEFA). There are other sources for fuel consumption data available, but most do not include fuel consumption data for LHVs. Average fuel consumption for each commodity is calculated using the average load for each commodity and assuming a linear relationship between the empty and full load consumption. As can be seen in Table 1, the LHVs have higher fuel consumption empty because of higher vehicle own weight (20 t vs. 15 t) and higher fuel consumption fully laden, because of higher gross vehicle weight (60 t vs. 44 t) [25].

3. Results

3.1. Freight Transport Profile in the United Kingdom and Finland

United Kingdom and Finland are very different countries in terms of population, the United Kingdom having 65.6 million people and Finland 5.5 million. The area is 242,000 km^2 for the United Kingdom and 338,000 km^2 for Finland, resulting in highly different population density; 271 versus 16 inhabitants/km^2. Also, in terms of economic structure, the countries differ, as agriculture, forestry, fishing, and industry represent 24% of Finnish gross value added (GVA), but only 14% in the United Kingdom, whereas in the United Kingdom the wholesale and retail, financial and scientific sectors constitute a larger share of total GVA than in Finland (Table 2).

Table 2. Sectoral gross value added and domestic freight transport in the United Kingdom and Finland in 2015 [25–27].

Gross Value Added in 2015	Finland		United Kingdom	
	Million €	%	Million €	%
Total	180,785	100%	2,299,669	100%
Agriculture, forestry, and fishing	4591	3%	14,981	1%
Industry (except construction)	37,341	21%	304,788	13%
Construction	11,552	6%	141,519	6%
Wholesale and retail trade, transport, accommodation and food service activities	28,770	16%	425,467	19%
Information and communication	10,303	6%	149,234	6%
Financial and insurance activities	5204	3%	166,698	7%
Real estate activities	22,814	13%	297,951	13%
Professional, scientific, and technical activities; administrative and support service activities	15,124	8%	282,934	12%
Public administration, defence, education, human health and social work activities	39,449	22%	425,328	18%
Arts, entertainment and recreation; other service activities; activities of household and extra-territorial organizations and bodies	5637	3%	90,769	4%
Domestic freight transport in 2015	Finland		United Kingdom	
	Road	Rail	Road	Rail
Freight transport by mode (million tonne-km)	24,488	8468	158,924	21,990
Total freight transport (million tonne-km)	32,956		180,914	
Transport intensity (tkm/€)	0.18		0.08	
CO$_2$ emissions (Mt)	2.9	0.06	19.6	0.57
CO$_2$ intensity (t/tkm)	0.118	0.007	0.121	0.026

The differences in economical, geographical, and demographical structures also affect the freight transport sector. The freight transport intensity, that is, the ratio of freight haulage in tkm to GVA in €, in Finland is more than twice that of United Kingdom, with 0.18 tkm/€ and 0.08 tkm/€, respectively.

The high transport intensity in Finland is largely because of the large forest industry sector in Finland. This can be seen from the breakdown of road freight transport by commodity (Figure 2). Forestry and logging is the largest commodity in Finland and constitutes about one-fifth of road tonne-kms in Finland. While the division to 20 NST2007 commodities gives valuable information, it is also necessary to further disaggregate the agricultural products into products of forestry and logging and the other food related agricultural products. Otherwise, the average load of this commodity would be very high in Finland and would overestimate in the United Kingdom the potential of LHVs in this commodity. In the United Kingdom, food products account for a quarter of tonne-kms. Mining and quarrying, which is mainly construction related soil, gravel, and sand transport are a large commodity in both countries, as are grouped goods, which are usually various types of palletized goods.

In terms of the types of vehicles used in road haulage, both countries have vast majority of tonne-kms produced with the heaviest vehicles, that is, 72% of tonne-kms with articulated vehicles over 33 t in the United Kingdom and 82% of tonne-kms with artics over 44 t in Finland. The share of heaviest vehicles is especially large for commodities such as coal, wood products, and chemical products. Rigid vehicles have a significant share in some commodities, such as mining and quarrying, textiles, transport equipment, furniture and household, and office removals.

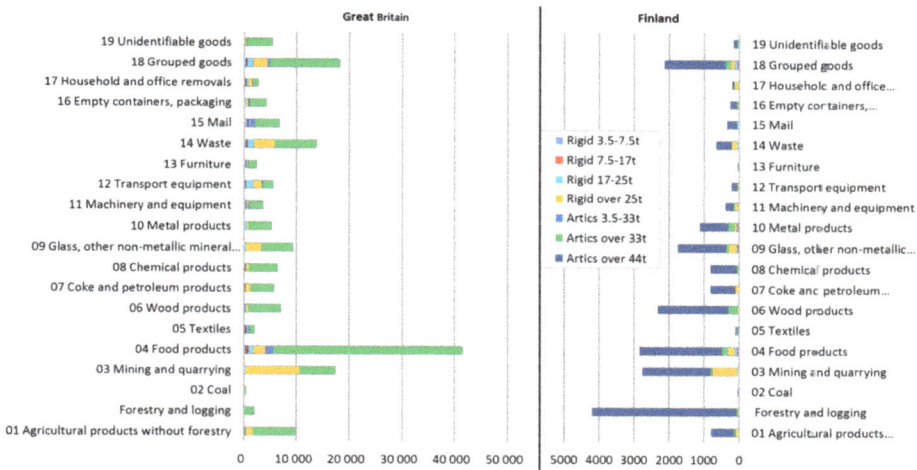

Figure 2. Tonne-kms (in million tkm) by commodity and type and weight of vehicle in Great Britain and Finland.

It can be seen from Figure 2, that vehicles with GVW of over 44 t (mostly 60 t) carry the vast majority of tonne-kms in Finland. These vehicles may carry up to 46% more payload (42.8 t vs. 29.3 t, [4]) than the British 44 t semitrailers. However, payload is usually restricted by other limitations than weight, that is, volume or cargo area. Hence, the maximum payload is actually rarely used for most commodities [28] and an increase in maximum payload cannot be used as such to estimate the potential for vehicle km savings from using LHVs. A change from 44 t and 16.5 m semitrailers to 60 t and 25.25 m vehicles would also increase the payload area and volume by about 46% (140 m^3 vs. 96 m^3 and 52 pallets vs. 36 pallets). Hence, the 46% increase in transport efficiency represents a theoretical maximum for the potential benefits of fully laden LHVs compared with fully laden HGVs. However, greater benefits may be possible if the use of LHVs also changes the logistics practices so that the utilization rate of vehicles increase. Such changes cannot be estimated and this must be acknowledged as a limitation of this study. This limitation could be addressed with more disaggregated commodity group data, but such data are not available for these countries and annual variation would also increase on a more disaggregate level.

There are six commodities in Table 3, in which the average load on laden trips are currently more than 46% greater in Finland than in the United Kingdom, namely forestry and logging, coal, wood products, chemical products, mail and household, and office removals. Although it is possible that the average load in the United Kingdom could increase to the current Finnish levels after the introduction of LHVs, this is considered unlikely, because it is more likely that the difference is due to differences in product mix, which would not change if LHVs would be introduced in the United Kingdom. Hence, the maximum increase in average payload on laden trips due to usage of LHVs has been limited to 46%. There are also two commodities, textiles and furniture, in which the current average load in Finland is lower than the current average load in the United Kingdom. These commodities have great annual variation in the Finnish data, so it is difficult to estimate the likely effect of LHVs in the United Kingdom. Hence, the average load is assumed to remain the same in the United Kingdom after the introduction of the LHVs. The resulting average payloads for calculating the benefits of LHVs are presented in the last column of Table 3.

Table 3. Current average loads by commodity on laden trips of over 33 t artics in the United Kingdom and Finland with estimated average loads in the United Kingdom if LHVs would be used.

Commodity and NST2007 Number	U.K. Artics Over 33 t Haulage (Million tkm)	U.K. Artics Over 33 t Mileage (Million km)	U.K. Artics Over 33 t Avg. Load (t)	FIN Artics Over 33 t Avg. Load (t)	Average Load Increase If Finnish Avg. Load Would Be Achieved	Estimated U.K. Artics Avg. load Using LHVs (t)
01 Agricultural products without forestry	8222	411	20.0	20.5	3%	20.5
Forestry and logging	1985	145	13.7	37.9	177%	20.0
02 Coal	471	19	24.8	39.7	60%	36.2
03 Mining and quarrying	6917	268	25.8	35.3	37%	35.3
04 Food products	35,970	2178	16.5	21.0	27%	21.0
05 Textiles	920	85	10.8	8.4	−23%	10.8
06 Wood products	6162	367	16.8	27.9	66%	24.5
07 Coke and petroleum products	4354	176	24.7	29.4	19%	29.4
08 Chemical products	5545	360	15.4	26.3	71%	22.5
09 Glass, other non-metallic mineral products	6092	286	21.3	26.9	26%	26.9
10 Metal products	4256	232	18.3	21.4	17%	21.4
11 Machinery and equipment	2541	209	12.2	16.7	37%	16.7
12 Transport equipment	1991	143	13.9	19.5	40%	19.5
13 Furniture	1484	142	10.5	7.9	−24%	10.5
14 Waste	8128	429	18.9	26.1	38%	26.1
15 Mail	4622	373	12.4	18.6	50%	18.1
16 Empty containers, packaging	3212	405	7.9	9.0	14%	9.0
17 Household and office removals	1169	96	12.2	18.3	50%	17.8
18 Grouped goods	13,420	899	14.9	19.1	28%	19.1
19 Unidentifiable goods	4858	270	18.0	20.7	15%	20.7
20 Other goods	198	12	16.5			24.1
Empty		2399				
All commodities	122,515	9904	16.3	25.4		

3.2. Maximum Potential of LHVs on the U.K. Road Freight Transport

Calculated based on the estimated average load by commodity on laden trips if LHVs would be used in the United Kingdom, Table 4 presents the maximum decrease of vehicle kms of over 33 t artics in the United Kingdom. The savings in vehicle kms range from 0% to 32% of the current vehicle kms.

The overall total saving in vehicle kms is 2.1 billion kms, which is 21% of current mileage with over 33 t artics and 11% of current total lorry mileage. Food products and empty running produce majority of the savings with 466 million kms and 510 million kms saved in these commodity groups, respectively. Wood products, chemical products, waste, mail, and grouped goods also provide estimated savings of more than 100 million kms each.

The commodity with the highest relative savings from current total vehicle kms are forestry and logging, coal, wood products, chemical products, and other goods. All of these are high density commodities, which currently have high percentage of vehicle kms driven with over 33 t lorries, and thus benefit from the extra weight capacity. However, also some commodities constrained by cargo area and volume, such as mail, grouped goods, and food products show savings of over 10% of current total lorry vehicle kilometres.

Table 5 presents the changes in fuel consumption. Total fuel savings are 178 million litres, which is 5% of the current fuel consumption of over 33 t artics in the United Kingdom. In terms of CO_2 emissions, the decrease is 0.5 Mt, which is 2.4% of total truck CO_2 emissions in the United Kingdom. Fuel consumption in terms of l/100 km increases because of increased payload and because LHVs have higher own weight and higher aerodynamic drag than HGVs. However, the decrease in vehicle fuel consumption due to decrease in vehicle kms is higher than the increase in fuel consumption per kilometre for all but two (agricultural products without forestry, empty containers and packaging) commodities. The negative fuel savings in these two commodities indicate that the average load using LHVs, which was based on the Finnish average load, does not increase the payload enough to decrease the mileage to outweigh the increase in fuel consumption per km. This is most likely due to national differences in the types of goods carried within the commodity group, but also because of annual variation in average load. In the following analysis, on transport costs, the same commodities show negative cost savings with two more commodities. Hence, it is necessary to analyse the issue further by taking into account the increase in cargo space capacity in addition to weight capacity.

Table 6 presents the transport cost savings if the average load for each commodity would increase with the implementation of LHVs. Total savings amount to £983 million annually, with food products and waste with saving of more than £150 million each. There are also four commodities, namely, agricultural products without forestry, metal products, empty containers and packaging, and unidentifiable goods, for which the cost saving is negative. For these commodities, the average load on laden trips is only slightly higher in Finland than in the United Kingdom, indicating that these commodities are constrained by cargo area or volume rather than by weight. For these commodities, the transport costs would increase because the decrease in vehicle kms is smaller than the increase in transport costs per kilometre if the cargo space would not be efficiently used. As it was discussed earlier with textiles and furniture, the increase in payload might be greater in the United Kingdom because of the additional area and volume capacity of LHVs than the comparison with Finnish average loads shows. Hence, it is justifiable to analyse the potential savings of these commodities if the average load in these commodities would increase by 46%, that is, by the theoretical maximum increase. This analysis gives an upper estimate to the potential savings.

Table 4. Decrease in vehicle kms in the United Kingdom if LHVs would be used.

Commodity and NST2007 Number	Total Current U.K. Mileage, All Vehicles (Million km)	U.K. Artics Over 33 t Mileage (Million km)	U.K. Artics Over 33 t Mileage With LHVs (Million km)	Decrease in U.K. Artics Over 33 t Mileage (Million km)	Decrease as % of Artics Over 33 t U.K. vkm	Decrease as % of Total U.K. vkm
01 Agricultural products without forestry	706	411	401	10	2%	1%
Forestry and logging	145	145	99	46	32%	32%
02 Coal	24	19	13	6	32%	25%
03 Mining and quarrying	806	268	196	72	27%	9%
04 Food products	3398	2178	1712	466	21%	14%
05 Textiles	427	85	85	0	0	0
06 Wood products	605	367	251	116	32%	19%
07 Coke and petroleum products	351	176	148	28	16%	8%
08 Chemical products	647	360	247	113	32%	18%
09 Glass, other non-metallic mineral products	639	286	226	60	21%	9%
10 Metal products	526	232	199	33	14%	6%
11 Machinery and equipment	593	209	153	56	27%	10%
12 Transport equipment	562	143	102	41	29%	7%
13 Furniture	503	142	142	0	0	0
14 Waste	1286	429	312	117	27%	9%
15 Mail	786	373	255	118	32%	15%
16 Empty containers, packaging	658	405	356	49	12%	7%
17 Household and office removals	429	96	66	30	32%	7%
18 Grouped goods	1764	899	704	195	22%	11%
19 Unidentifiable goods	386	270	234	36	13%	9%
20 Other goods	18	12	8	4	32%	21%
Empty	3974	2399	1889	510	21%	13%
All commodities	19,233	9904	7797	2107	21%	11%

Table 5. Fuel and CO_2 savings.

Commodity and NST2007 Number	HGV Fuel Consumption (l/100 km)	Current Fuel Consumption (million l)	LHV Fuel Consumption (l/100 km)	Fuel Consumption with LHVs (Million l)	Fuel Saving (Million l)	Fuel Saving as % of Current	CO_2 Saving (Mt)
01 Agricultural products without forestry	37.9	156	42.6	171	−15	−10 %	−0.04
Forestry and logging	34.3	50	42.4	42	8	15 %	0.02
02 Coal	40.6	8	49.4	6	1	17 %	0.00
03 Mining and quarrying	41.2	110	49.1	96	14	13 %	0.04
04 Food products	35.9	782	42.8	733	49	6 %	0.13
05 Textiles	32.7	28	38.4	29	0	0	0
06 Wood products	36.1	132	44.4	112	21	16 %	0.06
07 Coke and petroleum products	40.6	71	46.5	69	2	3 %	0.01
08 Chemical products	35.3	127	43.5	107	20	16 %	0.05
09 Glass, other non-metallic mineral products	38.6	110	45.4	103	8	7 %	0.02
10 Metal products	36.9	86	43.0	85	0	0 %	0.00
11 Machinery and equipment	33.5	70	40.9	62	7	11 %	0.02
12 Transport equipment	34.5	49	42.2	43	6	13 %	0.02
13 Furniture	32.5	46	38.2	48	0	0	0
14 Waste	37.3	160	45.0	140	20	12 %	0.05
15 Mail	33.6	125	41.6	106	19	15 %	0.05
16 Empty containers, packaging	31.1	126	37.6	134	−8	−7 %	−0.02
17 Household and office removals	33.5	32	41.4	27	5	15 %	0.01
18 Grouped goods	35.0	315	42.0	295	19	6 %	0.05
19 Unidentifiable goods	36.7	99	42.7	100	−1	−1 %	−0.00
20 Other goods	35.9	4	44.2	4	1	16 %	0.00
Empty	26.6	638	33.7	637	2	0 %	0.00
All commodities		**3325**		**3151**	**178**	**5 %**	**0.47**

Table 6. Transport cost savings.

Commodity and NST2007 Number	HGV Transport Costs (£/km)	Current Transport Costs (M£)	LHV Transport Costs (£/km)	Transport Costs with LHVs (M£)	Cost Saving (M£)	Cost Savings as % of Current
01 Agricultural products without forestry	0.97	397	1.14	457	−60	−15%
Forestry and logging	1.11	161	1.25	124	37	23%
02 Coal	1.25	24	1.37	18	6	25%
03 Mining and quarrying	1.75	468	1.85	363	105	22%
04 Food products	0.97	2104	1.14	1950	154	7%
05 Textiles	1.10	94	1.30	94	0	0
06 Wood products	0.97	355	1.14	286	68	19%
07 Coke and petroleum products	1.25	220	1.37	203	17	8%
08 Chemical products	1.25	450	1.37	337	113	25%
09 Glass, other non-metallic mineral products	0.97	276	1.14	258	19	7%
10 Metal products	0.97	224	1.14	226	-2	−1%
11 Machinery and equipment	1.10	230	1.30	199	32	14%
12 Transport equipment	1.10	157	1.30	133	24	16%
13 Furniture	1.10	156	1.30	156	0	0
14 Waste	1.75	749	1.85	578	172	23%
15 Mail	0.97	360	1.14	291	69	19%
16 Empty containers, packaging	1.10	446	1.30	464	−18	−4%
17 Household and office removals	1.10	106	1.30	86	20	19%
18 Grouped goods	0.97	869	1.14	802	67	8%
19 Unidentifiable goods	0.97	261	1.14	267	−6	−2%
20 Other goods	0.97	12	1.14	12	0	0
Empty	0.97	2318	1.14	2152	166	7%
All commodities		10,437		9455	983	9%

It was seen in Table 6 that there are six commodities for which the Finnish average load was lower or only slightly higher than the current average load in the United Kingdom, which resulted in negative transport cost savings for these commodities. Hence, an additional analysis (Table 7) is made in which the average load for these commodities is increased by 46% to reflect the change in cargo area and volume, which are likely constraints for these commodities instead of weight.

The additional analysis gives an upper estimate for the potential savings (Table 8). It could be argued that the upper estimate should be calculated by simply increasing the average load by 46% for all commodities. However, such an estimate is likely to unrealistic, because it is unlikely that all commodities would implement LHVs to the full as there are restrictions due to physical infrastructure and logistics practices. Hence, it can be concluded that implementing LHVs in the United Kingdom could reduce lorry vehicle kilometres by 2.1–2.6 billion km, which is 11–13% of current lorry mileage. This reduction would save 178–272 million litres of diesel and reduce CO_2 emissions by 0.5–0.7 Mt. The transport costs of over 33 t articulated lorries could decrease by £1.0–1.5 billion, which is 9–14% of current transport costs. It might also be argued that the effects of LHVs may be even greater and estimated assuming that the average load would increase by 46% for all commodities. However, this would give too high estimate on the effects of LHVs because it is unlikely that LHVs could be used on all journeys currently made with over 33 t articulated vehicle. Infrastructure may not enable use of LHVs everywhere and LHVs may also cause a modal shift from rail to road, which has an effect on the lower estimate of LHV benefits.

Sustainability **2018**, 10, 2754

Table 7. Savings in vehicle kms, transport costs and fuel using LHVs based on increased cargo space.

Commodity and NST2007 Number	Decrease in vkm Using LHVs Based on Weight (Million km)	Decrease in vkm Using LHVs Based on Cargo Space (Million km)	Cost Savings Using LHVs Based on Weight (M£)	Cost Savings Using LHVs Based on Cargo Space (M£)	Fuel Savings Using LHVs Based on Weight (Million l)	Fuel Savings Using LHVs Based on Cargo Space (Million l)
01 Agricultural products without forestry	10	129	−60	76	−4	35
05 Textiles	0	27	0	0	0	0
10 Metal products	33	73	−2	43	6	17
13 Furniture	0	45	0	0	0	0
16 Empty containers, packaging	49	128	−18	85	−4	22
19 Unidentifiable goods	36	85	−6	50	6	20
Additional savings based on cargo space		359		340		94

Table 8. Range of estimated savings in the United Kingdom using LHVs.

	Decrease in UK artics over 33 t vkm	Decrease as % of artics over 33 t UK vkm	Decrease as % of total UK vkm
Lower estimate	2107	21%	11%
Upper estimate	2581	26%	13%
	Fuel saving (Ml)	Fuel savings as % of current over 33 t artics fuel use	CO_2 saving (Mt)
Lower estimate	178	5%	0.47
Upper estimate	272	8%	0.72
	Cost saving (M£)	Cost savings as % of current over 33 t artics costs	
Lower estimate	983	9%	
Upper estimate	1454	14%	

3.3. Effects of Infrastructure and Modal Shift on the Lower Estimate of LHV Benefits

The estimated benefits presented in the previous section show the effects, which are internal in the road freight sector, but do not take into account possible modal shift or effects on road infrastructure. In order to estimate the net economic benefits on a national scale, two additional major factors must be taken into account, namely road infrastructure improvement costs and modal shift from rail to road. TRL [9] estimated significantly lower effects of LHVs in the United Kingdom than the estimates presented in the previous section. TRL estimated 0.4–1.3% reduction in vehicle kms, 1.4–3.6% reduction in transport costs and 0.5–1.4% increase in CO_2 emissions. The differences are due to significantly lower estimate on the potential use of LHVs and shift from rail to road. TRL [9] estimates that the take up rate of LHVs would be 5% as a low estimate and 10% as a high estimate, that is, 5–10% of articulated vehicle's tkm would be carried by LHVs. The low estimate was primarily due to restricting the use of LHVs on only certain types of commodities, which is not justified based on the Finnish example where all types of commodities are mainly transported using LHVs (Figure 2). Of course the situation in Finland is due to long term development, but in the long term, LHVs are likely to take over transport from HGVs in the United Kingdom too, simply because of the indisputably lower transport costs per unit transported.

TRL [9] also estimates the effects of possible route restrictions on LHV and shows that if LHVs would be allowed only on motorways and dual carriageways, the reduction in vehicle km would be 20% lower than in if LHVs would be allowed on all roads. Road infrastructure, especially some roundabouts in urban areas and bridges in rural areas, as well as docking areas in distribution centres, may not be able to accommodate LHVs because of the larger turning circle and GVW of LHVs. Hence, it is justified to take the estimate by TRL [9] to calculate a new lower estimate on the effect of LHVs in the United Kingdom. Applying the suggested 20% decrease in the lower estimate of the vehicle km savings result in a new estimate of 1.7 billion vkm (8.8% of total lorry vkm). In terms of fuel, CO_2, and cost savings, the savings are also reduced to 142 million l, 0.38 Mt, and £786 million, respectively.

Regarding the modal shift, ORR [29] reported that rail freight had 10% share (about 17 billion tkm) of domestic freight transport in the United Kingdom and 1.7 billion lorry kilometres would be required to transport the amount of freight moved by rail in 2015–2016. Finland (27%) and Sweden (29%) currently have considerable higher shares of rail freight of total tkms than United Kingdom [26]. However, it should be noted that majority of the freight transport on rail are due to extensive forest industry in these countries. In Finland, forestry and products of paper industry represent about 48% of total rail tkms [30], whereas in the United Kingdom, 40% of rail tkms are intermodal freight [29]. According to TRL [9], 13% shift from total rail haulage to road is a mid-range estimate, which would be about 2.2 billion tkm. Supposing this would shift from rail to road, using LHVs with an average load of 20 t and the share of empty running of 33% of total mileage (which is a high estimate for empty running, as the average for over 33 t artics in the United Kingdom is 24%), the resulting increase in road freight would be approximately 0.15 billion vehicle kms. The resulting lower estimate of the net change in lorry vkm would be a decrease of 1.5 billion km (8% of total lorry vkm) and resulting fuel, CO_2, and cost savings can be seen in Table 9.

Sustainability **2018**, 10, 2754

Table 9. Range of estimated savings in the United Kingdom using LHVs when taking infrastructure and modal shift into account.

	Decrease in U.K. artics over 33 t vkm	Decrease as % of artics over 33 t U.K. vkm	Decrease as % of total U.K. vkm
Lower estimate	1 535	15%	8%
Upper estimate	2 581	26%	13%
	Fuel saving (Ml)	Fuel savings as % of current over 33 t artics fuel use	CO_2 saving (Mt)
Lower estimate	129	4%	0.35
Upper estimate	272	8%	0.72
	Cost saving (M£)	Cost savings as % of current over 33 t artics costs	
Lower estimate	716	7%	
Upper estimate	1 454	14%	

As can be seen from Tables 8 and 9, the upper estimate was not changed as a result of infrastructure and modal shift. Regarding the modal shift, the upper limit does not take any change into account, reflecting the finding by Steer et al. [3] that empirical evidence on modal split shows lower effects than desk studies anticipate. Regarding infrastructure, implementing LHVs could also change the structure and operational practices of freight transport networks so that average payload increases more than estimated here when they are mostly used on trunk routes on motorways and dual carriageways between distribution centres, unlike HGVs currently are. However, it should be noted that there are currently about 250 'substandard' bridges on motorways and major A-roads and more than 3000 such bridges on council-maintained roads [31]. Some of these bridges may not be fit to carry even the current heaviest HGVs. Hence, it is likely that significant investments in the magnitude of hundreds of millions of pounds would be required to improve the bridges to allow 60 t GVW on major roads.

4. Discussion

The results of this study estimated that the savings of longer and heavier vehicles would be 1.5–2.6 billion vehicle kms (8–13% of total lorry mileage), £0.7–1.5 billion in transport costs, and 0.35–0.72 Mt in CO_2 emissions (1.8–3.7% of total lorry emissions). These findings are well in line with some estimates of other studies. McKinnon [14] reported that the previous increase in maximum weight in the United Kingdom, from 40 t to 44 t in 2001, saved 0.13 billion vehicle kms, £0.11 billion in transport costs, and 0.13 Mt in CO_2 emissions in 2003. As this was a 10% increase in payload weight and LHVs would increase the payload weight by 46%, as well as an increase in the cargo area and volume, the results are roughly comparable. Leach et al. [13] estimated that allowing longer (25.25 m) vehicles without increasing the maximum weight could have an effect on 15% of current articulated vehicle mileage and lead to transport cost savings of about £0.2 billion, with CO_2 emission reductions of 0.1–0.2 Mt, depending on the effects on rail freight transport. As Leach et al. [13] estimated that longer vehicles could affect 15% of articulated vehicle kms, and this analysis took into account 94% of articulated vehicle kms, the results are highly comparable.

Arki [32] reported that allowing LHVs in the whole of Europe would decrease the vehicle kms by 13%, transport costs by 6.8%, and CO_2 emissions by 3.6%. These figures are very similar with the results of this study, although the reduction of CO_2 emissions is lower than in this analysis. This could be partly due to the fact that LHVs are already in operation in some European countries. Vierth et al. [24] studied the opposite situation, that is, abandoning LHVs and using HGVs instead in Sweden, and found that the amount of vehicle kms would increase by 24%, transport costs by 7%, and CO_2 emissions by 6% if only HGVs were used, and there would be no shift from road to rail. Again, these results are well in line with the estimates of this study.

5. Conclusions

The potential effects of implementing longer and heavier vehicles (LHVs) in road freight transport have been studied in various countries nationally and internationally in Europe. These studies have focused on the implementation of LHVs on certain types of commodities and the experiences from countries like Finland and Sweden, which have a long tradition in using LHVs and in which LHVs used for all types of commodities have not been widely utilised. This study aimed to assess the impacts of long and heavy vehicles on various commodities in the United Kingdom based on the Finnish experiences in order to estimate the possible savings in road freight transport vehicle kilometres, costs, and CO_2 emissions in the United Kingdom if LHVs would be introduced and used similarly to Finland in the transport of various commodities.

The international commodity level approach induced some challenges to the analysis. Firstly, some commodities have very limited amount of data and there is large annual variation, which might lead to slightly different results if other years would have been chose. Hence, it is recommended that data from several years should be used in future commodity level studies. Secondly, international comparison even at the commodity level might not be sufficiently detailed to make assumptions on the utilisation of LHVs, and thus the changes in average loads moving from HGVs to LHVs. Even within the same commodity type, very different goods may be transported, as illustrated by the case of agricultural products and forestry in this study. Hence, international comparisons should aim to utilise the most disaggregated data available. However, there are variations between countries in the level of disaggregation, so the 20 commodity types used in NST2007 might remain the deepest reliable level of disaggregation, as European countries deliver this data annually [21].

The study shows that the savings of introducing longer and heavier vehicles in the United Kingdom would be 1.5–2.6 billion vehicle kms, £0.7–1.5 billion in transport costs, and 0.35–0.72 Mt in CO_2 emissions. These findings are well in line with previous findings in other countries. The results confirm that considerable savings in traffic volume and emissions can be achieved and the savings are very likely to outweigh possible effects of modal shift from rail to road. Previous research also somewhat agrees that LHVs are unlikely to cause negative effects on modal shift or transport safety. Hence, LHVs are likely to provide a viable solution in the United Kingdom to decrease greenhouse gas emissions from transport in order to mitigate climate change, although the effect on greenhouse gases is very limited.

However, in order to gain the maximum benefits of LHVs, investment requirements in infrastructure are likely to emerge. Hence, further research on the state of road infrastructure, especially regarding weight-restricted bridges and manoeuvrability issues, would be required in order to estimate the investment costs and compare them against the savings projected in this study. Alternatively, LHVs can be allowed only on certain types of roads to avoid infrastructure investments. Further research that takes into account routing and likely use of various road types by LHVs should be carried out to find the right balance between the LHV road network and investment costs. Such research can build on the recent study by Palmer et al. [33], which focused on the fast moving consumer goods (FMCG) sector.

Author Contributions: Conceptualization, H.L. and P.G.; Methodology, H.L.; Formal Analysis, H.L.; Data Curation, H.L. and P.D.; Writing—Original Draft Preparation, H.L.; Writing—Review & Editing, A.K.; Visualization, H.L.; Supervision, P.G.; Project Administration, H.L.; Funding Acquisition, H.L.

Funding: This work was supported by the Kone Foundation (grant number b4b919).

Conflicts of Interest: The authors declare no conflict of interest. The funders had no role in the design of the study; in the collection, analyses, or interpretation of data; in the writing of the manuscript; and in the decision to publish the results.

References

1. Liimatainen, H.; Kallionpää, E.; Pöllänen, M.; Stenholm, P.; Tapio, P.; McKinnon, A. Decarbonizing road freight in the future—Detailed scenarios of the carbon emissions of Finnish road freight transport in 2030 using a Delphi method approach. *Technol. Forecast. Soc. Chang.* **2014**, *81*, 177–191. [CrossRef]

2. Sims, R.; Schaeffer, R.; Creutzig, F.; Cruz-Núñez, X.; D'Agosto, M.; Dimitriu, D.; Figueroa Meza, M.J.; Fulton, L.; Kobayashi, S.; Lah, O.; et al. Transport. In *Climate Change 2014: Mitigation of Climate Change. Contribution of Working Group III to the Fifth Assessment Report of the Intergovernmental Panel on Climate Change*; Edenhofer, O., Pichs-Madruga, R., Sokona, Y., Farahani, E., Kadner, S., Seyboth, K., Adler, A., Baum, I., Brunner, S., Eickemeier, P., et al., Eds.; Cambridge University Press: Cambridge, UK; New York, NY, USA, 2014.

3. Steer, J.; Dionori, F.; Casullo, L.; Vollath, C.; Frisoni, R.; Carippo, F.; Ranghetti, D. A Review of Megatrucks—Major Issues and Case Studies. Available online: http://www.europarl.europa.eu/thinktank/en/document.html?reference=IPOL-TRAN_ET(2013)513971 (accessed on 31 July 2018).

4. Glaeser, K.P.; Ritzinger, A. Coparison of the performance of heavy vehicles. Results of the CECD study: 'Moving freight with better trucks'. *Procedia—Soc. Behav. Sci.* **2012**, *48*, 106–120. [CrossRef]

5. Statistics Finland. Goods Transport by Road. Official Statistics of Finland: Helsinki, Finland, 23 April 2017. Available online: http://stat.fi/til/kttav/2016/kttav_2016_2017-04-28_en.pdf (accessed on 31 July 2018).

6. Liimatainen, H.; Nykänen, L.; Pöllänen, M. Impacts of increasing maximum truck weight—Case Finland. C. *Stud. Transp. Policy* **2018**. under review.

7. Christidis, P.; Leduc, G. Longer and Heavier Vehicles for Freight Transport. European Commission Joint Research Centre Institute for Prospective Technological Studies. Available online: https://ec.europa.eu/jrc/en/publication/eur-scientific-and-technical-research-reports/longer-and-heavier-vehicles-freight-transport (accessed on 31 July 2018).

8. Fraunhofer ISI, TRT, NESTEAR. Long-Term Climate Impacts of the Introduction of Mega-Trucks. Study for the Community of European Railway and Infrastructure Companies (CER): Karlsruhe, Germany, 2009. Available online: http://www.cer.be/sites/default/files/publication/090512_cer_study_megatrucks.pdf (accessed on 31 July 2018).

9. TRL. Longer and/or Longer and Heavier Goods Vehicles (LHVs)—A Study of the Likely Effects if Permitted in the UK: Final Report. DfT: London, UK, 2008. Available online: https://www.google.com/url?sa=t&rct=j&q=&esrc=s&source=web&cd=1&ved=2ahUKEwiM7I-4tcvcAhWCZt4KHesBywQFjAAegQIBBAC&url=https%3A%2F%2Ftrl.co.uk%2Fsites%2Fdefault%2Ffiles%2FPPR285%25283%2529.pdf&usg=AOvVaw0nuLBY6YYLZkXnbqfuKUd- (accessed on 31 July 2018).

10. Meers, D.; van Lier, T.; Macharis, C. Longer and heavier vehicles in Belgium: A threat for the intermodal sector? *Transp. Res. Part D* **2016**, *61*, 459–470. [CrossRef]

11. Sanchez Rodrigues, V.; Piecyk, M.; Mason, R.; Boenders, T. The longer and heavier vehicle debate: A review of empirical evidence from Germany. *Transp. Res. Part D* **2015**, *40*, 114–131. [CrossRef]

12. Liljestrand, K. Improvement actions for reducing transport's impact on climate: A shipper's perspective. *Transp. Res. Part D* **2016**, *48*, 393–407. [CrossRef]

13. Leach, D.; Savage, C.; Maden, W. High-capacity vehicles: An investigation of their potential environmental, economic and practical impact if introduced to UK roads. *Int. J. Logist. Res. Appl.* **2013**, *16*, 461–481. [CrossRef]

14. McKinnon, A. The economic and environmental benefits of increasing maximum truck weight: The British experience. *Transp. Res. Part D* **2005**, *10*, 77–95. [CrossRef]

15. Ortega, A.; Vassallo, J.; Guzman, A.; Perez-Martinez, P. Are longer and heavier vehicles (LHVs) beneficial for society? A cost benefit analysis to evaluate their potential implementation in Spain. *Transp. Rev.* **2014**, *34*, 150–168. [CrossRef]

16. Ericson, J.; Lindberg, G.; Mellin, A.; Vierth, I. Co-modality—The Socio-economic Effects of Longer and/or Heavier Vehicles for Land-based Freight Transport. In Proceedings of the 12th WCTR, Lisbon, Potugal, 11–15 July 2010.

17. Varin, P.; Saarenketo, T. Effect of Axle and Tyre Configurations on Pavement Durability—A Prestudy. ROADEX Network Report. Available online: http://www.roadex.org/wp-content/uploads/2014/01/ROADEX_Axle_Tyre_Prestudy_15102014-Final.pdf (accessed on 31 July 2018).

18. Vierth, I.; Haraldsson, M. Socio-economic Effects of Longer and/or Heavier Road Transport Vehicles—The Swedish case. In Proceedings of the Setting Future Standards: International Symposium on Heavy Vehicle Transport Technology—HVTT12, Stockholm, Sweden, 16–19 September 2012.

19. Risk Solutions. The GB Longer Semi-Trailer Trial. 2016 Annual Report Summary. 2017. Available online: https://www.gov.uk/government/uploads/system/uploads/attachment_data/file/646226/longer-semi-trailer-trial-annual-report-2016-summary.pdf (accessed on 31 July 2018).

20. Bergqvist, R.; Behrends, S. Assessing the effects of longer vehicles: The case of pre- and post-haulage in intermodal transport chains. *Transp. Rev.* **2011**, *31*, 591–602. [CrossRef]

21. Regulation (EU) no 70/2012 on Statistical Returns in Respect of the Carriage of Goods by Road. Available online: http://eur-lex.europa.eu/legal-content/EN/TXT/PDF/?uri=CELEX:32012R0070&from=EN (accessed on 31 July 2018).

22. DfT. Road Freight: Domestic and International Statistics. 2017. Available online: https://www.gov.uk/government/collections/road-freight-domestic-and-international-statistics (accessed on 31 July 2018).

23. RHA. *Cost Table*; Road Haulage Association: Peterborough, UK, 2013.

24. Vierth, I.; Berell, H.; McDaniel, J.; Haraldsson, M.; Hammarström, U.; Yahya, M.-R.; Lindberg, G.; Carlsson, A.; Ögren, M.; Björketun, U. The Effects of Long and Heavy Trucks on the Transport System. *VTI rapport 605A*. Available online: https://www.diva-portal.org/smash/get/diva2:675341/FULLTEXT02.pdf (accessed on 31 July 2018).

25. VTT. LIPASTO—Transport Emissions Database. VTT Technological Research Centre of Finland, 2017. Available online: http://lipasto.vtt.fi/en/index.htm (accessed on 31 July 2018).

26. Eurostat. Eurostat Database. 2017. Available online: http://ec.europa.eu/eurostat/data/database (accessed on 31 July 2018).

27. National Statistics. Final UK Greenhouse Gas Emissions National Statistics: 1990–2015. 2017. Available online: https://www.gov.uk/government/statistics/final-uk-greenhouse-gas-emissions-national-statistics-1990-2015 (accessed on 31 July 2018).

28. Nykänen, L.; Liimatainen, H. Possible impacts of increasing maximum truck weight: Finland case study. In *Towards Innovative Freight and Logistics: Research for Innovative Transports Set*; Blanquart, C., Clausen, U., Jacob, B., Eds.; Wiley-ISTE: London, UK, 2016; Volume 2, pp. 121–133.

29. ORR. Freight Rail Usage. 2016-17 Q4 Statisticla Release. Office of Rail and Road, 2017. Available online: http://orr.gov.uk/__data/assets/pdf_file/0007/24892/freight-rail-usage-2016-17-quarter-4.pdf (accessed on 31 July 2018).

30. FTA. The Finnish Railway Statistics 2016. Statistics from the Finnish Transport Agency, 2017. Available online: https://julkaisut.liikennevirasto.fi/pdf8/lti_2017-09_rautatietilasto_2016_web.pdf (accessed on 31 July 2018).

31. RAC Foundation. Substandard Road Bridges. 2017. Available online: https://www.racfoundation.org/research/economy/substandard-road-bridges-foi-2017 (accessed on 31 July 2018).

32. Arki, H. European Study on Heavy Goods Vehicles' Weights and Dimensions. In Proceedings of the European Transport Conference 2009, Noordwijkerhout, The Netherlands, 5–7 October 2009.

33. Palmer, A.; Mortimer, P.; Greening, P.; Piecyk, M.; Dadhich, P. A cost and CO_2 comparison using trains and higher capacity trucks when UK FMCG companies collaborate. *Transp. Res. Part D* **2018**, *58*, 94–107. [CrossRef]

sustainability

MDPI

Article

Electric Road Systems: Strategic Stepping Stone on the Way towards Sustainable Freight Transport?

Jesko Schulte * and Henrik Ny

Department of Strategic Sustainable Development, Blekinge Institute of Technology, SE-37179 Karlskrona, Sweden; henrik.ny@bth.se
* Correspondence: jesko.schulte@bth.se; Tel.: +46-455-385-519

Received: 6 March 2018; Accepted: 4 April 2018; Published: 11 April 2018

Abstract: Electrification of the transport sector has been pointed out as a key factor for tackling some of today's main challenges, such as global warming, air pollution, and eco-system degradation. While numerous studies have investigated the potential of electrifying passenger transport, less focus has been on how road freight transport could be powered in a sustainable future. This study looks at Electric Road Systems (ERS) in comparison to the current diesel system. The Framework for Strategic Sustainable Development was used to assess whether ERS could be a stepping stone on the way towards sustainability. Strategic life-cycle assessment was applied, scanning each life-cycle phase for violations against basic sustainability principles. Resulting sustainability "hot spots" were quantified with traditional life-cycle assessment. The results show that, if powered by renewable energy, ERS have a potential to decrease the environmental impact of freight transport considerably. Environmental payback times of less than five years are achievable if freight traffic volumes are sufficiently high. However, some severe violations against sustainability principles were identified. Still, ERS could prove to be a valuable part of the solution, as they drastically decrease the need for large batteries with high cost and sustainability impact, thereby catalyzing electrification and the transition towards sustainable freight transport.

Keywords: Electric Road Systems; sustainable freight transport; life-cycle assessment; E-freight; Strategic Sustainable Development; electric vehicles

---

1. Introduction

Transportation is a necessity and facilitator for people to meet their needs in today's society. At the same time, side effects of the current, fossil-based transport system, such as emissions of carbon dioxide, particulate matters, nitrogen and sulfur oxides, undermine human health as well as eco-system quality [1]. In the EU, the transportation sector accounts for one third of the total energy use and one fifth of all greenhouse gas (GHG) emissions [2]. At the same time, living up to the Paris Agreement requires drastic emission reductions and Europe wants to be the leading region in the transition towards a sustainable society. Electrification of vehicles has been pointed out as a key factor for success, due to zero exhaust emissions in the use phase [3]. However, there are still sustainability constraints in other life-cycle phases [4]. So far, most attention has focused on electric vehicles (EVs) for passenger transport. Still, trucks account for 25% of GHG-emissions of EU's transport sector [5] and the number of heavy trucks, especially, is increasing more and more [6]. Battery electric vehicles are often regarded as the main solution and several fully electric, battery-powered trucks have been presented to the public, for example the Tesla Semi and the Nikola One. Enabling a heavy truck to drive 800 km on one charge, however, requires large batteries. Batteries have a substantial sustainability impact during their life-cycle, at least with current designs [7,8]. Also, the substitution of today's global truck fleet with battery-powered freight transport is limited by resource constraints, especially considering metals like cobalt and lithium [9,10]. In addition, the batteries account for a major part of the vehicle cost, which is one of the largest barriers for the introduction of EVs [11].

Electric Road Systems (ERS)—defined as roads that support dynamic power transfer from the road to vehicles while the vehicles are in motion—could be a supplement to overcoming some of the challenges of battery EVs [12]. Still, it is important to reflect on the original aim of pursuing EV technology, namely making the transition towards a sustainable transport system, and to investigate if and how ERS can contribute to reaching this aim [13]. Previous studies have so far investigated technical aspects of ERS or conducted environmental comparisons based on specific life-cycle stages, focusing on the potential for GHG emission reductions [14]. However, when focusing only on a concept's potential to simply decrease the sustainability impact of a system, a strategic perspective is missing. Hence, the new concept might be better than the existing solution, but may still be incapable of reaching all the way to sustainability and, thereby, prove to be a costly dead end. The purpose of this study is, therefore, to broaden the perspective to investigate the complete life-cycle of ERS infrastructure from a full socio-ecological, strategic sustainability perspective. More specifically, this study aims at providing insights to the following research questions:

1. What is the sustainability impact of overhead line ERS in comparison to the current fossil-powered system?
2. What is the relative importance of different life-cycle phases?
3. Could the introduction of ERS be a strategic stepping stone on the way towards sustainable transport?

By answering these questions, the contribution of this study is (i) the analysis of the complete life-cycle from raw material extraction to end-of-life; (ii) including both the ecological and social dimensions; and (iii) applying a long-term strategic planning perspective by using backcasting from basic principles for sustainability, which is further explained in Section 2.2.

2. Background

2.1. Electric Road Systems

ERS have emerged as one of few realistic solutions to make freight transport more energy efficient and sustainable [15]. According to the same study, there are two use case scenarios for ERS: they can either be used in closed systems, for example for bus routes or on mining sites, or in open systems (i.e., highways for the general traffic). By electrifying main roads, convenient long-distance transport would be possible, at the same time as allowing the battery size to be relatively small, delivering approximately 150 km of range depending on how much of the road network that is electrified [11]. Mainly three technical concepts exist, which have been described in more detail in previous studies [11,16]:

i   Conductive power supply through overhead lines, similar to trains;
ii Conductive power supply through an electric rail in the road, similar to some subways; and
iii Inductive power supply without any physical contact through electric coils in the road.

For all concepts, research is ongoing and test tracks exist at various places around the world, but they are still far from constituting large-scale commercial systems and have technology readiness levels between three and seven [16]. The concepts differ significantly from each other in many aspects including function, cost and environmental impact. Concept (i) can only be used by high vehicles, for example trucks and buses, while concepts (ii) and (iii) also could be used by passenger battery EVs. However, passenger cars usually travel much shorter distances per day and, therefore, have a much lower need for charging other than home charging. All three concepts require that trucks have an additional source for propulsion, for example, an internal combustion engine, fuel cell, or a small battery. This is necessary as it is only meaningful to electrify the parts of the road network that have high traffic flows. In addition, sections like bridges or interceptions might not be possible to electrify. Therefore, a hybrid solution is needed for driving on the non-electrified sections. As usually only one lane has access to the ERS, another power supply is also needed for overtaking on the non-electrified lane. In addition, not being completely dependent on electricity supply from the ERS increases the system's resilience to malfunction and power failure considerably.

So far, most scientific literature has focused on inductive power transfer (IPT), for example [13,17–22], and test tracks are in operation in, among other places, South Korea and Italy [23]. Opinions regarding costs vary, but in Europe it is considered to be the most expensive of the three concepts [11,24,25]. Although IPT has several advantages, there is also some uncertainty regarding health effects of electromagnetic fields [26,27] and as to whether the technology is robust enough for safe and reliable long-term operation in harsh climates, such as in northern Europe [24,28]. Conductive power transfer with electric rails in the road is being tested by Volvo, Elways and other actors [29]. This solution is estimated to be less expensive, but it is not yet sufficiently tested in regard to safety and functionality, especially because the electric rail is located in the road surface, where it is exposed to weather influences and potential objects in the rail [30]. Conductive power supply from overhead lines is a more proven technology, due to its similarities to railway and trolleybus systems. Siemens and Scania are two main actors involved in the development of this concept and test tracks have been built and are in operation in Germany and Sweden. Some disadvantages are that passenger cars cannot use this ERS type, that masts and overhead lines have a visual impact, and that the overhead lines pose a risk for accidents.

This study focuses on conductive ERS with overhead lines, because (i) this technology is more mature [11,12]; (ii) information and data is available; and (iii) a large innovation procurement focusing on conductive solutions is currently taking place in Sweden with the aim to validate and test different ERS concepts [31].

2.2. Strategic Planning towards Sustainability

The transport system is highly complex and interconnected with other systems, such as, for example, the energy system. This makes it challenging to identify the actions and technologies that are strategic stepping stones on the way towards sustainability and, respectively, the ones that may later turn out to be costly dead ends. Therefore, there is a need for a framework that provides an understanding of the full scope of the sustainability challenge that includes a structure and inter-relational model to distinguish goals, tools, guidelines, and processes, and that offers an operational definition of sustainability that can be used as a basis for strategic thinking. Broman and Robèrt [32] have summarized two decades of research on the Framework for Strategic Sustainable Development (FSSD), which is designed to be such a framework that can be used to plan for sustainability in complex systems. The FSSD consists of five levels: system, success, strategy, action, and tools. Central to the FSSD is a backcasting approach: instead of asking what is likely to happen based on today's trends, backcasting starts out from the looked-for goal and then questions are asked as to what has to happen today and tomorrow in order to reach that goal, in this case a sustainable society. Apparently, a future vision of success plays a central role in this approach. A detailed vision of a sustainable future is, however, inflexible and difficult for many people to agree upon. On the other hand, a definition that is too general and vague is not useful as guidance for innovation. Therefore, the following basic sustainability principles (SPs) are used as a science-based definition of success [33]:

"In a sustainable society, nature is not subject to *systematically increasing* . . .

1 . . . concentrations of substances extracted from the Earth's crust;

2 . . . concentrations of substances produced by society;

3 . . . degradation by physical means; and, in that society . . .

4 . . . people are not subject to conditions that systematically undermine their *capacity* to meet their needs."

These science-based, first-order principles essentially work as root causes for social and environmental issues. The social dimension of the SPs is currently under further development [34]. However, the above version of the principles was used as the new social SPs were not published and sufficiently tested at the time of the study. Rather than solving one current problem at a time and thereby risking running into costly dead ends as new unforeseen problems arise, the FSSD makes planning long term strategic by identifying and taking smart steps that lead towards compliance with the SPs [33].

3. Methods

This study used a combination of Strategic Life-Cycle Assessment (SLCA) and traditional Life-Cycle Assessment (LCA) with the aim to add a strategic planning perspective and to focus the LCA on the most important sustainability issues in the life-cycle (see Figure 1).

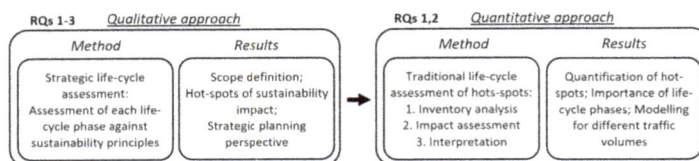

Figure 1. Schematic overview of the applied research methods and the results of each step.

3.1. Strategic Life-Cycle Assessment

Ny et al. [35] presented a method called Strategic Life-Cycle Assessment, which is used to identify a system's or product's violations against the SPs along the life-cycle. By doing so, "hot spots" of environmental and social impact are mapped in a qualitative way. These are further analyzed in a next step through quantitative LCA. That means that the results of the SLCA dictate the scope of the LCA, focusing it on the most important life-cycle stages and sustainability aspects. The SLCA in this study was conducted in multiple group sessions with several researchers and compared diesel-powered heavy trucks with corresponding trucks that are powered by ERS. In the beginning of the sessions, the team investigated the two systems and did a life-cycle mapping to get an overview of the most important processes, materials and energy flows. A template, cross referencing the life-cycle stages with the SPs was then used to systematically map SP violations throughout the life-cycles. Violations against SPs in each life-cycle are displayed in table format, in line with previous studies, combining text and a color scheme, ranging from 'neutral' over 'slightly negative' to 'negative', to indicate the magnitude of the violations [1,4]. In addition, one table describes how the systems would need to look like in order to fit into a sustainable future, or in other words, to comply with the SPs. A comparison between the "as is" and "to be" tables can give insight into whether it is realistically possible to achieve sustainability with these solutions. The most severe violations, that is, hot spots, were verified and partially quantified through LCA.

3.2. Life-Cycle Assessment

The LCA in this study followed the process of the ISO standard 14040 [36] but its scope was guided by the results of the SLCA as described above. The purpose was not to do a detailed and complete inventory and assessment of the systems. LCA was only used to quantify the most important environmental impacts of overhead line ERS powered trucks as compared to diesel trucks to be able to make a strategic assessment in line with the research questions.

The LCA software tool SimaPro 8.2 with the Ecoinvent 3.2 database [37] was used to model and compare the life-cycles. Model and results were verified by an independent LCA consultancy (Miljögiraff KB). The process started with a life-cycle inventory (LCI), in which inflows (e.g., raw materials and energy) and outflows (e.g., emissions) were mapped for all activities in the included life-cycle phases. Also, assumptions for the model were derived based on existing literature, statistics, etc., which is further explained in Section 4.2. The life-cycle impacts of ERS- and diesel-powered truck transport were compared per transported ton kilometer (tkm). The latter thereby represents the functional unit of the study. Next, a life-cycle impact assessment (LCIA) was performed, following the steps of the ISO 14040 standard, to analyze the impact of the components and activities and their related in- and outflows. This phase started with the selection of impact categories. For this study, ReCiPe (H) midpoint and endpoint [38] were used as impact assessment methods, because they link all basic principles for socio-ecological sustainability with LCA, as described by Borén and Ny [4]. According to their work, systematic increase of substances in nature as described by SP1–2 can be linked to effect indicators in the following way: (i) combustion of fossil fuels leads to increasing concentration of CO_2 in the atmosphere, which links to 'Climate Change' and 'Fossil Depletion'; (ii) usage of metals like copper, lead, nickel et cetera contribute to 'Metal Depletion' as long as they are not kept in closed loop systems; (iii) emissions of nitrogen oxides (NOx) from combustion of air and fuels link to 'Acidification', 'Eutrophication', 'Particulate Matter' and 'Photochemical oxidants'; and (iv) toxic and persistent chemicals such as dioxins and persistent organic pollutants contribute to 'Ecotoxicity' and 'Human Toxicity' categories. Systematic degradation of nature by physical means, SP3, for example, by open pit mining and landfills, links to the 'Land Use' categories. SP4, which defines social sustainability, is violated through negative health impacts from, for example, air pollution or radiation and is therefore linked to the categories 'Human Toxicity' and 'Ionizing Radiation'. Furthermore, the dissipate use of scarce resources, including fossil fuels and metals like copper, are a SP4 issue. Based on the ReCiPe method, LCI results were classified, meaning that all flows were assigned to one or multiple impact categories. In the following step, characterization, category indicator results were calculated with the help of characterization factors. To assess the magnitude of impacts and the relative importance of different impact categories, results were normalized by relating them to the yearly impact of an average citizen. Weighting, which is an optional step according to the ISO 14040 standard, was not performed in this study, as it is purely subjective. After the LCIA, the results were interpreted in relation to the research questions.

4. Results

4.1. Strategic Life-Cycle Assessment of ERS and Diesel Truck Transport

The SLCA mapped violations against SPs for truck transport on ERS- and diesel-powered trucks. The results are presented in one section for each SP in table format and the main impacts are discussed in text. Finally, a vision and requirement of a sustainable system is described as a definition of success for the specific case.

4.1.1. Assessment against Sustainability Principle 1

The biosphere and the lithosphere have always been connected to each other: substances from the lithosphere have entered the biosphere through, for example, volcanic eruptions. At the same time, substances from the biosphere have become part of the lithosphere as in the case of fossil fuels. Besides these natural flows, humans have since 200 years or so ago started to influence these flows to a considerable degree. However, this influence has almost entirely been in the form of an increased flow from the lithosphere to the biosphere. Consequences of such an increase and imbalance can have very negative implications both for humans and other living organisms: substances such as lead and cadmium, that normally only occur in very low concentrations in nature, can become a major health and ecosystem threat when their concentration increases systematically in nature as a result of, for example, leakage from mining. Even the increase of carbon dioxide in the atmosphere and its effects on global warming and ocean acidity are an example of a violation of SF1, as coal from the lithosphere is added to the biosphere much faster than it is removed from it [33].

When the case of ERS is assessed against SP1, Table 1, a considerable use of raw materials is identified, especially copper and steel, for catenaries, electric facilities, road barriers, catenary masts, etc. The extraction of these materials, as well as the production, cause spreading of heavy metals and other substances, which leads to increasing concentrations in the biosphere. In the use phase, SP1 is violated because of diffuse copper emissions from catenary friction and emissions from energy production, depending on the electricity mix. For the diesel system, production and combustion of the fuel clearly violates SP1. Common for both systems are diffuse emissions from, for example, road, tire, and break wear.

Table 1. Sustainability Principle 1 (SP1) strategic life-cycle assessment comparing Electric Road Systems (ERS) and diesel-powered freight transport.

Life-Cycle Phase	SP1 Effects of ERS-Powered Trucks	SP1 Effects of Diesel-Powered Trucks
Extraction	Heavy metals in components and processes. Emissions from fossil fuel usage.	Heavy metals in components and processes. Emissions from fossil fuel usage. Oil leakages, gas flaring.
Production	Heavy metals in components and production. Emissions from fossil fuel usage.	Heavy metals in components and production. Emissions from fossil fuel usage.
Distribution	Emissions from truck transports of infrastructure systems, vehicles.	Emissions from truck transports of infrastructure systems, vehicles and fuel.
Use	Copper emissions from catenary wire friction. Heavy metals in maintenance. Emissions from maintenance transport.	Combustion emissions. Heavy metals and fossil oil in maintenance. Emissions from maintenance transport.
Waste	Incomplete recycling of heavy metals and other materials related to SP1. Some cables and other components may be left in the ground and leak heavy metals. Emissions and leakages from recycling processes and landfills.	Incomplete recycling of heavy metals and other materials related to SP1. Emissions and leakages from recycling processes and landfills.

Red: negative sustainability impact; yellow: slightly negative sustainability impact.

4.1.2. Assessment against Sustainability Principle 2

Emissions of persistent chemicals or NO_X are examples of SP2 violations. When ERS is assessed against SP2, Table 2, it is found that the infrastructure includes many electric facilities and cables, which can contain plastic insulation with persistent additives. These chemicals can leak and accumulate in nature, for example when cables are left in the ground even after the ERS' end-of-life. Another main SP2 violation occurs when burning fossil fuels, both in vehicles and for electricity production (depending on the electricity mix), which causes NO_X emissions, contributing to eutrophication and acidification.

Table 2. SP2 strategic life-cycle assessment comparing ERS- and diesel-powered freight transport.

Life-Cycle Phase	SP2 Effects of ERS-Powered Trucks	SP2 Effects of Diesel-Powered Trucks
Extraction	NO_X emissions from combustion.	NO_X emissions from combustion.
Production	NO_X emissions from combustion. POP and Dioxin emissions.	NO_X emissions from combustion. POP and Dioxin emissions.
Distribution	NO_X emissions from truck transports of infrastructure systems and vehicles.	NO_X emissions from truck transports of infrastructure systems and vehicles.
Use	NO_X emissions from truck transports of infrastructure systems and maintenance vehicles. Leakage of persistent chemicals from electric components.	NO_X emissions from truck transports of infrastructure systems and maintenance vehicles. NO_X emissions from the vehicle's engine. Leakage of persistent chemicals from electric components.
Waste	Incomplete recycling of compounds related to SP2. Emissions and leakages from recycling processes and landfills.	Incomplete recycling of compounds related to SP2. Emissions and leakages from recycling processes and landfills.

Red: negative sustainability impact; yellow: slightly negative sustainability impact.

4.1.3. Assessment against Sustainability Principle 3

According to SP3, nature must not be degraded by physical means. In contrast to SP1–2, SP3 is not about systematically increasing concentrations but about destruction of nature through land use or mismanagement of ecosystems. Open pit mining and oil extraction are the most relevant violations of SP3 in the ERS case, Table 3, because they occupy and degrade large surface areas through leakages of hazardous substances, risking destruction of soil and water resources. Such areas are often not usable, neither by humans, animals nor plants. SP3 is here also violated by building infrastructure for electricity and by fuel production and distribution, because they hinder the use of productive surfaces and might contribute to deforestation and fragmentation. Finally, as some components are not recycled, they contribute to increasing landfill space.

Table 3. SP3 strategic life-cycle assessment comparing ERS- and diesel-powered freight transport.

Life-Cycle Phase	SP3 Effects of ERS-Powered Trucks	SP3 Effects of Diesel-Powered Trucks
Extraction	Open pit mining of metals.	Open pit mining of metals and other resources. Oil extraction.
Production		Contamination at refineries.
Distribution	Land use for roads and power grids.	Land use for roads and pipelines.
Use	Land use for roads.	Land use for roads.
Waste	Non-recycled materials to landfills.	Non-recycled materials to landfills.

Red: negative sustainability impact; yellow: slightly negative sustainability impact; blue: neutral.

4.1.4. Assessment against Sustainability Principle 4

While the first three principles focus on ecological sustainability, SP4 is about meeting human needs, which also is a requirement for a sustainable future. The assessment against SP4, Table 4, emphasizes the mining and fossil fuel industries that are plagued by conflicts, whose effects prevent people from meeting their needs, for example because they get wounded or are forced to flee their homes. These industries also cause ecosystem degradation by physical and chemical means, which undermines not only people's health but also their possibilities to make a living with agriculture, fishing or livestock farming. The same is true for the production phase and metal recovery in some countries that expose people and nature to hazardous emissions [39]. Another violation of SP4 may occur when scarce resources are extracted and used in a way that limits their availability for future generations. This is the case for fossil fuels and metals like copper [40,41] and platinum [42].

Table 4. SP4 strategic life-cycle assessment comparing ERS- and diesel-powered freight transport.

Life-Cycle Phase	SP4 effects of ERS-Powered Trucks	SP4 Effects of Diesel-Powered Trucks
Extraction	Use of scarce resources such as copper. Open pit mining causes negative health effects and forces people to move.	Use of scarce resources such as platinum. Open pit mining causes negative health effects and forces people to move.
Production	Negative health effects from emissions related to fossil fuel use and component production. Harmful job conditions at some places.	Negative health effects from emissions related to fossil fuel use and component production. Harmful job conditions at some places.
Distribution	Health effects from transport emissions.	Health effects from transport emissions.
Use	Health risks due to high voltage and overhead line accidents.	Negative health effects from emissions related to fossil fuel use.
Waste	Harmful emissions and working conditions in some countries.	Harmful emissions and working conditions in some countries.

Red: negative sustainability impact; yellow: slightly negative sustainability impact.

The introduction of ERS might also create positive effects like new jobs in all life-cycle phases. On the other hand, they might just replace jobs from other infrastructure systems, which leaves the net effect on job creation uncertain. However, an increased demand for renewable energy and a transition of the energy sector is expected to lead to a net increase of jobs in Europe [3,43,44].

4.1.5. Requirements for Sustainable Freight Transport Systems

When investments in new infrastructure are planned with the goal to make transportation more sustainable, it is important to investigate what would be required for the systems to be fully sustainable. Otherwise there is a risk of sub-optimization and a risk that society once again gets locked into an unsustainable system for half a century or longer. Especially in the case of ERS, which have a long life-time, one has to consider the question of whether the concept will provide a stepping stone towards a sustainable society and compliance with the SPs. The results of the SLCA revealed that the most severe sustainability challenges for ERS are (i) the extensive use of raw materials, including scarce metals like copper; (ii) diffuse emissions in the use phase, especially copper from catenary friction; (iii) environmental impact of the electricity used in the system; and (iv) use of fossil fuels for processes and activities throughout the life-cycle.

Table 5 shows the most important aspects of how the life-cycle would need to look like in order to not violate any SP. Of central importance for reaching compliance with the SPs are the strict application of best available technology (BAT) and precautionary and substitution principles in all life-cycle phases. As Robèrt et al. [33] emphasize, SP1 does not forbid all use of metals. Rather, they should be handled in closed loops so that the concentration in nature does not increase and the available resources are not depleted. However, it would be more favorable to design a system that is completely independent of rare or toxic materials and substances. Fossil oil should be avoided completely. Good, safe and just working conditions have to be guaranteed in all life-cycle phases. When these requirements are met, a sustainable transport solution can be a valuable, long-term satisfier of human transportation needs.

Table 5. Life-cycle requirements on ERS, viewed through the lens of the four sustainability principles.

Life-Cycle Phase	Requirements for Sustainable ERS
Extraction	Very limited extraction of new resources. Extraction with best available technology (BAT). Complete restauration of the site after operation. Respecting indigenous people's rights.
Production	Strict application of BAT, precautionary and substitution principle. Rare substances with high accumulation potential are kept in closed loops.
Distribution	Only using sustainable modes of transport powered by renewable energy.
Use	Ensuring a safe, comfortable and effective satisfaction of people's need for transportation *. Only renewable energy input and no emissions of critical substances from use and maintenance.
Waste	Optimized for following EU's waste hierarchy [45]: prevention, reuse, recycle, recovery; except there should be no need for landfilling. Circular material flow.

* A contribution to sustainable development. blue: neutral.

4.2. Life-Cycle Assessment of ERS and Current Fossil-Powered Truck Transport

As a result of the SLCA, the main hot spots identified were raw material extraction, production and use phases, which therefore were selected for quantification with LCA. Included processes and components are: (i) raw material extraction for roads, lorries, diesel, and road electrification; (ii) processes for turning the raw material into products; (iii) combustion of diesel, emissions from electricity generation, catenary friction, road, break, and tire wear emissions, and lorry maintenance. The other life-cycle phases were excluded from the LCA. The inventory analysis of the infrastructure utilized data from 1000 V railway systems, which use very similar components to overhead line ERS. This data was further adjusted with the help of a railway and ERS expert. For the components where there was no suitable data in Ecoinvent, simplified components were modelled (see Table 6).

Table 6. Simplified modelling of ERS components, using Ecoinvent data.

Component	Material Use per km	Technical Life Time, Years	Reference
Overhead lines (double)	Copper: 4800 kg	40	Swedish Transport Administration [46], Stripple and Uppenberg [47]
Catenary masts	Steel: 7752 kg PE: 122 kg Fiberglass: 152 kg	50	Swedish Transport Administration [46], Stripple and Uppenberg [47]
Semi-rigid roadside barriers	Concrete: 5500 kg Steel: 16,000 kg	20	Swedish Transport Administration [46], Stripple and Uppenberg [47]
Electrical equipment, protective relay, fuses	Steel: 57 kg	40	Uppenberg [48]
Cables, AXQJ 1 kV 3×70/21	Aluminium: 363 kg Copper: 121 kg PE: 216 kg	40	Uppenberg [48]
Distribution sheds	Steel, low-alloyed: 145 kg	50	Uppenberg [48]
Transformer 1/0,4 kV, 100 kVA	Steel: 3 kg Copper: 1 kg	40	Uppenberg [48]
Transformer 1/0,4 kV, 50 kVA	Steel: 89 kg Copper: 38 kg	40	Uppenberg [48]
Transformer 1/0,4 kV, 30 kVA	Steel: 3 kg Copper: 1 kg	40	Uppenberg [48]
Transformer 1/0,4 kV, 16 kVA	Steel: 3 kg Copper: 1 kg	40	Uppenberg [48]
Transformer 1/0,4 kV, 5 kVA	Steel: 3 kg Copper: 1 kg	40	Uppenberg [48]

Market processes and system model "allocation, cut-off by classification" were used for all materials and processes [49]. Europe was chosen as the geographical reference point. The value chain can, however, include materials and processes from other parts of the world. Four scenarios for electricity generation were applied, all based on Ecoinvent data: European mix, Nordic mix, wind-generated electricity, and a worst-case scenario that assumed coal-generated marginal electricity. The type of electricity for vehicle propulsion in the use phase is the only difference between the scenarios, thereby showing the sensitivity of the model for this key factor. Truck traffic volume was initially set to 1000 vehicles per direction and day, which corresponds to a major Swedish highway [30]. Trucks had a gross vehicle weight of 16–32 ton, with an average load factor of 5.79 ton [50]. Diesel trucks met Euro 6 emission limits. Diesel consumption was 0.037 L/tkm, which corresponds to 0.21 L/km (Ecoinvent 3.2 database). Electricity consumption was set to 0.17 kWh/tkm, which corresponds to the same amount of energy consumption as the diesel truck when calculating with diesel engine efficiency of 42%, electric engine efficiency of 95%, and electricity losses in the ERS of five percent. Catenary friction was estimated as 10 kg of copper per km and year, which is the same amount as for railway systems [51]. Without full scale tests it is, however, yet uncertain whether this data is fully valid for ERS due to differences in traffic intensity, speed, and movement of the pantograph-shaped pick-up.

Comparing characterization results of ERS- and diesel-powered trucks, Figure 2, revealed: (i) wind-powered ERS have lower environmental impact than the diesel system in 11 out of 18 impact categories; (ii) ERS powered by European electricity mix or marginal electricity have higher environmental impact in 12 and 13 out of 18 categories respectively, as compared to the current diesel system. That number is 10 for the Nordic electricity mix scenario; (iii) there are substantial differences in GHG emissions: ERS that use coal-based marginal electricity cause the highest emissions (229 g/tkm), followed by diesel (165 g/tkm), EU mix electricity (117 g/tkm), Nordic mix (41 g/tkm) and wind-generated electricity (31 g/tkm); and (iv) a closer look at ERS infrastructure reveals that most environmental impacts are tied to the three components of copper catenaries, catenary masts and road barriers. Other parts, such as converters or cables, play a minor role.

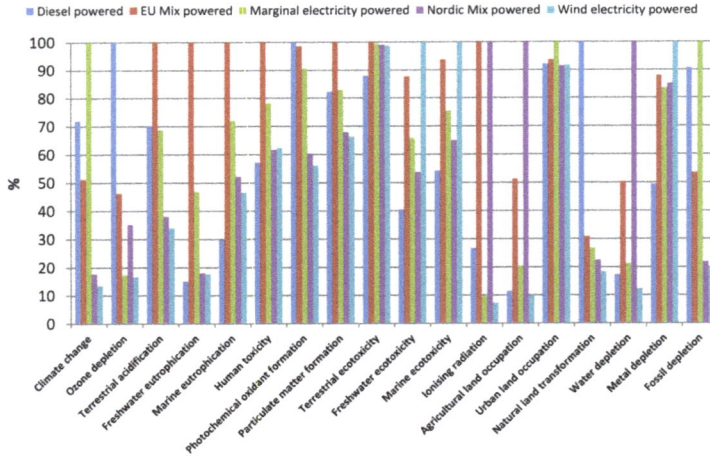

Figure 2. Characterization results with ReCiPe midpoint showing life-cycle environmental impact for diesel-powered freight transport and ERS. Whether ERS have lower life-cycle environmental impact is strongly dependent on how electricity is produced.

Normalization of the results shows that freshwater eutrophication, human toxicity, eco-toxicity and natural land transformation are the most relevant impact categories for the investigated systems. ERS, no matter how electricity is generated, have higher impact in the toxicity categories, which is largely due to the extensive use of copper and emissions from catenary friction. The fossil-powered system has a much higher impact on natural land transformation, mainly due to petroleum production.

A closer look at the role of different life-cycle phases for the climate change category, Figure 3, reveals that most emissions occur in the use phase if trucks are powered by diesel, marginal electricity or EU mix electricity. If electricity is produced in a less carbon intensive way, as is the case with Nordic mix and wind-power, extraction to distribution (E–D) phases constitute the main source of CO_2 equivalent (CO_2e) emissions. It is, however, not the ERS infrastructure that causes high emissions in E–D phases. Instead, road and lorry production cause about 20 g CO_2e emissions per tkm, which can be compared with 1.5 g/tkm for electrification of the road with ERS.

The amount of tkm transported on ERS is a central parameter: the more goods that are transported on an ERS, the less is the share of building the infrastructure, E–D phases, for each tkm. The number of tkm is in turn dependent on the number of trucks and their average load factor. In order to evaluate if electrification of a specific road is favorable from a sustainability perspective, it is necessary to adjust the assumptions of the LCA model. Figure 4 clarifies the dependence of life-cycle environmental impact and the number of tkm transported on ERS. It shows the environmental break-even times for the different scenarios, that is, the time until the environmental impact of the road electrification is compensated by lower impact in the use phase. ReCiPe endpoint was used to get one accumulated, comparable number for environmental impacts. The marginal electricity scenario is not displayed because it has higher life-cycle environmental impact per tkm than the diesel system, hence there is no break-even.

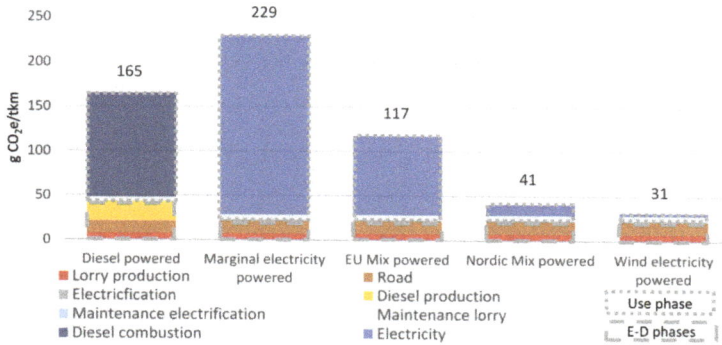

Figure 3. Contribution of extraction to distribution and use life-cycle stages to total climate change impact of diesel- and ERS-powered freight transport.

Generally, break-even times are short if Nordic mix or wind-power generated electricity are used: assuming 1000 trucks per day, break-even is only three to four years and with 500 trucks per day that time is below 10 years. The situation is different, though, if EU mix electricity is assumed: a minimum of about 700 trucks per day is required to achieve a break-even time of 10 years, and with 1000 heavy vehicles per day, that time is seven to eight years. A break-even time shorter than five years is only possible on roads with high traffic flow of more than 1400 trucks per direction and day. From a purely GHG emission perspective, break-even times are considerably (about 70%) shorter when assuming the same amount of freight transport per day. Climate change is, however, only one of many impacts that have to be considered.

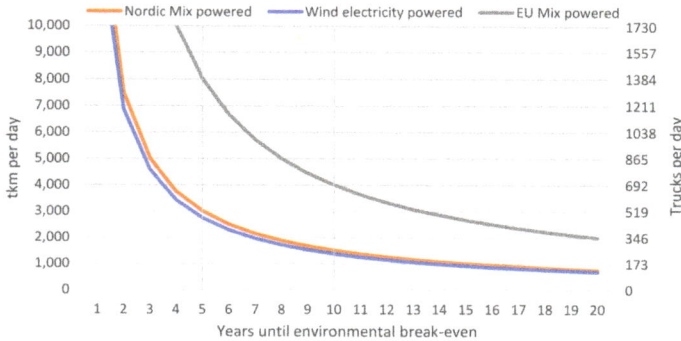

Figure 4. Environmental break-even times for ERS in relation to electric freight traffic amounts, that is, the time it takes until impact (ReCiPe endpoint) from building ERS is offset by lower emissions in the use phase, as compared to diesel-powered freight transport.

5. Concluding Discussion

5.1. What Is the Sustainability Impact of ERS in Comparison to the Current Fossil-Powered System?

The SLCA showed that both ERS and diesel freight transport have some severe sustainability impacts and violations against the SPs, especially in raw material extraction, production and use phases. For ERS, these are mostly due to usage of large amounts of copper and other raw materials, as well as impacts from electricity production and diffuse emissions in the use phase. For diesel transport, the value chain of oil and fuel combustion are main causes of the violations. Possibilities to reduce these SP violations were identified and are mostly about the application of 'best available technology', 'substitution' and 'precautionary' principles throughout the life-cycle as well as closing the loop of material flows and using sustainably harvested, renewable energy.

The more detailed LCA showed that GHG emissions can be decreased significantly as long as electricity generation is not coal-based. GHG payback times are five years for roads with more than 400 lorries per day and below two years if there are at least 1000 lorries per day. However, based on normalization results, the most relevant environmental impact categories are eco-toxicity, human toxicity, eutrophication and, especially for the diesel system, natural land transformation. These findings underline the importance of widening the view on sustainability beyond climate change and GHG emissions. In total, endpoint results show that transport on coal electricity powered ERS causes higher environmental impact than driving on diesel, while the impact is lower if EU mix, Nordic mix or wind-generated electricity is used. This emphasizes the fact that the sustainability performance of ERS is highly dependent on how electricity for the use phase is produced. Therefore, a transition of the transport sector to be powered by electricity has to be simultaneously accompanied by a transition of the energy system to 100% flow-based, renewable energy. As intrinsic to LCA, results are dependent on model assumptions, especially concerning the share of electrified roads, traffic volume and load factor, which is related to average lorry weight. Therefore, these parameters have to be adjusted in order to evaluate the environmental impact for specific cases.

5.2. What Is the Relative Importance of Different Life-Cycle Phases?

Through SLCA, the raw material extraction, production and use phases were identified as hot spots of sustainability impact. In general, most environmental impacts occur in the use phase, both for diesel and ERS systems, even if wind-generated electricity is used. This dominance of the use phase is a common result of LCAs for transport systems and many other products [1,4,47]. However, if the electricity for ERS is produced in a sustainable way, the impact of the infrastructure itself becomes more prominent, which was also apparent in Figure 3. The SLCA showed that ERS infrastructure still has major sustainability challenges, especially due to copper and steel use for catenaries, masts and road barriers. As these components constitute relatively pure fractions, there might be good possibilities for a high degree of material recycling. In this regard, diffuse emissions in the use phase, like copper from catenary friction, are more difficult to prevent and control. It is therefore important to include infrastructure in LCAs of transport systems. Sustainability challenges of the current diesel system are mainly characterized by the petroleum value chain, which includes extensive land use, and diesel combustion with a variety of emissions.

5.3. Is the Introduction of ERS a Strategic Stepping Stone on the Way towards Sustainable Transport and What Role Could They Play in That Transition?

5.3.1. Lock-In and Threshold Effects

The life-cycle inventory revealed that many ERS components have a long life-time. Even though that has some sustainability advantages, it also means that ERS as a transport solution has a long life-time. That fact in combination with high initial investment costs means that there might be a strong lock-in effect: once built and invested in, ERS would have to be used for a long time, presumably many decades, in order to be environmentally and economically viable, depending on how the infrastructure is financed. Thus, ERS is not a flexible solution as compared to some other infrastructure solutions, such as fast charging networks that relatively easily and quickly can be expanded, decreased or even phased out. On the other hand, this study indicated that environmental pay-off times are relatively short, if electricity is produced in a sustainable way and the freight transport volume is sufficiently high. A challenge that ERS share with several other systems, for example fast charging, battery swapping, and fuel cell stations, is the threshold effect: as long as there is no substantial network of infrastructure, it is not attractive for haulage contractors to invest in more expensive vehicles that are adjusted for using ERS. Today, it is largely unclear how strong such a threshold effect is for ERS. It depends largely on the share of electrified roads, business models, and to what degree a specific lorry travels on a fixed or flexible and constantly changing route.

5.3.2. Stepping Stones towards Sustainability

As a comparison of the future requirements for a sustainable infrastructure (Table 5) and the assessment of today's state (Tables 1–4) showed, there is a considerable discrepancy between 'as is' and 'to be'. That in itself is not a reason to dissuade from investments in ERS, as all alternatives, for example fuel cell or battery-powered systems, today have some violations against the SPs [4]. In addition, some violations might be possible to solve through smart design and technology development. However, the most important question to ask is which infrastructure system is the smartest "stepping stone" [32] on the way to a sustainable future. This means that it is not only important to find today's most sustainable solution, but also to find the technology platform with the largest potential to lead us on the right track and closer to reach full socio-ecological sustainability over time. From this perspective, ERS could prove to be a valuable part of the puzzle, mostly because they drastically decrease the need for large batteries, which results in lower vehicle cost and sustainability impact. Thereby, ERS could catalyze electrification and the transition towards sustainable freight transport. ERS are, however, at an early stage of development [16]. In the short term, the application of ERS for closed systems, for example transport of ore between a mine and a production facility or a harbor, is important for testing and further development of the concept. Closed systems have much lower complexity as fewer stakeholders are involved, the need for standardization is limited, and both economic cost–benefit and wider sustainability assessments are considerably simpler [15]. The application in closed systems would, however, only solve a small fraction of the freight transports' sustainability challenges and the total GHG reduction potential is rather low. Still, it is a valuable way to gather experience and knowledge on ERS before heavily investing in road electrification to overcome the threshold effect for ordinary freight transport.

5.4. Recommendations and Future Research

Based on the findings of this study, the following recommendations are made to accelerate the transition of road freight transport towards sustainability. Firstly, the ERS concept should be further explored and developed, specifically regarding business models for open systems and the dynamics between the amount of electrified road and the rationale for actors to invest in ERS-compatible vehicles. Secondly, test applications and demonstrators play a key role in the early phases of technology development but are in many cases dependent on public funding. Incentives for public funding could be strengthened by detailing the societal benefits of ERS, including savings in relation to the externalities of the current system. Thirdly, political will and a clear strategy are required to reduce uncertainty for private companies, making it more attractive to invest in ERS. Finally, and most importantly, strategic leadership is needed to guide the development of freight transport towards sustainability, without sub-optimizations in the transport sector and solutions in the transport sector that block sustainable solutions in other sectors. For this purpose, Robèrt et al. [52] and Borén et al. [53] presented an FSSD-based process model for cross-sector and cross-disciplinary cooperation, ensuring cohesive creativity across sectors and groups of experts as well as stakeholders. In the end, ERS also need to be compared with alternative technologies like battery electric trucks and fuel cell trucks. It is likely that a combination of technologies will exist in the future to fit different needs and contexts, even if some technology might dominate. Exactly which role ERS will play is yet to be seen and also dependent on progress with other technologies. For instance, the possibilities for a sustainable scale-up of battery electric trucks is largely dependent on breakthroughs in battery design. For fuel cell solutions, increased system efficiency and a substitution or limited use of platinum would be necessary. In any case, with the applying of a strategic perspective based on backcasting from a vision of full sustainability, policy and decision-makers can ensure that actions lead step-wise towards a sustainable society.

Acknowledgments: Sincere thanks to our colleague Sven Borén for valuable discussions on SLCA and LCA methodology and results. Also, Pär Lindman at the Swedish SimaPro dealer Miljögiraff KB contributed to the study by supporting and verifying the LCA model.

Author Contributions: Jesko Schulte collected the data, developed the LCA model, performed most of the analysis and led the writing process. Henrik Ny headed the research group, initiated the study, supported the life-cycle assessments, participated in the analysis, and revised the manuscript.

Conflicts of Interest: The authors declare no conflict of interest. The founding sponsors had no role in the design of the study; in the collection, analyses, or interpretation of data; in the writing of the manuscript, and in the decision to publish the results.

References

1. Nurhadi, L.; Borén, S.; Ny, H. Advancing from Efficiency to Sustainability in Swedish Medium-sized Cities: An Approach for Recommending Powertrains and Energy Carriers for Public Bus Transport Systems. *Procedia Soc. Behav. Sci.* **2014**, *111*, 586–595. [CrossRef]
2. European Commission. *EU Transport in Figures: Statistical Pocketbook 2015*; European Commission: Luxembourg, Luxembourg, 2015.
3. Teske, S.; Sawyer, S.; Schäfer, O.; Pregger, T.; Simon, S.; Naegler, T. *Energy [R]evolution—A Sustainable World Energy Outlook 2015*; Greenpeace International, Global Wind Energy Council, Solar Power Europe: Brussels, Belgium, 2015.
4. Borén, S.; Ny, H. A Strategic Sustainability and Life Cycle Analysis of Electric Vehicles in EU today and by 2050. In Proceedings of the 18th International Conference on Sustainable Urban Transport and Environment (ICSUTE), Madrid, Spain, 24–25 March 2016; Volume 10, pp. 256–264.
5. European Commission. *A Strategy for Reducing Heavy Duty Vehicles' Fuel Consumption and CO_2 Emissions*; European Commission: Brussels, Belgium, 2014.
6. ICCT. *European Vehicle Market Statistics—Pocketbook 2014*; ICCT: Berlin, Germany, 2014.
7. Notter, D.A.; Gauch, M.; Widmer, R.; Wäger, P.; Stamp, A.; Zah, R.; Althaus, H.-J. Contribution of Li-ion batteries to the environmental impact of electric vehicles. *Environ. Sci. Technol.* **2010**, *44*, 6550–6556. [CrossRef] [PubMed]
8. Hawkins, T.R.; Singh, B.; Majeau-Bettez, G.; Strømman, A.H. Comparative Environmental Life Cycle Assessment of Conventional and Electric Vehicles. *J. Ind. Ecol.* **2013**, *17*, 53–64. [CrossRef]
9. Sverdrup, H.U.; Ragnarsdottir, K.V.; Koca, D. An assessment of metal supply sustainability as an input to policy: Security of supply extraction rates, stocks-in-use, recycling, and risk of scarcity. *J. Clean. Prod.* **2017**, *140*, 359–372. [CrossRef]
10. Kushnir, D.; Sandén, B.A. The time dimension and lithium resource constraints for electric vehicles. *Resour. Policy* **2012**, *37*, 93–103. [CrossRef]
11. Connolly, D. Economic viability of electric roads compared to oil and batteries for all forms of road transport. *Energy Strateg. Rev.* **2017**, *18*, 235–249. [CrossRef]
12. Tongur, S.; Engwall, M. The business model dilemma of technology shifts. *Technovation* **2014**, *34*, 525–535. [CrossRef]
13. Chen, F.; Taylor, N.; Kringos, N. Electrification of roads: Opportunities and challenges. *Appl. Energy* **2015**, *150*, 109–119. [CrossRef]
14. Lennartsson, M. Elektriska Vägar, Miljöanalys (Electric Roads, Environmental Analysis). Available online: http://elvag.se/en/archive/2010-04-16/Elektriska-vagar-Miljoanalys.pdf (accessed on 7 July 2016).
15. Tongur, S.; Sundelin, H. The electric road system transition from a system to a system-of-systems. In Proceedings of the Asian Conference on Energy, Power and Transportation Electrification (ACEPT), Singapore, Singapore, 25–27 October 2016; pp. 1–8.
16. Sundelin, H.; Gustavsson, M.G.H.; Tongur, S. The maturity of electric road systems. In Proceedings of the International Conference on Electrical Systems for Aircraft, Railway, Ship Propulsion and Road Vehicles & International Transportation Electrification Conference (ESARS-ITEC), Toulouse, France, 2–4 November 2016.

17. Choi, S.Y.; Member, S.; Gu, B.W.; Member, S.; Jeong, S.Y.; Member, S.; Rim, C.T.; Member, S. Advances in Wireless Power Transfer Systems for Roadway-powered Electric Vehicles. *IEEE J. Emerg. Sel. Top. Power Electron.* **2015**, *3*, 18–36. [CrossRef]

18. Suh, N.P.; Cho, D.H.; Rim, C.T. Design of On-Line Electric Vehicle (OLEV). In *Global Product Development*; Springer: Berlin/Heidelberg, Germany, 2011; pp. 3–8.

19. Stamati, T.; Bauer, P. On-road charging of electric vehicles. In Proceedings of the 2013 IEEE Transportation Electrification Conference and Expo (ITEC), Detroit, MI, USA, 16–19 June 2013; pp. 1–8.

20. Sallan, J.; Villa, J.L.; Llombart, A.; Sanz, J.F. Optimal Design of ICPT Systems Applied to Electric Vehicle Battery Charge. *IEEE Trans. Ind. Electron.* **2009**, *56*, 2140–2149. [CrossRef]

21. Covic, G.A.; Boys, J.T. Modern Trends in Inductive Power Transfer for Transportation Applications. *IEEE J. Emerg. Sel. Top. Power Electron.* **2013**, *1*, 28–41. [CrossRef]

22. Gill, J.S.; Bhavsar, P.; Chowdhury, M.; Johnson, J.; Taiber, J.; Fries, R. Infrastructure cost issues related to inductively coupled power transfer for electric vehicles. *Procedia Comput. Sci.* **2014**, *32*, 545–552. [CrossRef]

23. FABRIC Test Sites. Available online: http://www.fabric-project.eu/index.php?option=com_k2&view=itemlist&layout=category&task=category&id=24&Itemid=214 (accessed on 13 February 2018).

24. Viktoria Swedish ICT. *Slide-in Electric Road System—Inductive Project Report*; Viktoria Swedish ICT: Gothenburg, Sweden, 2014.

25. Shin, J.; Shin, S.; Kim, Y.; Ahn, S.; Lee, S.; Jung, G.; Jeon, S.-J.; Cho, D.H. Design and Implementation of Shaped Magnetic Resonance Based Wireless Power Transfer System for Roadway-Powered Moving Electric Vehicles. *IEEE Trans. Ind. Electron.* **2014**, *61*, 1179–1192. [CrossRef]

26. International Commission on Non-Ionizing Radiation Protection. Guidelines for Limiting Exposure to Time-Varying Electric and Magnetic Fields (1 Hz to 100 kHz). *Health Phys.* **2010**, *99*, 818–836.

27. Council of Europe. *Resolution 1815—The Potential Dangers of Electromagnetic Fields and Their Effect on the Environment*; Council of Europe: Strasbourg, France, 2011; Volume 1815.

28. Covic, G.A.; Boys, J.T. Inductive power transfer. *Proc. IEEE* **2013**, *101*, 1276–1289. [CrossRef]

29. Emre, M.; Vermaat, P.; Naberezhnykh, D.; Damousuis, Y.; Theodoropoulos, T.; Cirimele, V.; Doni, A. FABRIC—Review of Existing Power Transfer. Available online: https://www.fabric-project.eu/images/Deliverables/FABRIC_D33.1_V1_20141215_Review_of_existing_solutions_PUBLIC.pdf (accessed on 7 July 2016).

30. Viktoria Swedish ICT. *Slide-in Electric Road System—Conductive Project Report*; Viktoria Swedish ICT: Gothenburg, Sweden, 2013.

31. Swedish Transport Agency First Electric Road in Sweden Inaugurated. Available online: http://www.trafikverket.se/en/startpage/about-us/news/2016/2016-06/first-electric-road-in-sweden-inaugurated/ (accessed on 13 February 2018).

32. Broman, G.I.; Robèrt, K.-H. A Framework for Strategic Sustainable Development. *J. Clean. Prod.* **2017**, *140*, 17–31. [CrossRef]

33. Robèrt, K.H.; Broman, G.; Waldron, D.; Ny, H.; Byggeth, S.; Cook, D.; Johansson, L.; Oldmark, J.; Basile, G.; Haraldsson, H.; et al. *Strategic Leadership towards Sustainability*, 6th ed.; Blekinge Institute of Technology: Karlskrona, Sweden, 2010.

34. Missimer, M. *Social Sustainability within the Framework for Strategic Sustainable Development*; Blekinge Institute of Technology: Karlskrona, Sweden, 2015.

35. Ny, H.; MacDonald, J.P.; Broman, G.; Yamamoto, R.; Robert, K.-H. Sustainability Constraints as System Boundaries: An Approach to Making Life-Cycle Management Strategic. *J. Ind. Ecol.* **2006**, *10*, 61–77. [CrossRef]

36. ISO. *ISO 14040:2006: Environmental Management, Life Cycle Assessment—Principles and Framework*; ISO: Geneva, Switzerland, 2006.

37. Wernet, G.; Bauer, C.; Steubing, B.; Reinhard, J.; Moreno-Ruiz, E.; Weidema, B. The ecoinvent database version 3 (part I): Overview and methodology. *Int. J. Life Cycle Assess.* **2016**, *21*, 1218–1230. [CrossRef]

38. Goedkoop, M.; Heijungs, R.; Huijbregts, M.; De Schryver, A.; Struijs, J.; van Zelm, R. *ReCiPe 2008: A life Cycle Impact Assessment Method Which Comprises Harmonised Category Indicators at the Midpoint and the Endpoint Level*; Ministry of Housing, Spatial Planning and Environment (VROM): Den Haag, The Netherlands, 2013.

39. Tsydenova, O.; Bengtsson, M. Chemical hazards associated with treatment of waste electrical and electronic equipment. *Waste Manag.* **2011**, *31*, 45–58. [CrossRef] [PubMed]

40. Northey, S.; Mohr, S.; Mudd, G.M.; Weng, Z.; Giurco, D. Modelling future copper ore grade decline based on a detailed assessment of copper resources and mining. *Resour. Conserv. Recycl.* **2014**, *83*, 190–201. [CrossRef]
41. Harmsen, J.H.M.; Roes, A.L.; Patel, M.K. The impact of copper scarcity on the efficiency of 2050 global renewable energy scenarios. *Energy* **2013**, *50*, 62–73. [CrossRef]
42. Elshkaki, A. An analysis of future platinum resources, emissions and waste streams using a system dynamic model of its intentional and non-intentional flows and stocks. *Resour. Policy* **2013**, *38*, 241–251. [CrossRef]
43. Blyth, W.; Gross, R.; Speirs, J.; Sorrell, S.; Nicholls, J.; Dorgan, A.; Hughes, N. *Low Carbon Jobs: The Evidence for Net Job Creation from Policy Support for Energy Efficiency and Renewable Energy*; UKERC: London, UK, 2014.
44. Cambridge Econometrics. *Employment Effects of Selected Scenarios from the Energy Roadmap 2050 Final Report for the European Commission (DG Energy)*; Cambridge Econometrics: Cambridge, UK, 2013.
45. European Commisssion Directive 2008/98/EC of the European Parliament and of the Council of 19 November 2008 on Waste and Repealing Certain Directives. 2008. Available online: http://eur-lex. europa.eu/legal-content/EN/TXT/?uri=CELEX:32008L0098 (accessed on 10 April 2018).
46. Swedish Transport Administration. *Swedish Transport Administration Climate Calculation—The Swedish Transport Administration's Model for Calculating the Climate Change Impact and Energy Use of Infrastructure from a Life Cycle Perspective*; Version: 3.0; Swedish Transport Administration: Borlänge, Sweden, 2015.
47. Stripple, H.; Uppenberg, S. *Life Cycle Assessment of Railways and Rail Transports*; IVL Report B1943; Swedish Transport Administration: Luleå, Sweden, 2010.
48. Uppenberg, S. *Comparative Life Cycle Assessment for Alternative Design of Electrification for Rail Systems [in Swedish: Jämförande Livscykelanalys för Alternativ Utformning av Hjälpkraft för Spåranläggning]*; Swedish Transport Administration: Luleå, Sweden, 2012.
49. Ecoinvent Allocation Cut-off by Classification. Available online: http://www.ecoinvent.org/database/ system-models-in-ecoinvent-3/cut-off-system-model/allocation-cut-off-by-classification.html (accessed on 7 July 2016).
50. Transport and Mobility Leuven TREMOVE Economic Transport and Emissions Model v.2.7b. Available online: http://www.tmleuven.be/methode/tremove/home.htm (accessed on 7 July 2016).
51. Gustafsson, M.; Blomqvist, G.; Håkansson, K.; Lindeberg, J.; Nilsson-Påledal, S. *Railway Pollution—Sources, Fate and Actions. [In Swedish: Järnvägens Föroreningar- källor, Spridning och Åtgärder]*; VTI: Linköping, Sweden, 2007.
52. Robèrt, K.H.; Borén, S.; Ny, H.; Broman, G. A strategic approach to sustainable transport system development—Part 1: Attempting a generic community planning process model. *J. Clean. Prod.* **2017**, *140*, 53–61. [CrossRef]
53. Borén, S.; Nurhadi, L.; Ny, H.; Robèrt, K.H.; Broman, G.; Trygg, L. A strategic approach to sustainable transport system development—Part 2: The case of a vision for electric vehicle systems in southeast Sweden. *J. Clean. Prod.* **2017**, *140*, 62–71. [CrossRef]

sustainability

MDPI

Article

Applying a Mesoscopic Transport Model to Analyse the Effects of Urban Freight Regulatory Measures on Transport Emissions—An Assessment

Jacek Oskarbski and Daniel Kaszubowski *

Faculty of Civil and Environmental Engineering, Gdansk University of Technology, 80-233 Gdańsk, Poland; jacek.oskarbski@pg.edu.pl
* Correspondence: daniel.kaszubowski@pg.edu.pl; Tel.: +48-692-478-220

Received: 9 May 2018; Accepted: 16 July 2018; Published: 18 July 2018

Abstract: Sustainable urban freight management is a growing challenge for local authorities due to social pressures and increasingly more stringent environmental protection requirements. Freight and its adverse impacts, which include emissions and noise, considerably influence the urban environment. This calls for a reliable assessment of what can be done to improve urban freight and meet stakeholders' requirements. While changes in a transport system can be simulated using models, urban freight models are quite rare compared to the tools available for analysing private and public transport. Therefore, this article looks at ways to extend Gdynia's existing mesoscopic transport model by adding data from delivery surveys and examines the city's capacity for reducing CO_2 emissions through the designation of dedicated delivery places. The results suggest that extending the existing model by including freight-specific data can be justified when basic regulatory measures are to be used to improve freight transport. There are, however, serious limitations when an exact representation of the urban supply chain structure is needed, an element which is required for modelling advanced measures.

Keywords: urban freight management; traffic modelling; dedicated delivery places; transport emissions

---

1. Introduction

Sustainable urban freight management is a serious challenge for decision-makers due to the increasingly more stringent environmental protection requirements. It involves being able to select the right solutions to reduce the adverse impact of freight activity on the environment and, most of all, to having a reliable quantitative evaluation of the results of the applied measures. This requires an analysis of how urban supply chains operate as a separate component of a city's transport system and the ability to verify their impact on this system in accordance with the adopted criteria. Despite the considerable interest in sustainable urban mobility, there are issues with selecting solutions that improve freight transport from the point of view of its environmental parameters [1,2]—particularly in the development of analytical methods and in multiple practical measures.

For cities, there are many challenges resulting from the complex definition of a sustainable urban freight system. Also, they must ensure access to all types of freight transport while at the same time reducing emissions of air pollutants and noise and maintaining the economic efficiency of this type of business [3]. It is difficult to take all of these assumptions into account because of the system's entity-related complexity [4], which is manifest in the simultaneous presence of many groups of participants and in the many ways transport activity is organised. In an urban environment, this results in a complex structure of relationships between the entities involved [5]. For example, in the public space, which is defined as an area of interest for city authorities, inhabitants and businesses, there are environmental issues associated with transport emissions. These complexities are also visible in the relationships between the city and the transport sector, often expressed in, for example, new environmental standards introduced for freight vehicles.

Despite the use of different analytical methods, it is clear that emission levels in freight transport are proportionately higher than its share of total transport activity in cities. It is estimated that, in Paris for example, freight emissions constitute 26% of CO_2 emissions with regards to vehicle-km [6] and from 15% to 26% of CO_2 emissions in Bordeaux, Dijon and Marseilles. In the same time, share of freight movements in vehicle-km in total transport was 13%, 13% and 19% respectively, [7]. In light of UK research, delivery vehicles (<3.5 t), which make up the majority of the fleet used in urban deliveries, are responsible for 14.5% of CO_2 emissions while constituting only 6.6% of total miles travelled [8]. There are various scenarios for what total CO_2 emissions will look like; they include, for example, technological changes or the structure of supply chains [9]. The entity-related complexity of urban freight transport is also accompanied by the presence of many possible instruments to influence this system [10,11]—which requires a qualitative assessment of possible results.

At the same time, cities have limited options for analysing the situation in practical terms and to respond to the need to limit transport emission levels. This pertains mainly to the analytical tools available to model transport, especially freight transport in cities and to the ability to provide them with the right inputs to include the environmental consequences of possible activities. To date, no universal freight transport modelling principles, which include the internal complexity of this phenomenon have been developed for cities. There are a number of models with various structures and levels of detail, most of which remain in the sphere of theoretical scientific analyses [12]. Among these, only a few are aimed at analysing the emission levels of CO_2 and other substances [13] in detail.

Two issues must be considered when searching for ways to model urban freight with a focus on environmental impacts. First, is how functional the available models are. Second, can the real requirements of the city, which is the actual user of the models, be realistically met. This includes the experience of urban freight management, availability of urban freight data and the city's transport policy and its goals. The earliest methods for tackling urban freight include classical four step models [14]. The main modelling unit is freight vehicle trips between origin and destination. Such models were first used in the 1970s [15,16] and by analogy to models of transport of people, these models estimate the number of trips generated by each zone, producing an O-D matrix [17]. The main data collection method was cordon line surveys. These surveys were specific to each city which limits their transferability and provides no information on the mode of how each trip is organised logistically. Another limitation of trip models is that they do not take into account the relation between trips of the same round and fail in the trip chain simulation [18] While models based on truck trips provide the basic functionality in terms of simulating the current scenario, they are not reliable when applied to forecasting [19]. As regards modelling of freight environmental impacts, these models come with a basic limitation. It is not possible to include in detail the factors that determine demand for freight. This includes how participants of a transport process influence its structure, e.g., by organising deliveries in a certain way. Even if a city has very limited requirements towards freight modelling, trip-based models are not a practical option.

To overcome problems with trip chain simulation, several models were developed [20,21]. Single trips can be combined into a tour using the savings function. This approach is implemented in the WIVER [22,23] software which was applied in several cities in Germany [18]. Despite their enhanced functionality this type of model has significant limitations as regards the practical requirements for implementation. Unlike simple cordon-line surveys they require extensive surveys of transport companies and logistics operators. Due to a fragmented private sector, this typically exceeds the capabilities of most local authorities. On the other hand, trip chain models can include a direct link between a receiver type and how a mode of transport is organised, which allows for better analysis of the environmental characteristics of the freight movements.

A prevalent approach in urban freight transport looks at the quantity of goods to be transported as the primary object of modelling. These models are known as commodity based [24–27]. They have a more complex structure and include three sub-models that give an estimate of vehicle O-D matrices. They are: the attraction model which provides commodity/quantity flows by each zone, the acquisition model which defines the zone from which the commodity flows originate and the quantity-to-vehicle model which converts commodity/quantity O-D and converts it into vehicles. While commodity based models are considered to be well evolved, few authors have proposed a complete modelling framework [19]. The transferability of selected models has been researched [28] but it has been proved that there are significant differences in the structure of goods distribution patterns making direct transfer of models difficult. These models provide more flexibility in terms of modelling environmental factors related to urban freight activity. It is because they directly link the type of receiver with the nature of generated transport activity, providing an opportunity to investigate how different measures may influence e.g., the utilisation level of different types of vehicles [29]. They also have high requirements in terms of data provision, including interviews with retailers and drivers and traffic counts of commercial and private vehicles [30]. These types of models use sophisticated mathematical methods to transfer commodity flows into vehicle flows, which increases their complexity and introduces another level of approximation.

In the last category of models under consideration, i.e., delivery-based models, a similar method of data provision is used. This approach focuses on deliveries which allows a direct link between generators and transport service providers. Movement of a vehicle may also be considered through road occupancy which ensures the measurability of modelling results and the possibility of merging them with other type of traffic flows [14,31]. The main example of this class of models is Freturb, developed in France [22,32]. It is the only functional urban freight transport model with a dedicated environmental assessment module available to local authorities [7]. However, this module cannot be used as a standalone solution as it is feeds on data from other modules of the Freturb model.

From the perspective of local authorities managing a city's transportation system, there is a gap between model functionality and the environmental impact of urban freight. This is partly due to the inconsistency in how urban freight and passenger transport models are developed. Despite a large number of studies on urban freight simulation, the proposed approaches have not yet been fully validated and have often failed to provide the expected results [33]. At the same time, a number of cities have developed advanced passenger transport models that fulfil the need to manage this type of flow. Some cities have also surveyed urban freight activity in order to understand the underlying problems and develop local solutions [34,35], validate demand reduction measures such as consolidation centres [36], or to assess the environmental impact of inner city deliveries [37].

The identified gap may be addressed by an investigation into whether the urban transport models already in use can be populated with inputs specific for urban freight transport in order to improve their usability in analysing environmental impacts, especially CO_2 emissions. Development of the classical four step models have resulted in a comprehensive solution that integrates different levels of detail in transport system analysis. They are supported by the availability of dedicated software which makes adaptation to the new challenges more feasible. The multi-layer structure of modern transport models developed in cities opens new areas of research on how urban freight transport could be included in transport planning practices with regards to specific issues such as emission reduction. It must be stressed that the intention is not to substitute dedicated urban freight models, but to provide local authorities with the possibility of addressing selected issues related to urban freight without setting up a complex freight modelling framework.

To achieve this goal, the main objective of this research was to assess the possibility of reducing CO_2 emissions by introducing dedicated delivery places in downtown Gdynia based on data about the structure of deliveries in downtown Gdynia obtained during the URBACT Freight TAILS project. The source data were used to feed a mesoscopic transport model for the City of Gdynia developed within the CIVITAS DYN@MO project and to evaluate how the expected reduction in the inconvenience caused by freight vehicles stopping on the road can have a positive effect on traffic conditions in the analysed streets and on the related reduction in CO_2 emissions.

2. Materials and Methods

2.1. Method Applied to Estimate Urban Emissions

The implementation of measures regarding an internally complex urban freight transport system requires an analysis of any possible effects that its individual constituents may have. An assessment of how the planned measures will influence the current condition of a transport system is meant to verify the possibility of sustainable urban development in economic, social and environmental terms [38].

An analysis of the emissions of air pollutants and green-house gases (GHG) is one of the most frequently used assessment indicators for urban transport sustainability [39]. From the point of view of assessing urban freight transport's impact on the environment, choosing such indicators makes it possible to include its most important features and related negative impacts [40]. At the same time, it ensures that requirements for their most important features are met, including target relevance, validity, measurability and sensitivity [41]. Due to their high practical significance, a number of models used to evaluate urban traffic related emissions have been designed. They can be classified into the following categories: average speed models, traffic situation models, traffic variable models, cycle variable models and modal models [42].

Emissions of harmful pollution from road traffic are becoming an increasingly challenging problem for engineers, planners and politicians and, above all, for urban residents. Assessing emissions is also an extremely complicated process, both in terms of actual emissions (e.g., fluctuations between vehicles) and their final dissipation. According to [16] in order to estimate emission levels, emission models have to be combined with vehicle flow estimations, either macro- or microscopic, depending on the characteristics of the emission model. Transport models help to show the movement of persons and goods in the transport network in a designated area with specific socio-economic characteristics and land use [43,44]. Models provide a tool that helps to illustrate the behaviour of the urban transport system and its users over time. They take into account changes in supply and demand in transport, both current and those included in forecasts. In the second half of the last century, extensive research on mathematical modelling of trips and vehicle flows was carried out [45], developing a series of software packages that help to build transport system models for an area or a road and to forecast traffic.

Models may differ in the scope they cover. Gdynia has developed an integrated multi-level model (MST) based on the London model [46–48]. Gdynia chose a three-tier structure with macro, meso and microscopic layers. While the scale of cities varies considerably, the overall concept of developing and using models is similar. The actual method of modelling and the way of using different levels depends on the approach to a given transport issue. The multi-level model supports a flexible approach to the analytical process. Land use models and planning of public and private transport networks are dominated by macroscopic modelling. Where more detail is needed, meso and microscopic models are necessary because they help to simulate road traffic along with its profiles, platoon dispersion, averaged traffic parameters such as queues and delays, saturation flow, etc., (mesoscopic models) and behaviour and interactions between individual road users (microscopic models). The macro and mesoscopic approach makes it possible to estimate typical transport network indicators (total travel time of vehicles in the network, vehicle kilometres or passenger kilometres, average speed, traffic assignment in the network and traffic volumes resulting from the assignment). Microscopic traffic simulators aim to realistically emulate the flow of individual vehicles in the road network. They are capable of replicating complex dynamic traffic systems that are difficult or impossible to simulate using traditional mathematical models.

Thanks to its multi-level structure, the model in Gdynia supports strategic, tactical and operational analyses. Because the multi-level model is hierarchical, specific models can feed data to one another which ensures transparency of the outcomes, regardless of the level of modelling [49]. In the case of the analysis of the impact of delivery vehicles on the obstruction of the flow of vehicles in the road lanes and the impact of these limitations on the environment, a mesoscopic model was used. The tool used in the analysis was the SATURN tool package. The macroscopic model supported with the PTV

VISUM software package was the basis for analysing changes in transport demand, modal split and modal shift of road users changing e.g., from a car to public transport as a result of improvements. The macroscopic model was also the basis for early analysis of traffic distribution and trips by public transport vehicles and by car. The trip matrices estimated in the macroscopic model form the basis for traffic distribution in the mesoscopic model (distribution adapted to the details of traffic organization).

The mesoscopic model helps to identify critical elements of the transport network, such as junction entry queues, including blocking back, which is blocking of previous intersections of streets by queues of vehicles. It also reflects the influence of traffic organization elements, such as types of intersections, the organization of traffic at intersections and traffic control, and takes into account queues at intersections or blocking back and the resulting delays. Analyses include the impact of changes or disturbances in the organization of traffic on the entire transport network or a selected area of the city (in the case of the analyses developed for the purpose of this paper, the downtown of Gdynia was selected). The multi-level transport model is applied on three levels which provides modelling options for various degrees of detail, currently mostly for private vehicles and public transport. The current structure of the model may be improved by providing additional data which makes analysis of freight traffic in Gdynia possible, taking into account operational factors of freight vehicles such as [50]:

- Delivery time and location availability (particularly in the city centre)
- Delivery access in terms of vehicle carriage capacity
- Indication of parking options for delivery vehicles in the city centre
- Option for permanent or temporary closing of specified street sections to traffic/parking of delivery vehicles
- Developing an information system for organisers of delivery traffic.

However, the analyses presented in the paper do not include complex freight transport modelling. They only look at the impact of delivery vehicles on traffic conditions and the accompanying increase in exhaust emissions on the basis of the selected indicators listed above.

In forecasting emissions, exogenous data influencing the dispersion of emissions such as meteorological data were not taken into account. The results presented in this article relate to data on quantitative emissions for vehicles in a transport network. The SATURN package contains internal procedures for the estimation and display of standard pollutants: carbon monoxide, carbon dioxide, hydrocarbons, nitrogen oxides. In the emission assessment procedure, a linear model for all types of pollutants with explanatory variables of cruise travel time on the link, time spent "idling" in queues at junctions, distance, number of primary and secondary stops per vehicle and vehicle volume was used. Emissions were calculated for each connection in the transport network and summed up for the whole city area and separately for the Gdynia city centre area. This example applies to "traffic variable" models in which emission factors are defined by traffic variables, such as average speed, traffic density or queue length. TEE model (Traffic Emissions and Energetics) [51] or the Matzoros model [52] are examples of this category. The results of the analyses are presented in Section 3.

2.2. Case Selection

The choice of dedicated delivery places as the subject of analysis was determined by a number of factors. They are one of the more common solutions used to facilitate the utilisation of urban transport infrastructure [53]. In combination with temporary regulations on the access to selected urban areas, they are often a city's first step in managing freight transport. This was also the case in Gdynita, where delivery places were introduced as a result of freight movements being included for the first time in a strategic document known as the 2016–2025 Sustainable Urban Mobility Plan [54].

The evaluation of the environmental aspects of an urban freight transport system focuses on three main areas [55]; an economic evaluation, social assessments and environmental assessment. Environmental assessment is usually based on expressing distance by means of the emissions of selected substances, usually CO_2, NO_x, and Pm_{10}, using direct emission models expressed as a function of vehicle type and its velocity [7,56–58]. There have also been attempts to assess emissions as a function of vehicle speed and acceleration using GPS data [59]. Therefore, reference to basic parameters of vehicle movement in a road network, such as speed, is confirmed in the research performed to date and in the capabilities of the available analytical instruments.

An incorrectly designed delivery place location may have an adverse effect on traffic structure and on the safety of users, including pedestrians [60,61]. Therefore, there have been attempts to provide methods for designating delivery areas to better utilise their potential as a constituent of an urban transport system [62–64]. However, their applicability is limited by the local context of available data and by the use of specific modelling techniques which may be difficult to duplicate in other cities. An introduction of advanced systems to increase the utilisation efficiency of delivery areas by their users has also been considered. They include delivery area booking systems [65,66] which make use of information technologies and mobile communication devices. These provide potentially significant options for increasing the operational efficiency of transport, also in the context of reducing total transport and the related CO_2 emissions. The influence of deliveries made directly on the roadway, without the application of delivery areas, on traffic parameters has also been studied [67].

Therefore, for many cities, an analysis of how dedicated delivery places operate may be the first stage in competence building for active freight transport management. Efforts should be made to find ways to include this matter in the practice of urban transport system planning, using extended available solutions to include aspects specific to freight transport.

2.3. Delivery Survey in Gdynia

A study of the delivery structure in Gdynia was performed in Q2, 2017 and included three downtown streets (Starowiejska, Świętojańska and Abrahama). In this way, the analysis covered Gdynia's most important high streets. The study was performed by means of a direct business survey and a day-long visual observation of freight vehicles activity. A previous French study was adapted to identify the principles for the study and to select the dedicated delivery areas [68]. In the area under analysis, 506 active businesses were identified on the ground floor of buildings. In total, 12 categories of receivers were used in the survey: services, convenient goods and groceries—independent retailer, convenient goods and groceries—chain retailer, clothing, restaurants and bars, hotels, electronic appliances, decoration and furniture, pharmacies, banks and financial services, other retailers and services, and public services. This classification reflects the retail and service character of the area under analysis. The survey had 337 participating businesses, which constituted 66% of the total number. Therefore, it is possible to reliably generalise the results for the whole area under consideration. The structure of the survey and its key results are presented in Table 1.

Receiver surveys made it possible to collect key information to characterise freight movements [69] in the area under analysis, especially because it was the first study dedicated to this subject in Gdynia. They were modelled on the French experience with the Freturb model. Due to the method of use (designation of dedicated delivery areas), they did not cover delivery driver behaviours included therein [70]. Despite certain limitations [71,72], sometimes resulting from low interest in the transport-related aspect of delivery organisation on the part of receivers, the surveys have significant potential as a component of a data collection and processing system, for the purposes of urban freight transport planning [71].

Based on the collected information, a recommended number of delivery places was calculated for the three streets under analysis. Their planned distribution and number were identified in consultation with the businesses in each street. The final number of physically feasible spaces was limited by infrastructure-related factors and by the availability of general-use parking spaces. Therefore, the aim of the consultations was to identify a location where deliveries could be made using an internal courtyard or access from parallel streets. In this way, the initial number of 49 recommended delivery areas was reduced to 29, out of which 11 were selected for pilot implementation. This was preceded by extensive consultation with the Municipal Police to establish practical enforcement rules. As local and national traffic regulations do not refer directly to delivery spaces, a mix of existing regulations regarding public parking spaces, parking time limits and respective signage and markings has been adopted. Moreover, the Municipal Police suggested the use of CCTV cameras at selected spots to monitor delivery spaces. In further analysis, the pilot project (11 areas) is treated as Variant 1 (V1), while the 29 target areas are treated as Variant 2 (V2).

Table 1. Overview of the delivery survey in Gdynia.

Category	Result
Type of receiver	12 categories
Average no. of employees	3
Type of transport service	75% logistics operators, 25% own transport
Number of received deliveries	3 deliveries daily per receiver (working days)
Day of delivery	Even distribution of deliveries during working days, slight domination of deliveries on Tuesday
Time of delivery	70% of deliveries between 10.00–16.00, peak at 11.00–12.00
Duration of delivery	10 min. (median)
Place of vehicle stop	30% roadside, 22% pavement, 36% public parking space, 19% premises' courtyard
Type of vehicle	93% < 3.5 t

Note: Average number of deliveries: pharmacies 5, electronic appliances/home equipment 4, bars and restaurants 4, convenience stores 3, clothing 2, services 2, other 2.

3. Results

The application of MST on a larger scale as part of the SUMP (Sustainable Urban Mobility Planning) process was carried out in Gdynia, which was the first city in Poland to develop and implement a multi-level model for the needs of the SUMP process.

The model covers all of Gdynia which is divided into 173 transport zones. Each is described with trip-generating variables such as, population, jobs, places of education, size of buildings divided by function and others. Gdynia's street network is represented in its entirety with links categorised based on technical class, cross-section, capacity and free-flow speed. The whole transport network is made up of more than 5500 links and more than 2100 nodes. To calculate the demand for trips and residents' transport behaviour in the 4-step model, functions were used that were calibrated on the basis of comprehensive traffic surveys conducted in the city of Gdańsk in 2009, and a 2013 survey of resident preferences and transport behaviour.

The macroscopic model is the basis for analysing changes in transport demand (new land use rules, new socio-economic and demographic data), modal split and modal shift with road users changing between public transport and the car users, O-D matrices as a result of changes in traffic organisation and control or other measures including development of transport network. The macroscopic model is also the basis for early analysis of traffic distribution and trips by public transport and car users. Four classes of car users have been considered: passenger cars, vans, heavy vehicles without trailers and heavy vehicles with trailers. O-D trip matrices calculated in the macroscopic model are the basis for traffic distribution in the mesoscopic model, where distribution is adjusted for details of traffic organisation.

The mesoscopic model helps to identify the critical elements of the transport network such as junction entry queues including blocking back (blocking of previous intersections of streets by queues of vehicles). It also helps to analyse ways to improve traffic efficiency, such as lanes at junction entry, corrections to signalization programmes, use of different junction types, adding traffic signals, etc., which may improve traffic conditions. The analyses look at the effects a measure has on the city's entire transport network.

For the assignment of traffic in a street network a stochastic method of load balancing (Stochastic User Equilibrium Assignment) was used, which includes the dependence of travel time on the size of the traffic flow. Iterative algorithms have been applied here. In this method, total traffic flow on the "source-to-target" is divided among a number of routes. The basic assumption of this method is that traffic arranges itself on congested networks such that the routes chosen by individual drivers are those with the minimum perceived cost; routes with perceived costs in excess of the minima are not used. A stochastic model is set up by assuming that the cost (travel time have been considered as the cost of trip) as defined by the model is the average cost but that there is a distribution about the average as perceived by individuals. The perceived cost of a route may therefore be simulated by selecting a cost at random from the perceived distribution of costs on each link. Algorithms designed to reflect the resulting flows are referred to as Burrell assignment models. This is how key network performance indicators can be obtained.

The model has been calibrated taking into account the stationary configuration of flows (traffic volumes on individual sections of streets and at junctions for turning flows for the state of the network without delivery vehicles blocking traffic lanes). Values of the parameters for the choice of link cost distributions (KOB and SUET), cumulative density function (KOB) and the generation of random numbers (KORN for initial seed value) were applied to calibrate the modelled flow to observed flow. The normal distribution of link cost has been chosen with a value of KOB = 2 and value of SUET = 0.2 with KORN = 1 for random numbers.

Models were validated with the use of a control group of traffic volume values (volumes from a random typical day for the period of morning and afternoon peak hours without delivery vehicles blocking traffic lanes). A regression analysis and correlation between the values of traffic volumes (modelled and observed) were developed. It has been estimated that a strong linear correlation should occur between this value with the angular coefficient of the regression function which equals 1.00, and the free expression of the function equal to 0.00. The calculations for the measurement points (individual street sections and turns at intersections) found that for the model, the coefficient of determination R2 (for the function Y = X) accepted values range from 0.78 (peak morning) to 0.83 (peak afternoon). Due to the large number of field measurements of volume used in the model of the city, it turned out to be extremely difficult to calibrate the model to achieve a satisfactory convergence of all measuring points. However, the results indicate that in this example about 85% of the volume is explained by the model. For each of the models statistical analyses were also performed, allowing a more accurate comparison of the measured intensity values obtained from the model. For this purpose, an analytical method was used, which is a form of statistics χ^2 (GEH) that takes into account both the relative and absolute error. The results of the analyses of statistics for GEH, which analyses the model of the network of Gdynia, are presented in Table 2 for the afternoon peak hour.

Table 2. Statistical results of compliance of the volumes observed and obtained from the city model for the afternoon rush hour.

Statistics	Value of the Statistics
The volumes measured in the range of <700 (passenger cars/hour) (Relative change compared to the volume of a model <100 passenger cars/hour)	85.3%
The volumes measured in the range 700 (passenger cars/hour) (Absolute change compared to the volumes of the model <15%)	82.5%
Statistics GEH < 5	64.2%
Statistics GEH < 10	83.4%
The average value of statistics GEH	8.64
Coefficient of determination R^2	83.5%

Based on the delivery survey described in Section 2.3, analyses were performed using the mesoscopic model described in Section 2.1 and in this section. Simulations were carried out for three variants of designating dedicated delivery places on Starowiejska and Świętojańska Streets, as shown in Figure 1. Variant 0 (V0—base variant) represents the current situation, in which delivery vehicles randomly block traffic lanes in places identified in delivery surveys. The pilot project (11 areas) will be treated as Variant 1 (V1), while the 29 areas represent a full-scale implementation as Variant 2 (V2). When street lanes are blocked by delivery vehicles, traffic is disrupted by periodic and random queues of vehicles at the delivery points, increase in the number of vehicle stops, change of route, and dangerous manoeuvres of vehicles. This contributes to an increase in exhaust emissions. The implementation of dedicated delivery places in locations that do not block traffic lanes have been included in Variant 1 (with fewer delivery points) and Variant 2 (greater range of improvements), respectively. Analyses using the mesoscopic model were carried out for the morning and afternoon peak hours, for a one-hour period during each peak.

In the base variant, lanes are simultaneously blocked in 4–5 places identified in the surveys of Gdynia centre (Starowiejska and Świętojańska Streets); this lasts for 15 min in the afternoon peak hour and 15 min in the morning peak hour in 1–2 places simultaneously within 15 min period. The modelling includes places that effectively contribute to the development of traffic disruptions. Within an hour, the total time of blocking the lane in various places is 60 min in the morning peak hour and 180 min in the afternoon peak hour. Throughout the day, the scale of blocking lanes is much greater but traffic disruptions are not as severe as in peak hours because there is less traffic. The survey results indicate that if the proposed measures are implemented, dedicated delivery locations will help to reduce the time of blocking lanes by 40% in Variant 1 and by 100% in Variant 2.

The size of the traffic disruptions depends on the place of blocking the lane in the road network. In the modelling process of the base variant, the existing situation was accurately mapped for temporary location of delivery vehicles blocking traffic lanes. Delivery vehicles may block the lane in two ways. In the first case, the lane is fully closed, without the possibility of bypassing the delivery vehicle blocking the lane. In this case, it was assumed that there is no possibility of passing during the blocking of the lane and the capacity of the street section is reduced to 0 vehicles during 15 min period of simulation. Other vehicles change their route as a result of closing the street section. In the second case, it is possible to bypass the delivery vehicle by using the traffic lane in the opposite direction (the capacity is dependent on the traffic volume and speed in the opposite direction, which determines the possibility of bypassing the delivery vehicle blocking the lane). The observations carried out in the field study showed that at a rate of 300–500 vehicles/hour on the opposite direction lane the capacity of the blocked lane is equal to 10–15 vehicles/15 min. The model assumed a capacity of 15 vehicles/15 min (omitting the delivery vehicle in traffic gaps from the opposite direction). The mesoscopic approach does not take into account the delays of vehicles from the opposite direction that let vehicles move to bypass the delivery vehicle, which is a simplification and affects the accuracy of the results. While the simulation element in SATURN does not exactly model the exact progression of each vehicle when they move down the link, it is possible to deduce certain properties of their progression. The distinction between the two forms of stop (primary and secondary) is essentially as follows. Imagine a small arm at the intersection with the "stop" sign at the end. Every vehicle approaching this intersection must come to a complete stand still either at the stop line (if there is no queue) or behind the last vehicle in the queue; this is a primary stop. If there is a queue and vehicles leave from the beginning of the queue, the vehicles go forward, accelerating and then slowing down to a stationary position; these are secondary stops. This two-way split does not exactly represent all possible vehicle movements in a queue but it is probably sufficient for estimating secondary parameters such as fuel consumption or emissions and for providing a very broad description of the state of a junction or any other place in the street network (e.g., a bottleneck). The rules for estimating primary and secondary stops are, like their definitions, somewhat arbitrary (mesoscopic approach). Thus, for minor arms at priority junctions all arriving traffic must make a primary stop if its turn is over capacity or if the queue per lane is greater than 2 vehicles. If the queue per lane is (in the limit) zero the probability of a primary stop is equal to the calculated probability of there being no gap. For queues per lane of between 0 and 2 vehicles, a linear relationship is assumed. Secondary stops are calculated by assuming that all primary stops make a further number of secondary stops equal to the queue length per lane divided by the number of vehicles that can depart from the stop line in a platoon once a gap occurs (assumed equal to one over the probability of a gap). For major priority arms, secondary stops are ignored and a primary stop only occurs if the arm is over capacity or if, at the moment of arrival, the expected queue length per lane is greater than 1 vehicle.

Figure 1. Change in CO_2 emissions in variant 2 compared to the baseline in Gdynia centre.

In order to obtain more reliable results in the process of stochastic traffic assignment in the network, a quasi-dynamic model was used in which over-capacity queues are passed between time periods. Thus, the model takes into account the dynamics of changes in blocking lanes in particular places (15 min periods were assumed) and traffic parameters changes spread over time periods. Sample calculations of some indicator values at 15 min intervals for the afternoon peak hour are shown in Figure 2.

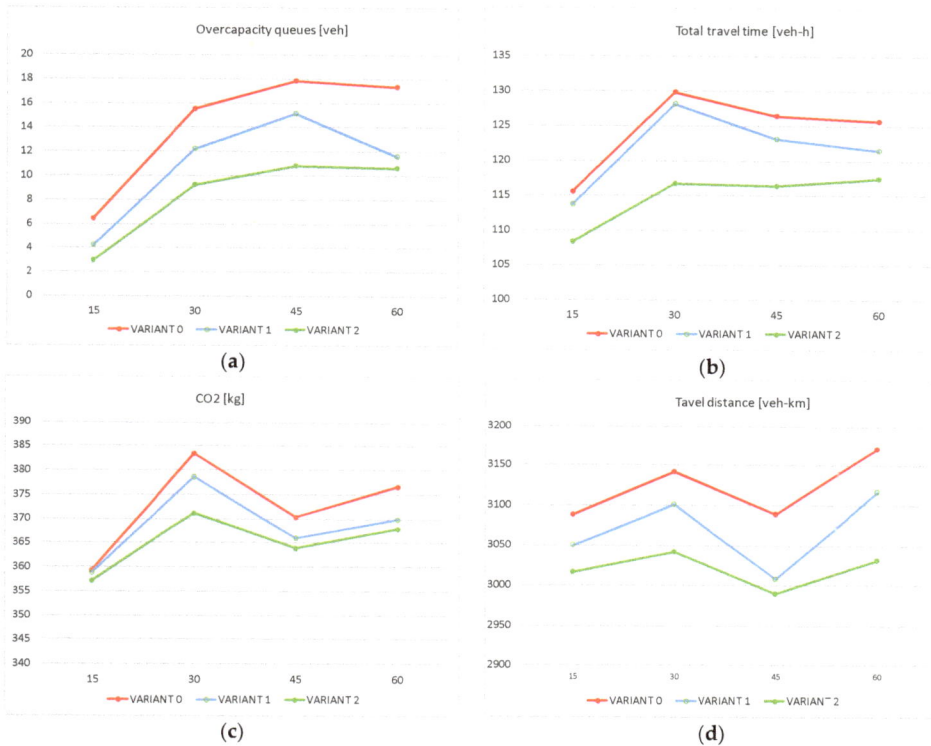

Figure 2. Selected indicator values for afternoon peak hour in the area of Gdynia centre. (**a**) overcapacity queues; (**b**) total travel time; (**c**) CO_2 emission; (**d**) travel distance (veh-km travelled).

Fluctuations of indicators observed in particular 15-min periods are a result of a different number of blocked lanes in these periods and their different location. In the case of lanes blocked along Świętojańska Street we can observe a greater deterioration of transport network indicators because of higher traffic volumes than along Starowiejska Street. In some more congested periods, drivers change their route by selecting nearby alternative routes, which contributes to an increase in distance travelled by vehicles and results in improved traffic conditions in the road network of the city centre (reduction in dynamics of growth of queue lengths, delays, travel time). The most stable traffic conditions can be observed in the case of implementing Variant 2 improvements (no cases of blocking traffic lanes by commercial vehicles).

The value of carbon dioxide depends on the travel time of all vehicles in the road network taking into account time spent in queues at junctions, travelled distance, number of primary and secondary stops per vehicle and vehicle volume at particular sections of network. If the above parameters deteriorate, there is an increase in exhaust emissions and fuel consumption. Examples of cumulative values of some indicators for a one-hour period are presented in Table 3. The equations for calculating emissions and default parameter values for the pollutants are presented below. The model still requires calibration according to the structure of the vehicle fleet in Gdynia. Therefore, the results are estimates, but prove the practical potential of the selected approach with regards to emission estimation.

The basic equation for the emission of CO_2 (E^i_{CO2}) from a link is:

$$E^i_{CO2} = (70d + 1200t_q + 16s_1 + 5s_2)V$$

E^i_{CO2}—emission of CO_2 at link i
d—link distance
t_q—time spent idling in queues at junctions
s_1—number of primary stops per vehicle
s_2—number of secondary stops per vehicle
V—vehicle volume at link i

The basic equation for the emission of NO_x(E^i_{NOx}) from a link is:

$$E^i_{NOx} = (103t_c + 1.8t_q + 0.42s_1 + 0.09s_2)V$$

E^i_{NOx}—emission of NO_x at link i
t_c—average cruise travel time on the link
t_q—time spent idling in queues at junctions
s_1—number of primary stops per vehicle
s_2—number of secondary stops per vehicle
V—vehicle volume at link i

Table 3. Example of simulation results from the mesoscopic model for variants of dedicated delivery spots—effect for the area of the city centre, morning and afternoon peak hour.

Variant	Veh-km Travelled	Travel Time	Fuel Consumption	Mean Speed	Number of Stops	CO₂ Emission	NOx Emission
	(veh-km)	(veh-h)	(L)	(km/h)	(-)	(kg)	(kg)
Morning peak (1 h period)							
V0	2166.64	87.03	231.28	29.87	15,238.30	231.81	7.00
V1	2160.86	86.73	230.73	29.90	15,182.73	230.47	6.98
V2	2139.64	84.41	226.46	30.42	14,896.88	227.06	6.88
% change (V1-V0)/V0	−0.3	−0.3	−0.2	0.1	−0.4	−0.6	−0.3
% change (V2-V0)/V0	−1.2	−3.0	−2.1	1.8	−2.2	−2.1	−1.7
Afternoon peak hour (1 h period)							
V0	3111.88	147.68	367.80	25.34	27,172.21	372.27	10.52
V1	3060.33	144.66	365.24	25.44	27,019.81	368.97	10.45
V2	3017.34	136.22	354.50	26.62	26,532.14	361.69	10.27
% change (V1-V0)/V0	−1.7	−2.0	−0.7	0.4	−0.6	−0.9	−0.7
% change (V2-V0)/V0	−3.0	−7.8	−3.6	5.0	−2.4	−2.8	−2.4

More vehicles in the network and higher traffic volumes occur during the afternoon peak hour than during the morning peak. The effects of planned changes are more visible for the period in which the transport network is more congested. Analyses have shown a positive effect with exhaust emissions reduced both in the morning and afternoon peak periods, primarily in the case of Variant 2. Taking into account that only two peak hours were analysed, accumulated gains in the long term could be significant.

Figure 1 shows changes in CO_2 emissions in individual sections of the transport system network in the centre of Gdynia. If implemented, the delivery vehicle scheme with dedicated parking spaces will contribute to reducing emissions primarily on the streets covered by the project.

4. Discussion

The aim of this paper was to verify whether the mesoscopic urban transport model already in use in Gdynia can be populated with inputs specific to urban freight transport in order to improve the ability to assess the environmental impact of freight transport. The analysis was executed by feeding the model with delivery data from downtown Gdynia's to assess whether dedicated delivery areas could be a way to reduce CO_2 emissions there. The objective was to evaluate how the expected reduction in the inconvenience caused by freight vehicles stopping on the road can have a positive effect on traffic conditions in the analysed streets and on the related reduction in CO_2 emissions.

The transport model proved to be useful in understanding the issue of dedicated delivery places. It was able to capture changes in CO_2 and NO_x emissions. Dedicated delivery places were implemented primarily for their effect on access to selected downtown streets and the nuisance caused by delivery vehicles when they stop, e.g., on the pavement. If, however, they can also help to reduce emissions, the scheme can be rolled out in as a comprehensive measure in all of the downtown area. As a result, the benefits would can be more evident.

In studying dedicated delivery places, use was made of the model's capacity to represent road traffic parameters and the potential effects on traffic in the area under analysis. Delivery places are a relatively simple regulatory solution which local authorities can use as a point of departure to more advanced urban freight management measures. Whether it is this scheme or other regulatory tools, such as time windows for accessing selected areas, standard transport models developed with a sufficient level of detail are able to assess the effects of measures being proposed with an acceptable level of detail.

If the effects on urban freight-related emissions are to be substantial, more needs to be done. Rather than focus on infrastructure only, the measures must be related to the structure of supply chains in urban areas. To that end, the structure of the model must account for a much broader scope of parameters to identify freight vehicle activity and the characteristics of demand for delivery and the variety of receivers using the service. They would meet the analytical complexity requirements of available urban freight optimisation solutions. Due to these challenges, the transport model used in this analysis has a major functionality deficit when compared to dedicated freight transport models such as France's Freturb [14,31] and Germany's Viver [22,23], as presented in Table 4.

As it can be noticed, even comprehensive models designed to analyse freight offer varying functionalities which translates into how well we can study the effects of selected solutions on emission levels. While the existing models based on a four-stage approach could theoretically be extended to cover the specificity of a study problem, the costs and workload of doing that remain an issue. In addition, it is likely that modified models will be difficult to calibrate due to lack of data which the majority of cities do not collect as a standard transport management practice. Regardless of the approach, the process of introducing a model is anything but simple because of the complexity and scarcity of established practices to use as a point of reference.

While the development of more comprehensive freight transport models based on data obtained from carriers is justified, it is extremely difficult due to the reluctance of carriers to share information. Reliable statistics on shipments create the opportunity to develop better quality models for freight transport demand, and on this basis, to estimate the trip matrix of freight vehicles (this approach is justified in modelling heavy traffic—through freight traffic, industry and sea port activities), when shipment dynamics are difficult to research (although this depends on the frequency of data provided by carriers, industry and seaport operators). The implementation of a reliable model also requires a regular and comprehensive study of traffic, including heavy goods vehicles and businesses (the present legislation does not make such studies mandatory, and the high costs discourage cities from conducting them). Such research should be carried out every 5 years and would provide valuable input and validation data for freight transport models. In the case of commercial distribution, commercial services, e-commerce and express courier municipal services (waste disposal/maintenance of roads) it is important to apply the approach based on cruise route modelling (rounds/travel chains). This allows the model to include fixed routes.

Table 4. Comparison of selected transport models' functionality for urban freight analysis.

Indicator	Freturb	Wiver	Gdynia's Multi-Level Transport Model
Distance covered during deliveries	Can define travel distribution between transport zones including deliveries combined in rounds	Can define travel distribution between transport zones including deliveries combined in rounds	Can define direct travel distribution between transport zones excluding deliveries combined in rounds. While technically the model can be extended, this would require a lot of simplification due to poor capacity to represent receiver distribution
Total freight vehicle working time	Cannot be defined directly	Cannot be defined directly	Only includes driving time (can be extended to define time delays to match delivery time)
Duration of a single delivery	The parameter helps to define total time of infrastructure use in connection with kerb-side delivery, parking spaces, etc.	Not used as model parameter	Not used as model parameter (can be extended by adding time lost and the effects of lane blockage)
Number of deliveries in a round	Model's basic parameter	Model's basic parameter	Not used as model parameter (the model can be extended with some necessary simplification)
Number of direct deliveries	Model's basic parameter	Model's basic parameter	Not used as model parameter (can be extended just as with the previous *indicator*, with the same limitations)
Number of deliveries per 1 employee	The employment by receiver category is one of the basic parameters decisive for delivery generation which is the basis of the model	Takes account of an area's trip generation, including the value of production expressed with e.g., employment	Not used as model parameter (can be extended with travel matrix—need to build dedicated delivery traffic matrices including supply chain structure)
Kilometres travelled	Can be calculated based on generated number and structure of freight vehicle movements	Can be calculated based on generated number and structure of freight vehicle movements	Can be calculated based on generated number and structure of freight vehicle movements, using limited generation methods and direct trips only
Number and type of freight vehicles	Three types of freight vehicles ($\leq$3.5 t < 40 t) and three types of transport services (transport by receiver, sender, logistic operator)	Number and structure of deliveries given per type of freight vehicle	Number and structure of freight vehicles referred to types of freight vehicles in traffic (the model can include different types of freight vehicles but linking them to types of business is limited)
Load capacity rate	Not used in the model, the model is based on the number of deliveries rather than on freight and freight vehicle-related parameters	Not used in the model, the model is based on the number of deliveries rather than on freight and freight vehicle-related parameters	Not used as model parameter
Temporal distribution of freight vehicle traffic	Temporal distribution of traffic can be presented using input data from receiver surveys	Temporal distribution of traffic not included in the model	Temporal distribution of traffic can be presented using input data from traffic volume surveys
Number and type of vehicles required to provide the service to the area's receivers	Can be calculated using the rate of receiver generated deliveries and the corresponding ways to organise a transport service	Can be calculated after a balancing out of trip generation and trip absorption in specific transport areas	Not used as model parameter (can be added just as in Viver)
Share of freight vehicles in road traffic	Share of freight vehicles can be presented using the PCU (*Passenger Car Unit*)	Total structure of passenger car and commercial vehicle traffic can be presented using VISSUM	Can identify the share of freight vehicles per each section of transport network by vehicle category

Sustainability **2018**, *10*, 2515

The next steps will focus on verifying whether the existing transport model could meet the requirements of urban freight management set by the Sustainable Urban Mobility Plan in Gdynia. This would include detailed parametrisation of the objectives to identify the data requirements for their analysis and how they could be met by the existing modelling framework. Also the most promising areas of MST implementation would be marked, as well as demand for additional data. Also, the relation between policy objectives and potential measures will be under investigation in terms of the model functionality to find the limits of the existing model application. These limits in terms of the relation between required inputs and possible outcomes frame the potential of MST for urban freight modelling and define where the implementation of dedicated urban freight model should begin. Another issue which should be considered is how to utilise existing systems such as Weigh-in-Motion [73] to improve data availability.

Understanding urban freight should be a consistent and gradual process and one that must be conducted in order to identify the scale of the challenges and equip decision-makers with the knowledge they need for planning. Even if no decisions are taken to introduce advanced analytical tools, the search for the right solutions can still continue so that they match the available resources and for private sector partner engagement.

Author Contributions: D.K. designed the methodology of the presented paper and its conceptualisation, conducted all calculations regarding delivery structure in the downtown of Gdynia and elaborated the number of dedicated delivery bays and their location. J.O. fine-tuned the modelling methodology and conducted all necessary calculation and visualisation of the transport model, as well as analysed results. D.K. wrote Sections 1, 2.2 and 2.3 and co-authored the conclusions. J.O. wrote Sections 2.1 and 3 and co-authored the conclusions.

Funding: This research received no external funding.

Acknowledgments: The study is based on the results of the URBACT Freight TAILS project implemented from 2016 to 2018. The transport model developed within the CIVTAS DYN@AMO project implemented under CIVITAS II PLUS was also used for traffic modelling purposes. No funding was obtained in the project to cover the costs of Open Access publication.

Conflicts of Interest: The authors declare no conflict of interest. The founding sponsors had no role in the design of the study; in the collection, analyses, or interpretation of data; in the writing of the manuscript, and in the decision to publish the results.

References

1. Wefering, F.; Rupprecht, S.; Bührmann, S.; Böhler-Baedeker, S. *Guidelines. Developing and Implementing a Sustainable Urban Mobility Plan*; Rupprecht Consult: Köln, Germany, 2014.
2. Hickman, R.; Hall, P.; Banister, D. Planning more for sustainable mobility. *J. Transp. Geogr.* **2013**, *33*, 210–219. [CrossRef]
3. Behrends, S.; Lindholm, M.; Woxenius, J. The Impact of Urban Freight Transport: A Definition of Sustainability from an Actor's Perspective. *Transp. Plan. Technol.* **2008**, *31*, 693–713. [CrossRef]
4. Dablanc, L. Goods transport in large European cities: Difficult to organize, difficult to modernize. *Transp. Res. Part A Policy Pract.* **2007**, *41*, 280–285. [CrossRef]
5. Macharis, C.; Verlinde, S. Sharing Urban Space: A story of Stakeholder Support. In *Urban Freight for Livable Cities. How to Deal with Collaboration and Trade-Offs*; Wolmar, C., Ed.; The Volvo Research and Educational Foundations, VREF: Nairobi, Kenya, 2012.
6. Dablanc, L.; Lozano, A. *Commercial Goods Transport in Paris*; United Nations Habitat: Nairobi, Kenya, 2013.
7. Segalou, E.; Ambrosini, C.; Routhier, J.-L. The environmental assessment of urban goods movement. In *Logistics Systems for Sustainable Cities*; Emerald Group Publishing Limited: Beck Lane, UK, 2004; pp. 15–207.
8. Mckinnon, A. CO_2 Emissions from Freight Transport: An Analysis of UK Data. In *Logistic Research Network 2007*; The Chartered Institute of Logistics and Transport: Corby, UK, 2007.
9. Piecyk, M.I.; McKinnon, A.C. Forecasting the carbon footprint of road freight transport in 2020. *Int. J. Prod. Econ.* **2010**, *128*, 31–42. [CrossRef]
10. Russo, F.; Comi, A. A classification of city logistics measures and connected impacts. *Procedia Soc. Behav. Sci.* **2010**, *2*, 6355–6365. [CrossRef]

11. Allen, J.; Anderson, S.; Browne, M.; Jones, P. *A Framework for Considering Policies to Encourage Sustainable Urban Freight Traffic and Goods/Service Flows. Report 1*; University of Westminster: London, UK, 2000.

12. Comi, A.; Donnelly, R.; Russo, F. *Urban Freight Models*; Elsevier Inc.: New York, NY, USA, 2013.

13. Anand, N.; van Duin, R.; Quak, H.; Tavasszy, L. Relevance of City Logistics Modelling Efforts: A Review. *Transp. Rev.* **2015**, *35*, 701–719. [CrossRef]

14. Bonnafous, A.; Gonzalez-Feliu, J.; Routhier, J.-L. An alternative UGM paradigm to O-D matrices: The Freturb model. In Proceedings of the 13th World Conference on Transport Research (13th WCTR), Rio de Janeiro, Brazil, 15–18 July 2013.

15. Slavin, H. Demand for urban goods vehicle trips. *Transp. Res. Rec.* **1976**, *591*, 32–37.

16. Ogden, K.W. *Modelling Urban Freight Generation*; Traffic Engineering Control; Hemming Group, Limited: London, UK, 1977; Volume 18.

17. Gentile, G.; Vigo, D. Movement generation and trip distribution for freight demand modelling applied to city logistics. *Eur. Transp. Trasp. Eur.* **2013**, *54*, 1–27.

18. Gonzalez-Feliu, J.; Routhier, J.-L. Modeling Urban Goods Movement: How to be Oriented with so Many Approaches? *Procedia Soc. Behav. Sci.* **2012**, *39*, 89–100. [CrossRef]

19. Comi, A.; Site, P.D.; Filippi, F.; Nuzzolo, A. Urban freight transport demand modelling: A state of the art. *Eur. Transp. Trasp. Eur.* **2012**, *51*, 1–17.

20. Sonntag, H. A computer model of urban commercial traffic—Analysis, basic concept and application. *Transp. Policy Decis. Mak.* **1985**, *3*, 171–180.

21. Janssen, T.; Vollmer, R. Development of a urban commercial transport model for smaller areas. In Proceedings of the German Society for Geography Annual Meeting, Berlin, Germany, 2005.

22. Ambrosini, C.; Meimbresse, B.; Routhier, J.-L.; Sonntag, H. Urban freight policy-oriented modelling in Europe. *Innov. City Logist.* **2008**, *2*, 197–201.

23. Sonntag, H.; Meimbresse, B. Modelling urban commercial traffic with model WIVER. In *L'intégration Des Marchandises Dans le Système Des Déplacements Urbains*; Patier, D., Ed.; Laboratoire d'Economie des Transports: Lyon, France, 2001; pp. 93–106.

24. Boerkamps, J.; van Binsbergen, A. GoodTrip—A new approach for modelling and evaluation of urban goods distribution. In *City Logistics I*; Taniguchi, E., Thompson, R.G., Eds.; Institute for City Logistics: Kyoto, Japan, 1999; pp. 175–196.

25. Wisetjindawat, W.; Kazushi, S.; Matsumoto, S. Supply chain simulation for modeling the interactions in freight movement. *J. East. Asia Soc. Transp. Stud.* **2005**, *6*, 2991–3004.

26. Wisetjindawat, W.; Sano, K. A Behavioral Modeling in Micro-Simulation for Urban. *J. East. Asia Soc. Transp. Stud.* **2003**, *5*, 2193–2208.

27. Russo, F.; Comi, A. A model system for the ex-ante assessment of city logistics measures. *Res. Transp. Econ.* **2011**, *31*, 81–87. [CrossRef]

28. Ibeas, A.; Moura, J.L.; Nuzzolo, A.; Comi, A. Urban Freight Transport Demand: Transferability of Survey Results Analysis and Models. *Procedia Soc. Behav. Sci.* **2012**, *54*, 1068–1079. [CrossRef]

29. Russo, F.; Comi, A. Urban freight transport planning towards green goals: Synthetic environmental evidence from tested results. *Sustainability* **2016**, *8*, 381. [CrossRef]

30. Nuzzolo, A.; Comi, A. Urban freight transport policies in Rome: lessons learned and the road ahead. *J. Urban. Int. Res. Placemak. Urban Sustain.* **2014**, *8*, 1–15. [CrossRef]

31. Routhier, J.; Toilier, F. FRETURB V3, A Policy Oriented Software of Modelling Urban Goods Movement. In Proceedings of the 11th Word Conference on Transport Research, Berkeley, CA, USA, 24–28 June 2007; p. 23.

32. Toilier, F.; Alligier, L.; Patier, D.; Routhier, J. *Vers un Modèle global de la Simulation de la Logistique Urbaine: FRETURB*; Version 2 Rapport Final; Laboratoire d' Economie des Transport: Lyon, France, 2005.

33. Nuzzolo, A.; Coppola, P.; Comi, A. Freight Transport Modeling: Review and Future Challenges. *Int. J. Transp. Econ.* **2013**, *40*, 183–206.

34. Cherrett, T.; Allen, J.; McLeod, F.; Maynard, S.; Hickford, A.; Browne, M. Understanding urban freight activity—Key issues for freight planning. *J. Transp. Geogr.* **2012**, *24*, 22–32. [CrossRef]

35. Browne, M.; Allen, J.; Steele, S.; Cherrett, T.; McLeod, F. Analysing the results of UK urban freight studies. *Procedia Soc. Behav. Sci.* **2010**, *2*, 5956–5966. [CrossRef]

36. City Ports. *City Ports Project Interim Report*; City Ports: Bologna, Italy, 2005.

37. Kijewska, K.; Iwan, S. Analysis of the Functioning of Urban Deliveries in the City Centre and Its Environmental Impact Based on Szczecin Example. *Transp. Res. Procedia* **2016**, *2015*, 739–749. [CrossRef]
38. Muñuzuri, J.; Cortés, P.; Onieva, L.; Guadix, J. Application of supply chain considerations to estimate urban freight emissions. *Ecol. Indic.* **2018**, *86*, 35–44. [CrossRef]
39. Haghshenas, H.; Vaziri, M. Urban sustainable transportation indicators for global comparison. *Ecol. Indic.* **2012**, *15*, 115–121. [CrossRef]
40. Browne, M.; Allen, J.; Nemoto, T.; Patier, D.; Visser, J. Reducing Social and Environmental Impacts of Urban Freight Transport: A Review of Some Major Cities. *Procedia Soc. Behav. Sci.* **2012**, *39*, 19–33. [CrossRef]
41. Joumard, R.; Gudmundsson, H.; Folkeson, L. Framework for Assessing Indicators of Environmental Impacts in the Transport Sector. *Transp. Res. Rec. J. Transp. Res. Board* **2011**, *2242*, 55–63. [CrossRef]
42. Smit, R.; Ntziachristos, L.; Boulter, P. Validation of road vehicle and traffic emission models—A review and meta-analysis. *Atmos. Environ.* **2010**, *44*, 2943–2953. [CrossRef]
43. The Use of Transport Models in Transport Planning and Project Appraisal, JASPERS Appraisal Guidance (Transport). Available online: http://kc-sump.eu/wordpress/wp-content/uploads/2015/04/Upotreba-Modela-u-prometnom-planiranju_JASPERS_kolovoz-2014.pdf (accessed on 7 April 2018).
44. Sivakumar, A. *Modelling Transport: A Synthesis of Transport Modelling Methodologies*; Imperial College: London, UK, 2007.
45. Travel Model Improvement Program—TMIP. *Model Validation and Reasonableness Checking Manual*; Federal Highway Administration: Washingdon, DC, USA, 1997.
46. Smith, J.; Blewitt, R. *Traffic Modelling Guidelines TfL Traffic Manager and Network Performance Best Practice*; Transport for London: London, UK, 2010.
47. Dimitriou, H.; Thompson, R. *Strategic Planning for Regional Development in the UK*; Routledge: Abingdon-on, UK, 2008.
48. Bliemer, C.J.; Mulley, C.; Moutou, C. *Handbook on Transport and Urban Planning in the Developed World*; Edward Elgar Publishing: Cheltenham, UK, 2016.
49. Okraszewska, R.; Romanowska, A.; Wołek, M.; Oskarbski, J.; Birr, K.; Jamroz, K. Integration of a Multilevel Transport System Model into Sustainable Urban Mobility Planning. *Sustainability* **2018**, *10*, 479. [CrossRef]
50. Oskarbski, J.; Kaszubowski, D. Potential for ITS/ICT Solutions in Urban Freight Management. *Transp. Res. Procedia* **2016**, *16*, 433–448. [CrossRef]
51. Negrenti, E. TEE: The ENEA traffic emissions and energetics model micro-scale applications. *Sci. Total Environ.* **1996**, *189–190*, 167–174. [CrossRef]
52. Matzoros, A.; van Vliet, D. A model of air pollution from road traffic, based on the characteristics of interrupted flow and junction control: Part I—Model description. *Transp. Res. Part A Policy Pract.* **1992**, *26*, 315–330. [CrossRef]
53. Muñuzuri, J.; Larrañeta, J.; Onieva, L.; Cortés, P. Solutions applicable by local administrations for urban logistics improvement. *Cities* **2005**, *22*, 15–28. [CrossRef]
54. Kaszubowski, D. Recommendations for Urban Freight Policy Development in Gdynia. *Transp. Res. Procedia* **2016**, *12*, 886–899. [CrossRef]
55. Gonzalez-Feliu, J. *Sustainable Urban Logistics: Planning and Evaluation*; Wiley: Hoboken, NJ, USA, 2018.
56. Gonzalez-Feliu, J.; Ambrosini, C.; Pluvinet, P.; Toilier, F.; Routhier, J.L. A simulation framework for evaluating the impacts of urban goods transport in terms of road occupancy. *J. Comput. Sci.* **2012**, *3*, 206–215. [CrossRef]
57. Fu, J.; Jenelius, E. Transport efficiency of off-peak urban goods deliveries: A Stockholm pilot study. *Case Stud. Transp. Policy* **2018**, *6*, 156–166. [CrossRef]
58. Holguín-Veras, J.; Encarnación, T.; González-Calderón, C.A.; Winebrake, J.; Wang, C.; Kyle, S.; Herazo-Padilla, N.; Kalahasthi, L.; Adarme, W.; Cantillo, V.; et al. Direct impacts of off-hour deliveries on urban freight emissions. *Transp. Res. Part D Transp. Environ.* **2018**, *61*, 84–103. [CrossRef]
59. Pluvinet, P.; Gonzalez-Feliu, J.; Ambrosini, C. GPS Data Analysis for Understanding Urban Goods Movement. *Procedia Soc. Behav. Sci.* **2012**, *39*, 450–462. [CrossRef]
60. Aiura, N.; Taniguchi, E. Planning On-Street Loading-Unloading Spaces Considering the Behaviour of Pickup-Delivery Vehicles and Parking Enforcement. In Proceedings of the 4th International Conference City Logistics, Langkawi, Malaysia, 12–14 July 2005; Volume 6, pp. 107–116.
61. Delaître, L.; Routhier, J.L. Mixing two French tools for delivery areas scheme decision making. *Procedia Soc. Behav. Sci.* **2010**, *2*, 6274–6285. [CrossRef]

62. Gardrat, M.; Serouge, M. Modeling Delivery Spaces Schemes: Is the Space Properly used in Cities Regarding Delivery Practices? *Transp. Res. Procedia* **2016**, *12*, 436–449. [CrossRef]

63. Alho, A.R.; Silva, J.D.E. Analyzing the relation between land-use/urban freight operations and the need for dedicated infrastructure/enforcement—Application to the city of Lisbon. *Res. Transp. Bus. Manag.* **2014**, *11*, 85–97. [CrossRef]

64. Dezi, G.; Dondi, G.; Sangiorgi, C. Urban freight transport in Bologna: Planning commercial vehicle loading/unloading zones. *Procedia Soc. Behav. Sci.* **2010**, *2*, 5990–6001. [CrossRef]

65. Gonzalez-Feliu, J.; Arndt, W.-H.; Beckmann, K.J.; Gies, J. The deployment of urban logistics solutions from research, development and pilot results. Lessons from the FREILOT Project. In *Stadtischer Wirtschaftsverkehr—Commercial/Goods Transportation in Urban Areas—Transports Commerciaux/Marchandises en Ville. Dokumentation der Internationalen Konferenz 2012 in Berlin*; Arndt, W., Beckmann, K., Gies, J., Gonzalez-Feliu, J., Eds.; Deutsches Institut fur Urbanistik: Berlin, Germany, 2013; pp. 104–121.

66. Patier, D.; David, B.; Chalon, R.; Deslandres, V. A New Concept for Urban Logistics Delivery Area Booking. *Procedia Soc. Behav. Sci.* **2014**, *125*, 99–110. [CrossRef]

67. Lopez, C.; Gonzalez-Feliu, J.; Chiabaut, N.; Leclerq, L. Assessing the impacts of goods deliveries' double line parking on the overall traffic under realistic conditions. In Proceedings of the 6th International Conference Information Systems, Logistics, and Supply Chain, Bordeaux, France, 1–4 June 2016; pp. 1–7.

68. Paris City Council. *Technical Guide to Delivery Areas for the City of Paris*; Paris City Council: Paris, Germany, 2005.

69. Allen, J.; Browne, M.; Cherrett, T.; McLeod, F. *Review of UK Urban Freight Studies*; Green Logistics Project Work Module 9; University of Westminster and University of Southhampton: Westminster/Southampton, UK, 2008.

70. Ambrosini, C.; Patier, D.; Routhier, J.L. Urban freight establishment and tour based surveys for policy oriented modelling. *Procedia Soc. Behav. Sci.* **2010**, *2*, 6013–6026. [CrossRef]

71. Holguín-Veras, J.; Jaller, M. Comprehensive Freight Demand Data Collection Framework for Large Urban Areas. In *Sustainable Urban Logistics: Concepts, Methods and Information Systems*; Gonzalez-Feliu, J., Semet, F., Routhier, J.-L., Eds.; Springer: Berlin/Heidelberg, Germany, 2014; pp. 91–112.

72. Allen, J.; Browne, M. *Survey Forms Used in Urban Freight Studies. Transport Studies Group*; University of Westminster: London, UK, 2008.

73. Oskarbski, J.; Kaszubowski, D. Implementation of Weigh-in-Motion System in Freight Traffic Management in Urban Areas. *Transp. Res. Procedia* **2016**, *16*, 449–463. [CrossRef]

![sustainability logo] *sustainability*

MDPI

Article

Decarbonisation of Urban Freight Transport Using Electric Vehicles and Opportunity Charging

Tharsis Teoh [1,*], Oliver Kunze [2], Chee-Chong Teo [3] and Yiik Diew Wong [3]

[1]   Civil Engineering and Geosciences, Delft University of Technology, 2628 CN Delft, The Netherlands
[2]   Resource and Risk Management, Neu-Ulm University of Applied Sciences, 89231 Neu-Ulm, Germany;
     oliver.kunze@hs-neu-ulm.de
[3]   School of Civil & Environmental Engineering, Nanyang Technological University,
     Singapore 639798, Singapore; teocc@ntu.edu.sg (C.-C.T.); cydwong@ntu.edu.sg (Y.D.W.)
*   Correspondence: t.g.h.teoh@tudelft.nl; Tel.: +31-6260-24238

Received: 16 June 2018; Accepted: 7 September 2018; Published: 12 September 2018

Abstract: The high costs of using electric vehicles (EVs) is hindering wide-spread adoption of an EV-centric decarbonisation strategy for urban freight transport. Four opportunity charging (OC) strategies—during breaks and shift changes, during loading activity, during unloading activity, or while driving on highways—are evaluated towards reducing EV costs. The study investigates the effect of OC on the lifecycle costs and carbon dioxide emissions of four cases of different urban freight transport operations. Using a parametric vehicle model, the weight and battery capacity of operationally suitable fleets were calculated for ten scenarios (i.e., one diesel vehicle scenario, two EV scenarios without OC, and seven EV scenarios with four OC strategies and two charging technology types). A linearized energy consumption model sensitive to vehicle load was used to calculate the fuel and energy used by fleets for the transport operations. OC was found to significantly reduce lifecycle costs, and without any strong negative influence on carbon dioxide emissions. Other strong influences on lifecycle costs are the use of inductive technology, extension of service lifetime, and reduction of battery price. Other strong influences on carbon dioxide emissions are the use of inductive technology and the emissions factors of electricity production.

Keywords: urban freight transport; battery electric vehicle; opportunity charging; carbon dioxide emissions; lifecycle costs; parametric vehicle model; evaluation framework

---

1. Introduction

International commitments to reduce carbon dioxide (CO_2) emissions—the most common and pervasive greenhouse gas—has fuelled efforts to *decarbonize* the freight transport sector. For long-distance transport, such as intercity, regional, national or international transport, efforts to reduce CO_2 emissions focus more on the shift to rail or waterways. Nevertheless, alternatives for urban freight transport (UFT) remain limited. One option, the use of battery electric vehicles (BEVs) in UFT is still lagging behind [1], despite its advantages in eliminating local air pollution [2], its relatively quiet [3] and more energy efficient [4] operations, and its capability to use renewable energy sources [5]. Furthermore, recent studies have demonstrated the effectiveness of the BEV-based freight transport to reduce CO_2 emissions, even while accounting for different energy production methods [6].

A wide-spread adoption of BEVs for freight transport faces technical and market-related challenges. Currently, the battery is seen as the limiting factor, linked to tightly constrained operational performance—due to a mix of limited driving distance and slow recharging time—and the high cost of the vehicle [7–9]. Besides the reduced driving distance compared to internal combustion engine vehicles, the addition of the battery also reduces its payload capacity, constrained by a fixed upper weight limit [10]. Further, the ecosystem that supports electric vehicles, such as maintenance and refuelling stations, is absent in many cities (and countries) that would otherwise be conducive for BEV operations [11]. While some governments have succeeded in incentivizing BEV adoption through subsidies for purchases, fiscal measures on fuel, sponsoring BEV trials, and penalizing conventional vehicles [12–14], these measures mainly affect the economic calculation for vehicle choice. They do not affect its operational capabilities. Coping with operational limitations is left to the logistics companies to manage. They have devised a range of strategies to compensate for the shortcomings of BEVs, as shall be explained next.

Fleet managers can deal with the operational limitations of the BEV in four ways: (1) reduce their scope of services, (2) modify transport operations, (3) modify vehicle, and (4) use opportunity charging (OC). Table 1 summarizes the specific measures and selected references to recent studies analysing or discussing them.

Table 1. Strategy to overcome operational limitations of battery electric vehicles (BEVs).

Strategy	Measures	References
Reduce scope of services	Reduce size of area served	-
	Reduce number of customers served	[7,15]
Modify transport operations	Optimize routes and schedules	[15,16]
	Use an urban consolidation centre	[15]
	Increase fleet size	[17,18]
Modify vehicles	Mix the fleet with conventional vehicles	[14]
	Increasing battery capacity of the BEV	[2,17]
	On-board power generators to supplement EVs	[10]
	Other efficiency measures (i.e., lightweighting, aerodynamics)	[16]
Use opportunity charging (OC)	Public charging infrastructure	[14,16,19]
	Semi-public charging infrastructure	[2,14,17]
	Dynamic charging	[17,20,21]
	Battery swap	[15,16]

The first and second strategy works within the limitations of the BEV. Reducing the scope of services aims at eliminating unprofitable routes or operations. The business, as a whole, may suffer, as revenues are expected to reduce along with the services provided. The same set of customers is served in the second strategy, but with significant changes with respect to how the vehicles are used. The third strategy adapts the vehicle's capability to the operational demands, in some cases compromising its pure electric operation. Retrofitted vehicles make use of modularity of their battery systems to provide their operators with the battery capacity they need. However, increasing the battery capacity significantly increases the overall purchase price of the BEV and reduces the payload capacity. The fourth strategy, using OC, integrates quick recharging events during working hours. This contrasts with the conventional time for charging, i.e., at night-time, outside of working hours. OC reduces the need for a large on-board battery, by increasing the dependence on external charging infrastructure. It effectively reduces the driving range requirement *from the daily driving distance to the distances between the locations of two planned charging activities*. The next opportunity for the recharging activity depends on the extent and availability of charging infrastructure, the type of equipment needed on the vehicle, and the pattern of vehicle usage (in time and within the transport network).

In comparison to other strategies, OC maintains the transport service capability, preserves the benefits of the pure electric drive, reduces the purchase cost of the BEV, maintains the operational capability (i.e., driving range and payload capacity), does not disrupt the existing operation schedule, and does not require additional logistics facilities. In general, the downsides of OC are dependence on availability of charging infrastructure and upgrades of electrical infrastructure to support fast charging, faster degradation of the battery, lower overall energy efficiency, and higher CO_2 emissions.

Evaluation studies currently do not consider the wide-range of possibilities to integrate fast-charging into BEV operations. This is regrettable, as different types of OC—depending also on the specifics of where and how they are incorporated—will have different compatibility with different UFT types. Companies willing to experiment with OC are therefore currently still left without comprehensive academic studies in support or in opposition to these options.

Hence, this paper aims to fill this gap by systematically deriving a set of OC strategies and technologies for supporting the use of BEVs in UFT and by evaluating the application of OC in consideration of financial and environmental criteria. The research questions are thus formulated as follows:

(1) To what extent does OC improve the BEV business case for UFT operators?
(2) To what extent does OC affect the decarbonisation benefits of the BEV for UFT operations?

The approach is applied to four different cases of UFT operations modelled according to real-world company data [22]. In the evaluation, the scenarios using OC are compared to scenarios using diesel vehicles, and to scenarios using BEVs but without the use of OC, thus providing evidence on the utility of OC in comparison to just enhancements to the vehicle or battery technology.

The next section is devoted to describing the methodology of the study: the case study descriptions, vehicle usage model, electric mobility system model, and indicator calculation. In Section 3, the results of the case study are presented: the modelled vehicle usage, the electric mobility system specifications, and calculated indicators representing the business case and the decarbonisation benefits. In Section 4, methodology and results are discussed critically in the broader context of BEV studies. Section 5 provides the general conclusions of the investigation, and recommendations for further research.

2. Methodology

Existing studies evaluating BEVs for UFT usually follow three main approaches, each at different levels of detail and emphasis: evaluation of vehicle class [6,23–25], operation-type [2,18], and detailed vehicle usage [15,17,26]. In the evaluation of vehicle class, the BEV is evaluated on the basis of a reference distance of the target vehicle class, such as daily distances of "48–6 km" for a medium-duty vehicle [6]. In the operation-type evaluation, the BEV is evaluated on the basis of simple transport operation scenarios, such as a simplified intermodal truck transport [2]. The detailed vehicle usage approach is evaluated according to micro-level usage of the vehicle, typically using an operations research model [15,17].

This study follows the detailed vehicle usage approach, which consists of the following sequence of steps:

(1) Define urban logistics scenario;
(2) Model vehicle movement for a representative time-period;
(3) Calculate energy consumption for vehicle operation;
(4) Calculate key performance indicators; and
(5) Evaluate indicators according to objectives.

2.1. Case Studies of Urban Freight Transport

A case study approach, in which the UFT activities of singular cases are modelled, was chosen because it would allow for a more specific look at how the characteristics of UFT operations influence their compatibility with BEVs [27]. The four case studies that were selected for the evaluation are summarized in Table 2. Each case is operated on the main island of Singapore.

For each case, a logistics planner (or equivalent role) was interviewed to collect data used to model their UFT operations for a single day. The most detailed data obtained was for Case A, which provided itineraries of deliveries and collections performed by their fleet for one day. When addresses could not be obtained from the interviewee or websites, a randomized selection was performed using QGIS' built-in random selection tool in order to emulate a realistic transport demand.

Table 2. Case study description according to industry sector, product type, and tour structure, as well as data obtained.

Cases	Industry Sector	Product Type	Tour Structure	Data Obtained
Case A	Courier-Express-Parcel	Mail, parcels	1 depot (and many cross-docking locations) to many addresses (delivery & collection)	Sample of itinerary, with addresses, shipment sizes, and service areas. Payload capacity.
Case B	Courier-Express-Parcel	Mail, parcels	3 depots to many addresses (delivery & collection)	Averages of schedule; service area description. Addresses from random selection. Fleet size. Payload capacity.
Case C	Furniture retail chain	Containerized furniture	1 depot to 1 store (7 shuttle trips of about 65 km each)	General schedule, fleet size. Addresses from website. Payload estimated.
Case D	Furniture retail chain	Containerized furniture	1 depot to 1 store (7 shuttle trips of about 16 km each)	General schedule, fleet size. Addresses from website. Payload estimated.

2.2. Vehicle Usage Model

Based on the information obtained, a full work-day vehicle usage schedule was modelled for each case study. A vehicle's usage mirrors the activity of the drivers assigned to it. The vehicle usage model shows the sequence of activities that the driver carries out while driving the vehicle (see Figure 1), with corresponding duration and distance travelled. The vehicle usage model has two main parts: route creation and assignment of routes to each vehicle in the fleet.

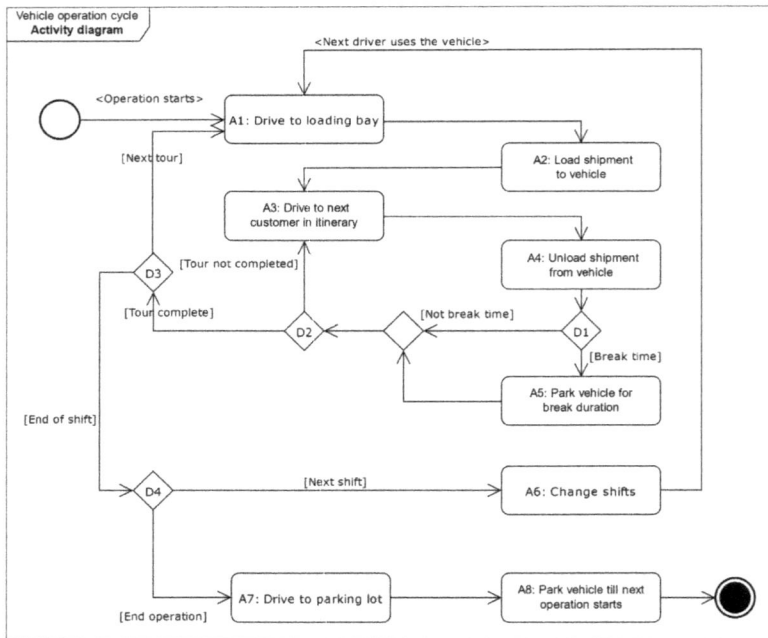

Figure 1. Activity diagram of vehicle operation cycle.

Route creation was performed based on a Vehicle Routing Problem model, implemented in the software XCargo by the company LOCOM GmbH, Karlsruhe, Germany. The software used map data of Singapore to calculate distances and synthetic shipment orders (created for each case using the data obtained from interviews, websites, and background literature) to calculate a set of routes that reduces the overall distance travelled. The number of routes created are determined by service area size and spread, vehicle fleet and number of routes of each vehicle in a day.

The routes are then assigned to individual vehicles in the fleet in a way to balance the total assigned route duration of each vehicle. The distance of each route leg is converted into duration based on constant vehicle speeds. The duration of each route is summed from the driving duration of each route leg and the estimated duration for loading and unloading activities.

The route assignment procedure is:

(1) Assign to each vehicle a route starting from the route with the longest duration;
(2) Assign to the vehicle with the lowest total route duration, the next longest duration route; and
(3) Repeat Step 2, until all routes are assigned or if each vehicle has been assigned the maximum number of routes.

The outcomes of the procedure are the average speed- and payload-time profiles of each vehicle in the fleet, throughout its operation. Note that this procedure can be replaced by any other modelling procedure (e.g., agent-based or operations research models) or simply by reproducing the speed- and payload-time profiles, such as by using GPS tracks in combination with vehicle-diaries.

2.3. Model of the Electric Mobility System

There are two technical subsystems of the electric mobility system: the BEV and the charging system. Cost-efficient BEV parameters shall be identified that can fulfil the travel capability requirements vis-à-vis the energy requirements of the battery and the weight dimensions of the vehicle. The BEV parameters are determined under influence of charging scenarios: a combination of the charging system and strategy.

The following sections describe the development of charging scenarios, the calculation method of the BEV parameters under different scenarios, and the calculation of energy usage at the vehicle and charging system level.

2.3.1. Charging Strategy

A key element of the study is to evaluate the effect of OC as affecting the suitability of BEV. Five OC strategies are evaluated:

- "no OC";
- "OC during break and shift change";
- "OC during loading activity";
- "OC during unloading activity"; and
- "OC while driving on highway".

The first serves as merely a BEV baseline. The BEV is only charged night-time in activity A8. The next three strategies are executed, while the vehicle is stationary, in activities (see Figure 1) A5 & A6, A2, and A4, respectively. The final strategy is performed, while the vehicle is driving on a highway. Note that these OC strategies complement overnight charging, which is assumed in each scenario.

2.3.2. Charging Technology

By considering the energy transfer method (whether conductive or inductive) and in-charging state of motion of the vehicle (whether stationary or dynamic), four general types of charging systems emerge [5]:

- Stationary conductive charging system;
- Dynamic conductive charging system;
- Stationary inductive charging system; and
- Dynamic inductive charging system.

Except for "dynamic conductive charging system", the other charging systems are evaluated in this study. Conductive charging while the vehicle is moving can work via an overhead catenary system or via a third-rail system. While both are commonly applied in rail, the former is also applied in trolley bus or truck systems. The eHighway program by Siemens is, to date, the only known trial of the trolley-truck concept for general cargo [28]. However, the systems have only been designed for large trucks. One can hypothesize that the fixed height of the catenary system would not be suitable for low vehicles, such as vans and smaller trucks. A third-rail system on the other hand is fairly unexplored as an option, except for a recently initiated project eRoadArlanda by the Swedish Transport Administration [29]. Still, little is known about the technical feasibility of that concept. These dynamic conductive charging systems are thus not considered because of interoperability concerns and current lacklustre support for the concepts.

2.3.3. Vehicle and Charging Scenarios

Given the five charging strategies and available charging systems, nine BEV scenarios are evaluated (see Table 3). In S0, the characteristics of the diesel vehicle (DV) is used. S0 serves as a comparison with the other scenarios. S1 and S2 are scenarios without OC. The BEVs are charged overnight using either the conductive or inductive charging systems.

Table 3. Scenarios investigated in the study composed of vehicle type, charging strategy and charging technology.

Scenario ID	Vehicle Type	Charging Strategy	Charging Technology
S0	DV	-	-
S1		no OC	Stationary conductive charging system
S2			Stationary inductive charging system
S3		OC during break and shift change	Stationary conductive charging system
S4	BEV		Stationary inductive charging system
S5		OC during loading activity	Stationary conductive charging system
S6			Stationary inductive charging system
S7		OC during unloading activity	Stationary conductive charging system
S8			Stationary inductive charging system
S9		OC while driving on highway	Dynamic inductive charging system

2.3.4. Parametric BEV Model

In contrast with previous studies that evaluate existing vehicles in the market, the BEVs in this study are adapted to the specified operational requirements of each UFT scenario, i.e., sufficient payload capacity and driving range. The full specifications of the BEV are defined by the gross vehicle weight (GVW), payload capacity, empty weight, battery capacity, and electric motor power. For a given vehicle usage, the amount charged using OC reduces the required battery capacity to fulfil the required driving range. The weight of the battery capacity is calculated by dividing the required battery capacity with the specific energy of 0.14 kWh/kg [30]. The battery weight influences the weight of the rest of BEV, which in turn influences its energy consumption rate while being driven. This circularity requires that the weight, energy consumption and battery capacity be determined simultaneously. The key components on the BEV model, the energy consumption model and the battery capacity estimation model, are discussed next.

2.3.5. Energy Consumption Model

In the vehicle, energy is consumed in three ways. First, energy is consumed when moving. Second, energy is consumed by idling engines. Third, energy is consumed to power up logistics-related equipment, such as refrigeration. For the cases being presented here, the vehicles are neither idle nor do they require additional logistics equipment. The assumption of zero idling energy can be justified in the Singapore's context, where switching off the engines is required by law, and a failure to do so is punished with a fine [31].

The energy consumption is calculated by multiplying the energy consumption rate at the route leg with the distance of the route leg. The rate varies according to the GVW and the current weight in each route. This rate is calculated using FASTSIM, an energy consumption simulation implemented in Excel created by Argonne National Laboratory. It incorporates factors such as vehicle weight, frontal area, length dimensions, driving profile, powertrain components, and regenerative braking [32] in its energy consumption model. Using FASTSIM on a set of dimensions of real-world vehicles, four linear models representing full and empty, diesel and electric vehicles were created (see Figure 2). The model uses the Heavy Duty Urban Dynamometer Driving Schedule as the driving profile.

To estimate the energy consumption rate for a vehicle, triangulate the weight of the vehicle at the route leg, using the GVW and empty weight and their corresponding energy consumption rates. This is calculated simultaneously with other BEV parameters.

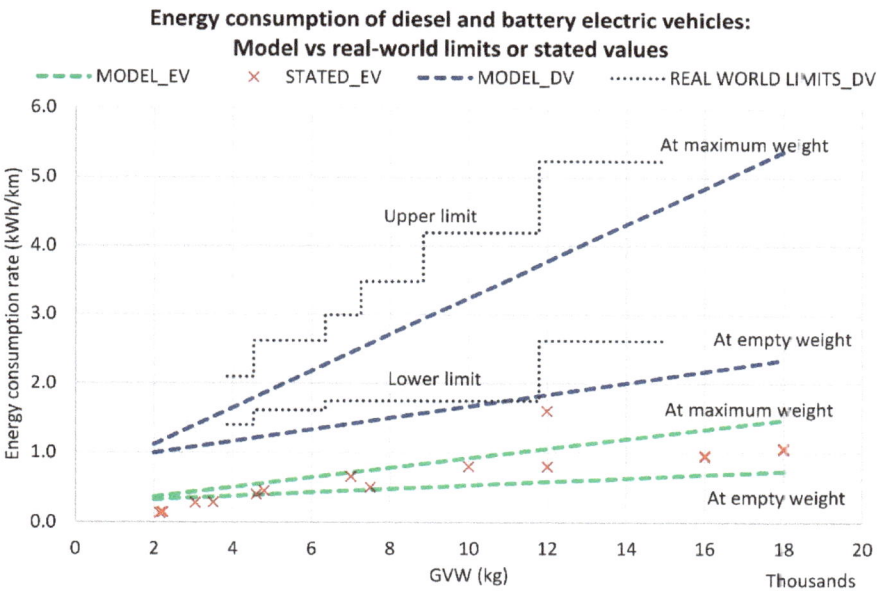

Figure 2. Comparison between energy consumption rates of diesel vehicle (DV) and BEV and the real-world limits for DVs and stated values of manufacturers of real-world BEVs.

The models show a reasonable correspondence to external values, such as the minimum and maximum limits of energy consumption from real-world testing of DVs [33], and the stated values of manufacturers of BEVs [34–43].

2.3.6. Battery Capacity Estimation Depending on Charging Strategy

The use of OC alters the critical energy capacity required of the on-board battery because the energy can be topped-up during the next OC event. In the absence of OC, the battery must last for the whole day till the vehicle returns to the depot at the end of the operations. It is assumed that each vehicle in the fleet uses the same battery capacity. This makes the vehicle with the most intensive "energy critical segment" the limiting vehicle. The battery capacity estimation is derived from the energy capacity of the limiting vehicle's energy critical segment.

Table 4 displays how the critical battery capacity is estimated. When the battery is being used, the battery level reduces, until the charging event. The charging event lasts the duration of the corresponding activity unless the battery depletes its capacity. Energy critical segments occur in between charging. The calculation of the required battery capacity considers the energy critical segments of all the vehicles in the fleet, to ensure that the battery level does not fall below 20% [44].

Table 4. Influence of charging strategy on battery estimation.

Charging Strategy	Charging Event [1]	Energy Critical Segment(s)	Important Determinant for Battery
No OC	A8	From A1 to A8	Vehicle with most energy intensive work load in the day.
OC during break and shift change	A5 or A6	From A1 to A5; from A5 to next A5 or A6	Duration of segmented operating hours
OC during loading activity	A2	From A2 to next A2	Longest route in fleet
OC during unloading activity	A4	From A1 to A4; from A4 to next A4; from A4 to A8	Longest distance from depot to first or last unloading stop
OC while driving on highway	During A1, A3 and A7, on highways only	Driving on urban roads	Longest route only on urban road

[1] The charging events correspond to the activities illustrated in Figure 1.

The vehicle's GVW, empty weight, and battery capacity are set simultaneously, as are the energy consumption and energy charged during each vehicle's route leg, route, and total operation. Other vehicle components are sized based on these parameters. The remaining necessary parameters are calculated as follows:

- Electric motor power: calculated based on a linear model, with the total vehicle weight as the dependent variable.
- Overnight charging power: calculated based on the battery capacity divided by the duration of overnight parking (see A8 from Figure 1).
- Battery replacement cycle in years: calculated based on a fixed charging cycle limit of 3000 cycles [30] and the energy usage of the fleet.

2.3.7. Usage of the Charging System

Efficiency of charging depends on the type of charging system used. The values used in this study are presented in Table 5. OC uses fast charging of either stationary Level 3 or dynamic fast charging systems.

Table 5. Efficiency of charging.

Charging System	Efficiency of Charging (%)	
	Conductive	Inductive
Stationary Level 1	85.8 [45]	78.4 [1]
Stationary Level 2	90.2 [45]	82.3 [46]
Stationary Level 3	88.7 [47]	81.0 [1]
Dynamic fast charging	-	75.0 [2]

[1] Efficiency values for inductive charging stationary level 1 and level 3 were estimated based on the differences in Levels 1, 2 and 3 of conductive charging; [2] Efficiency values for inductive dynamic fast charging were not found in literature but taken as 75%.

2.4. Indicator Calculation

In the comparison between DVs and BEVs, the most important indicators are presented in Table 6. Each indicator has a specific impact scale [48] and relevance to the vehicle types. If a category is found irrelevant to a vehicle type, the value of the indicator is zero. The table also presents the main input variable affecting the quantity of the indicator. The study only focuses on the key indicators, which are relevant to both vehicle types, and whose calculation would not significantly overlap. These are the costs incurred to the fleet owner and the emissions of CO_2.

Table 6. Indicator relevance to DVs and BEVs, in terms of its source and influence.

Categories	Indicators	Impact Scale	DV	BEV	Main Input Variable
Costs incurred to fleet owner	Vehicle cost (and charging system)	Individual	Yes	Yes	Fleet size
	Energy/fuel cost	Individual	Yes	Yes	Energy used
	Maintenance cost	Individual	Yes	Yes	Distance travelled
	Taxation and subsidies	Individual	Yes	Yes	Fleet size
Air and noise pollution	Nitrogen oxides emissions	Local	Yes	No	Energy used
	Volatile organic compounds emissions	Local	Yes	No	Energy used
	Particulate matter emissions	Local	Yes	No	Energy used
	Sulphur oxides emissions	Local	Yes	No	Energy used
	Ozone concentration	Local	Yes	No	Energy used
	Noise exposure	Local	Yes	Yes	Vehicle speed in sensitive area
Energy security and climate change	Efficiency of energy consumption	National	Yes	Yes	Energy used and power mix
	Efficiency of vehicle fuel/energy consumption	National	Yes	Yes	Energy used
	Use of renewable energy sources	Global	No	Yes	Power mix
	CO_2 emissions and other greenhouse gases	Global	Yes	Yes	Energy used and power mix

The study does not include several indicators for the following reasons. BEVs, because of its electric powertrain, do not produce air pollution at the location where the effects of air pollution are detrimental. Instead it is emitted, usually at the outskirts, where the power plants are located. Hence, *local* air pollution produced by BEVs is zero. Also, though noise exposure is an important advantage of the BEV, the calculation is not possible using the methods and data collected in this study, as it requires a full traffic model and population density model [49]. Nevertheless, BEVs are significantly quieter at speeds of less than 30 km/h, just quieter at speeds less than 50 km/h, and non-distinguishable from DVs at speeds above 50 km/h [3], thus excluding them from the study will not be detrimental. Finally, the study evaluates implicitly the energy efficiencies and use of renewables in the evaluation of CO_2 emissions.

The next sections present the procedures to calculate the costs using the lifecycle cost analysis method and the CO_2 emissions.

2.4.1. Lifecycle Cost

The costs incurred to the fleet owner is calculated using the lifecycle cost analysis, which "focuses primarily on capital or fixed assets", emphasizes "purchase price of the asset", and the costs "to use, maintain and dispose of that asset during its lifetime" [50]. The costs incurred throughout the lifecycle are adjusted to the current day value using a discount factor, and finally aggregated into a single indicator, the Net Present Value (NPV) [51]. As per the observable behaviour of vehicle owners in Singapore, the NPV is calculated for the lifecycle period of 10, 15, and 20 years.

The calculated costs are presented in Table 7, together with the cost schedule and relevance to different vehicle types. The selection of cost categories is an important step. Some costs, such as parking and road pricing costs have been excluded, because of zero difference between the DV and BEV.

The discount rate implies that transactions occurring in the future have less worth, although the currency value may be completely the same. This is based on the concept of time preference in micro-economics. The discount factor used in the study is based on a discount rate of 5%, though other studies have used values ranging from 5% to 15% [2,6,15,18,23,52]. The change of the NPV of the BEV scenarios in comparison to the DV scenarios are presented in percentages.

Table 7. Overview of costs calculated per vehicle in the lifecycle cost analysis.

Cost Categories	Cost Schedule	Relevant Factors According to Vehicle Type	
		DV	BEV
Vehicle purchase price	Beginning of lifecycle		
Vehicle base price		Vehicle size	Vehicle size
Battery cost		NA	Size of battery
Electric motor cost		NA	Size of electric motor
Charging receiver		NA	Charging system type
Vehicle purchase cost	Beginning of lifecycle		
Certificate of entitlement		Certificate of entitlement cost	Certificate of entitlement cost
Vehicle registration fees		Vehicle type	Vehicle type
Charging system cost	Beginning of lifecycle		
Charging system price		NA	Charging system type
Installation costs		NA	Charging system type
Battery replacement cost	According to battery replacement cycle	NA	Battery cost in year of replacement
Renewal of certificate of entitlement	In year 10, if the lifetime is extended.	Extension period	Extension period
Road tax	Annually	Vehicle type, size, age	Vehicle type, size, age,
Vehicle insurance	Annually	Vehicle purchase price	Vehicle purchase price
Salary	Annually	Vehicle size	Vehicle size
Maintenance cost	Annually	Total distance travelled and vehicle type and size	Total distance travelled and vehicle type and size
Energy cost	Annually	Fuel prices and total energy consumed.	Eectricity prices, total energy consumed and opportunity charging strategy
Resale of vehicle	End of lifecycle	Vehicle price	Vehicle price

2.4.2. Carbon Dioxide Emissions

The CO_2 emitted in each scenario are estimated based on the energy produced by the power plant for the BEV and the fuel used by the DV. For the BEV, the fuel is burned at the power plant with an emission factor ε_{co_2} of 0.4332 kg CO_2/kWh [53] (as of 2014), with a transmission loss factor of 1.0383 [54]. For the DV, the emission factor $\varepsilon_{co_2,DV}$ of 0.2677 kg CO_2/kWh is used [55]. Note that the DV consumes more energy in kilowatt-hours than the BEV per distance travelled (see Figure 2), so the lower value here is not indicative of lower CO_2 emissions. The change of the CO_2 emissions of the BEV scenarios in comparison to the DV scenarios are presented in percentages.

3. Results

3.1. Vehicle Usage

The routes of Cases A and B (see Table 2) are depicted in Figure 3. Case A (Figure 3a) has a distribution centre in the east and various cross-docking locations scattered around the rest of the island. Case B (Figure 3b) has three distribution centres in Singapore, serving the three different regions. The density of the stops is high and require multiple loading of the vehicles in the day. The routes of Cases C and D are not presented here, because they have only a single delivery location each.

A detailed look at the modelling of fleet's distance travelled is presented in Table 8. The distance categories are chosen as it mirrors the expected critical distances for various OC strategies. Generally, the vehicle in Case C is very intensively used, about 4 times the usage in Case D, over 4 times the average distance travelled in Case B, and over 6 times the average distance travelled in Case A.

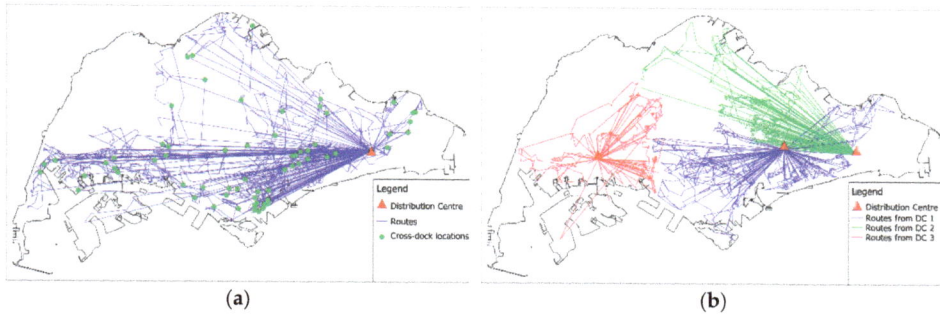

Figure 3. Routes for cases (**a**) A and (**b**) B with transhipment points.

Both Cases A and B show a high discrepancy between the mean and maximum values for the various distance categories. Case B has a slightly lower discrepancy than Case A. This might be attributed to the use of three distribution centres in the latter, compared to the use of a single depot and multiple cross-docks. Note however that the route and schedule planning did not aim to balance the distances, and that this is not a general observation about multiple crossdocking.

Table 8. Route description according to various distance categories.

Case	Case A		Case B		Case C		Case D	
Fleet size	64		53		1			
Total distance	4683		5230		453		114	
Distance statistics	Mean	Max	Mean	Max	Mean	Max	Mean	Max
Distance driven per vehicle	73.2	149.3	98.7	170.8	453.1		114.3	
Distance per schedule segment	34.2	113.4	38.7	87.6	151.0	194.2	38.1	49.0
Distance per route	34.2	113.4	38.7	87.6	64.7		16.3	
Distance per leg	2.3	47.6	1.7	31.4	32.4	33.0	8.2	8.8
Urban roads distance per vehicle	41.1	87.4	85.3	155.5	67.7		73.4	

High discrepancies for the distances mean that the battery capacity for the fleet will likely be oversized, because it is based on the requirement of the limiting vehicle. This leads to carrying additional, expensive and heavy battery in vehicles, which are mostly underused.

3.2. Vehicle System Specification

The battery capacity of BEVs is modified to meet the energy requirements of the operations, according to the different OC strategies. The percentage of energy transferred via OC in each charging scenario is presented in Table 9. The addition of battery to vehicles impacts the total vehicle weight significantly. For instance, the weight increase for S1 and S2 of Case C is 5800 kg for a 594 kWh battery, which is 45% of the weight of the DV. However, the use of OC has a strong impact on the required battery capacity, reducing it down to 29 kWh for S9 of Case C. This reduction varies from case to case, which implies varying suitability to the OC strategies.

The energy transferred via OC shows how much the BEV relied on the external charging network in the scenarios. The increased reliance on the OC network also implies that overnight charging infrastructure can be reduced. As the table shows, the reduction of the battery capacity does not strictly increase with the dependence on OC, although a logical relation can be assumed. More importantly is "when" the OC takes place, as is exemplified in comparing the required battery capacity of S3 and S4 with S9 in Case D. The energy transferred via OC is about the same, but the battery capacity of S9 is at least a quarter for S3 and S4.

Table 9. Vehicle system and effectiveness of opportunity charging (OC).

Case	Scenarios	GVW (kg)	Battery Capacity (kWh)	Energy Transferred Via OC (%)
Case A	S0	2400	-	-
	S1, S2	3100	78	-
	S3, S4	2900	58	68%
	S5, S6	2900	58	46%
	S7, S8	2600	27	73%
	S9	2600	27	79%
Case B	S0	2400	-	-
	S1, S2	3200	88	-
	S3, S4	2700	37	73%
	S5, S6	2800	47	75%
	S7, S8	2500	17	72%
	S9	2700	37	67%
Case C	S0	13,000	-	-
	S1, S2	18,800	594	-
	S3, S4	16,200	332	43%
	S5, S6	14,700	180	74%
	S7, S8	16,200	332	43%
	S9	13,200	29	100%
Case D	S0	13,000	-	-
	S1, S2	14,400	150	-
	S3, S4	13,800	90	57%
	S5, S6	13,200	29	95%
	S7, S8	13,200	29	86%
	S9	13,100	19	56%

3.3. Indicators

To illustrate the changes accrued by different OC scenarios, the indicators were compared with that of the DV scenarios. The change in NPV according to the respective service lifetimes are presented in Figure 4a–c. The changes in CO_2 emissions are presented in Figure 4d.

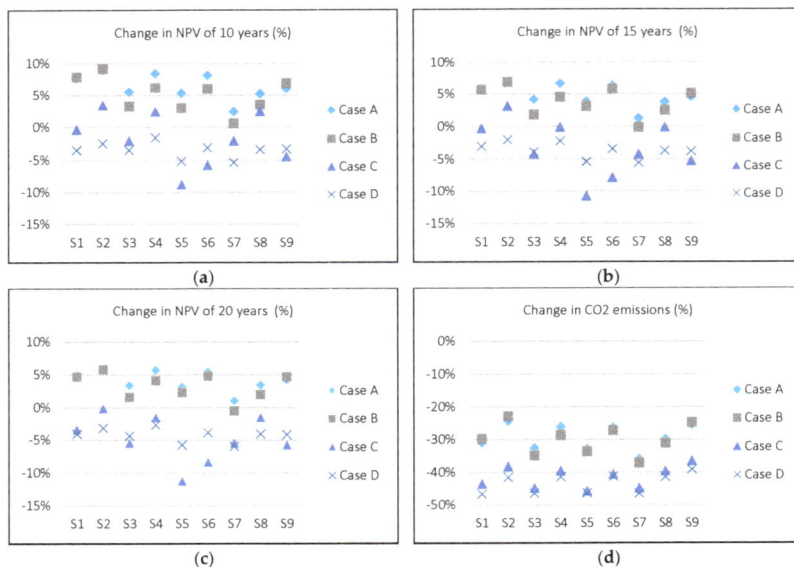

Figure 4. Percentage change of net present value (NPV) for service lifetime (**a**) 10, (**b**) 15 and (**c**) 20 years, and (**d**) CO_2 emissions.

The comparisons of the NPV show that BEV scenarios for the courier transport operations (Cases A and B) perform financially worse than for the furniture full-container-load transports (Cases C and D). More specifically, the BEV-based courier transports are generally not financially viable (i.e.. positive change in NPV). The scenarios in Case C are mostly financially viable. The scenarios also show stark reactions to the OC scenarios and to the use of inductive charging. All scenarios in Case D are fully financially viable, even without the use of OC. They also display a moderate reaction to OC scenarios.

Based on the change in CO_2 emissions, the potential reduction for courier transports (Cases A and B) are systematically less than for furniture full-container-load transports (Cases C and D). Strikingly, the reactions to the charging strategies are similar between the pairs Cases A and B and Cases C and D. There is also a clear increase of CO_2 emissions in inductive charging scenarios.

4. Discussions

With reference to the two research questions, the extent to which OC supports the business case or affects the decarbonisation benefits of using BEVs are discussed.

4.1. Role of Opportunity Charging to Reduce Carbon Dioxide Emissions

The use of OC results in a reduction of CO_2 emissions compared to the scenario without OC (see Rows 1–4, Table 10), except for OC during highway driving for Case C and all the scenarios in Case D. Each case reacts differently to the OC types (i.e., during breaks and shift changes, during loading activity, during unloading activity, or while driving on highways).

To put the size of the impacts into perspective, a ceteris paribus sensitivity analysis was performed testing the influence of charging technology, battery specific energy, and emissions of electricity production (Rows 5–11, Table 10).

4.1.1. Role of Charging Technology

The calculated values for OC (Rows 1–4, Table 10) are based on conductive charging technology. Moving from conductive charging to inductive charging (which could simplify operations) will significantly add to the CO_2 emissions in all case studies as analysed (see Row 5, Table 10)—almost always negating the CO_2 emission benefits of OC. Note that dynamic charging was performed using only inductive technology (Scenario S9) in this study, thus it is always accompanied by an increase in the CO_2 emissions by a large margin. The efficiency of inductive charging should therefore be improved as an enabler of dynamic charging.

4.1.2. Role of Battery Energy Density

The outcome of the sensitivity analysis on the specific energy (Rows 6–7, Table 10) agrees with the literature that its influence on CO_2 emissions is only slight [44]. Unexpectedly, the results do not show that the influence is larger for BEVs with larger batteries, such as in Cases C and D.

4.1.3. Role of Electricity Production Emission Factors

In this study, the role of emissions during electricity production was not analysed in greater detail. An average value for emissions factors based on the electricity production in Singapore of year 2014 [53] was used as the basis for the calculation. Generally, these emissions factors in Singapore could be expected to reduce with renewable energy, improved power plant technology and the import of energy from neighbouring countries [56]. However, the use of static averaged values might also mask the temporal changes of the emissions factors. For instance, Finenko and Cheah [57] showed that in Singapore, real-world emissions factors are only close to the averaged values in the early mornings on weekdays and Saturdays, and generally throughout Sundays and public holidays. The marginal emissions factors can vary up to double the averaged values [57].

A brief sensitivity analysis of the emissions factor (see Rows 8–11, Table 10) show that whether in the positive or negative direction, the value of the emissions factor has a significant impact on the benefits of BEVs, much larger than provided by OC.

Table 10. Changes to the CO_2 emissions due to scenario modifications.

Modifications to Scenario	A	B	C	D
OC during break or shift change	−1.8%	−5.7%	−1.3%	0.2%
OC during loading activity	−2.0%	−4.3%	−2.3%	0.3%
OC during unloading activity	−5.5%	−8.2%	−1.3%	0.2%
OC while driving on highway	−1.0%	−1.9%	1.8%	2.5%
Inductive technology	6.0%	5.8%	5.8%	5.6%
10% specific energy	−0.8%	−1.1%	−0.3%	−0.1%
20% specific energy	−1.6%	−1.9%	−0.5%	−0.1%
+10% emissions factor	6.9%	7.0%	5.6%	5.3%
+20% emissions factor	13.8%	14.1%	11.3%	10.7%
−10% emissions factor	−6.9%	−7.0%	−5.6%	−5.3%
−20% emissions factor	−13.8%	−14.1%	−11.3%	−10.7%

4.2. Role of Opportunity Charging to Improve the Financial Business Case

OC's main role is to reduce the operational limitation, while improving the financial attractiveness of BEVs. To analyse the influence of OC on the lifecycle costs, the changes between inductive charging scenarios and the DV scenario were calculated for each case (see Figures 5 and 6). The inductive charging scenarios (S2, S4, S6, S8, and S9) were used, since the "OC while driving on highway" strategy was calculated only with inductive charging technology. This isolates the influence of charging technology to focus solely on the difference caused by each OC.

The financial cost categories in the analysis are the same as introduced in Table 7, except for the "Misc. finances" category, which includes all taxes and registration fees, and "Vehicle purchase minus resale (minus battery)", which is self-explanatory. The battery costs are considered separately since it is a major cost component and to compare it with the battery replacement costs. A positive value in the figure implies an increase in the cost compared to the DV scenario, and a negative value implies a benefit.

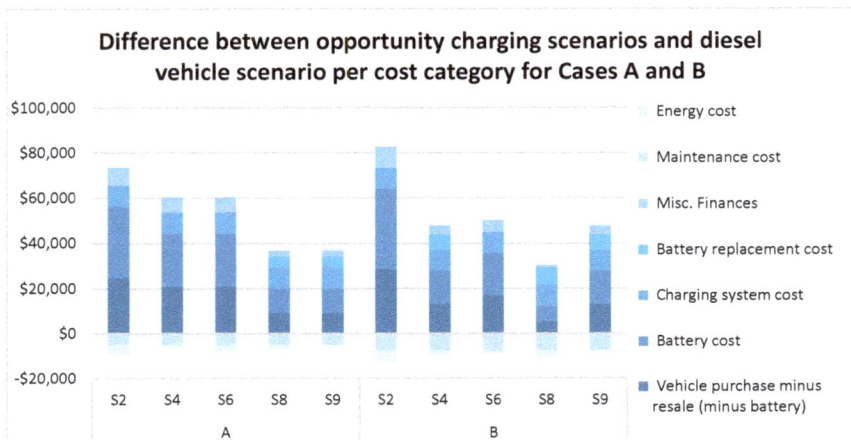

Figure 5. Cost difference breakdown of Cases A and B.

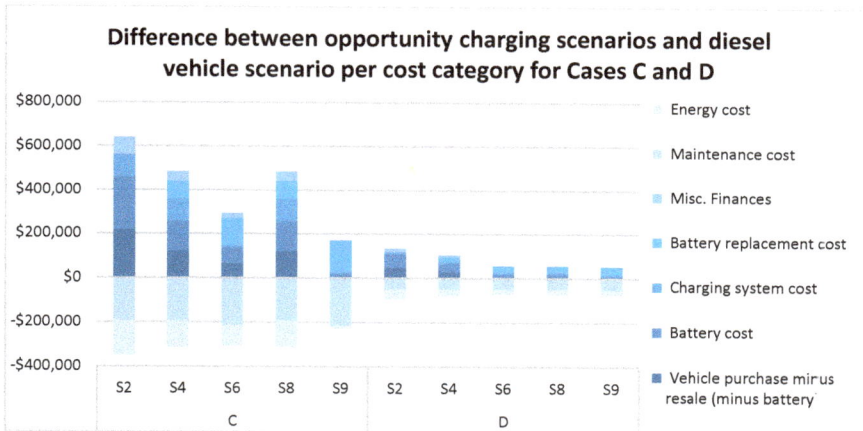

Figure 6. Cost difference breakdown of Cases C and D.

The results show that OC reduces the magnitude of both cost and benefit, for all cases Strikingly, however, is that the costs are reduced substantially more than the benefits, particularly in the purchase costs (vehicle, charging system and battery). On the other hand, battery replacement costs increase slightly. The benefit of lower energy costs also reduces with lower purchase costs. Maintenance costs (according to vehicle model) does not reduce, although it could be expected with lighter vehicles.

In summary, OC improves the business case, although not always sufficiently (i.e., negative total difference). The magnitude of the influence of OC on the NPV is compared with other factors (charging technology, service lifetime, battery specific energy, battery unit price, and electricity prices) using a sensitivity analysis and presented in Table 11.

Table 11. Changes to the net present value.

Modifications to Scenario	A	B	C	D
OC during break or shift change	−0.6%	−3.0%	−1.0%	0.9%
OC during loading activity	−0.9%	−3.2%	−9.2%	−0.6%
OC during unloading activity	−3.8%	−5.7%	−1.0%	−0.9%
OC while driving on highway	−3.0%	−2.3%	−7.9%	−0.8%
Inductive technology	2.5%	2.5%	4.0%	1.8%
+5 years' service lifetime	−1.6%	−1.3%	−1.6%	−0.2%
+10 years' service lifetime	−2.3%	−1.9%	−3.1%	−0.8%
10% energy density	−0.2%	−0.4%	−0.3%	−0.1%
20% energy density	−0.5%	−0.5%	−0.5%	−0.2%
−20% battery price	−2.1%	−2.3%	−3.7%	−1.6%
−10% battery price	−1.0%	−1.1%	−1.9%	−0.8%
10% electricity prices	0.3%	0.4%	1.5%	0.6%
20% electricity prices	0.6%	0.8%	2.9%	1.2%
−20% electricity prices	−0.6%	−0.8%	−2.9%	−1.2%
−10% electricity prices	−0.3%	−0.4%	−1.5%	−0.6%

4.2.1. Role of Charging Technology

Like the effects on CO_2 emissions, the use of inductive charging reduces the benefits of OC (see Row 5, Table 11). The magnitude is greater than most reductions using OC in all the cases, with some exceptions, like in Case C. Part of the reason for the negative influence of inductive charging is the additional costs of the systems. However, as it also substantially increased CO_2 emissions, another reason would be the loss of energy efficiency caused by the systems. As the technology is still at the developmental stage, manufacturers will need to devise smarter ways to reduce the inefficiencies [58,59].

4.2.2. Role of Service Lifetime

Extension of service lifetime is generally accepted to improve the business case of BEVs [23,24], as is shown in our study (Rows 6–7, Table 11). The reasoning is that the high purchase cost of the vehicle and charging system can be potentially off-set by the relatively lower operating and maintenance costs. However, as battery replacement is considered after a fixed set of charging cycles, the potential savings may differ [52]. Further, the study did not consider degradation of the battery over time, assuming that the capacity fade is minimal at 3000 cycles [60].

4.2.3. Role of Battery Specific Energy and Price and Electricity Prices

Improving the battery specific energy (Rows 8–9, Table 11), without reducing the price of battery (Rows 10–11, Table 11) is not effective to reduce the NPV. The decreasing prices for battery replacement based on a variety of factors [61]—from production processes to market forces—was already considered in the main study. Further reduction would have a direct bearing on the purchase price of the vehicle.

Also, unsurprisingly the business case depends on the variation of the electricity prices (Rows 12–15, Table 11). The strongest effect is found in Case C, which also has highest energy usage and cost compared to the other cases (compare Figures 5 and 6). This implies that operational characteristics resulting in high energy cost savings take precedence in improving the business case before the reduction of electricity prices.

4.3. Unused Battery Capacity of the Fleets

The puzzle remains as to why Cases A and B performed poorly financially compared to Cases C and D. One potential reason is the unused battery capacity of the fleet. In the study, the fleets for Cases A and B were taken to be homogeneous in terms of battery capacity and vehicle weight. The battery capacity was sized according to the need of the limiting vehicle, which had the highest workload measured in energy consumption. This meant that those BEVs with a lower workload did not fully utilize the potential cost savings from the lower energy and maintenance costs associated with driving distance range.

If this reason holds true, the effect of changes to the workload over time should also be investigated. The study tried to recreate the vehicle usage for a single day, using the available data. While the study assumed unchanging routes over time, daily transport operations happen within a more complex context, where new routes could be added, old routes modified, and re-routing occur on the fly. This too could result in unrealized savings potential.

As noted previously, OC can help with reducing the battery capacity needed on the vehicles, thus reducing waste. However, in addition, two solutions already identified in literature can also help deal with the expected variability of operational requirements in the fleet and in the future: modularity of the battery system and the use of BEV-suitable routing and scheduling decision support systems (DSS). In the first solution, within the fleet, each BEV can be fitted with the battery capacity it needs. This can be changed in the future, though probably not regularly, when the operational requirements change. In the second solution, the use of a DSS that balances energy expenditure, rather than distance or duration, would reduce energy requirement variability within the fleet or that accounts for mixed fleets with different driving ranges [15,62].

4.4. Availability of Charging Infrastructure

As previously stated, the study assumed installation of charging infrastructure at charging locations and that it is not owned by the freight carrier. Since it is not owned by the user, it is not included into the lifecycle cost analysis, although the price of electricity at different charging locations were varied to reflect different costs. The study also assumed 100% availability of the charging infrastructure at the time needed by freight carrier. These are two strong assumptions about ideal conditions for OC. But, as one can argue, these assumptions do not detract from the utility of the study, rather they highlight the importance of further research.

The need for public charging infrastructure is a common issue [11,63]. In the study, the energy unit cost for all OC was assumed to be higher than for overnight charging by more than 33% [64]. This was used to account for the commercial case of public charging services. Existing literature do not currently discuss business models of charging services for commercial vehicles. However, in comparison to charging stations for passenger vehicles, the business case for providing these services to commercial vehicles are better for the following reasons:

- BEVs for commercial trips do not occupy a parking-cum-charging slot for a long period compared to passenger vehicles, thus the turnover rate for that slot is higher.
- Related to that, existing IT-based management and booking of loading bays can help to ensure availability and high utilization of charging slots for BEVs.
- BEV drivers, which depend on OC to extend their journey, would be more willing to pay the additional premium on the charging bill.

These a priori reasons provide a basis for further research into the business models of charging services for commercial vehicles. Some interested parties could be: utility providers, who have an interest in increasing electricity usage; logistics facilities owners, who can increase revenue sustainably; or vehicle manufacturers, who have an interest in supporting its own products [65].

5. Conclusions

The study argued for the utility of four different OC strategies, particularly from the perspective of lifecycle costs and of the decarbonisation benefits. The BEV scenarios reduced CO_2 emissions by at least 23%, up to at least 39% specifically for full-container-load transport cases (i.e., Cases C and D). Stakeholders, who desire to see CO_2 emissions reduce in the road transport sector will find OC a good approach for most cases. In general, OC was found to reduce lifecycle costs, without a significant trade-off of the decarbonisation benefits. One notes that despite a general reduction of lifecycle costs, none of the scenarios of Case A were financially suitable. Other solutions from the fleet managers' perspective that can be used (see Table 1) must instead be considered.

The study highlighted other potential optimal technological and operational conditions that work together with OC to reduce costs and CO_2 emissions, such as restricting the use of inductive charging, increasing the service lifetime, and reducing the battery and electricity prices. Policy makers can make use of the results, particularly in supporting the business models of charging service providers, reducing regulations that limit the service lifetime of BEVs and promoting the reduction of battery and electricity prices.

Further research in this field could consider a more complete coverage of UFT operations, perhaps using agent-based models that can recreate vehicle usage at an operational level. There is also a need for further understanding the charging service ecosystem and how land use and transport policy can be co-opted to support its development in the commercial vehicle segment. Finally, future work should integrate the plethora of strategies outlined in Table 1 to find optimal bundles of solutions that can push for BEV use in urban freight.

Author Contributions: The paper is based on the Ph.D. dissertation of T.T. Conceptualizatior, T.T., O.K., and C.-C.T.; methodology, T.T. and O.K.; investigation, T.T.; writing—original draft preparation, T.T. and O.K.; writing—review and editing, T.T., O.K., C.-C.T. and Y.D.W., supervision, O.K., C.-C.T. and Y.D.W.

Funding: This work was financially supported by the Singapore's National Research Foundation under its Campus for Research Excellence and Technological Enterprise (CREATE) program. This open access publication is funded by TU Delft.

Acknowledgments: The research study reported here is based on the first author's Ph.D. dissertation under the Joint-PhD program of Technical University Munich and Nanyang Technological University Singapore.

Conflicts of Interest: The authors declare no conflict of interest. The founding sponsors had no role in the design of the study; in the collection, analyses, or interpretation of data; in the writing of the manuscript, and in the decision to publish the results.

References

1. Thomas, M.; Ellingsen, L.A.-W.; Hung, C.R. *Research for TRAN Committee—Battery-Powered Electric Vehicles: Market Development and Lifecycle Emissions: Study*; European Parliament: Brussels, Belgium, 2018.
2. Macharis, C.; van Mierlo, J.; van Den Bossche, P. Combining Intermodal Transport With Electric Vehicles: Towards More Sustainable Solutions. *Transp. Plan. Technol.* **2007**, *30*, 311–323. [CrossRef]
3. Marbjerg, G. *Noise from Electric Vehicles—A Literature Survey: COMPETT (WP3)*; Transportøkonomisk Institutt: Oslo, Norway, 2013.
4. Tie, S.F.; Tan, C.W. A review of energy sources and energy management system in electric vehicles. *Renew. Sustain. Energy Rev.* **2013**, *20*, 82–102. [CrossRef]
5. Sandén, B. (Ed.) *Systems Perspectives on Electromobility*; Chalmers University of Technology: Göteborg, Sweden, 2013.
6. Lee, D.-Y.; Thomas, V.M.; Brown, M.A. Electric Urban Delivery Trucks: Energy Use, Greenhouse Gas Emissions, and Cost-Effectiveness. *Environ. Sci. Technol.* **2013**, *47*, 8022–8030. [CrossRef] [PubMed]
7. Lebeau, P.; de Cauwer, C.; van Mierlo, J.; Macharis, C.; Verbeke, W.; Coosemans, T. Conventional, Hybrid, or Electric Vehicles: Which Technology for an Urban Distribution Centre? *Sci. World J.* **2015**, *2015*, 302867. [CrossRef] [PubMed]
8. Yilmaz, M.; Krein, P.T. Review of Battery Charger Topologies, Charging Power Levels, and Infrastructure for Plug-In Electric and Hybrid Vehicles. *IEEE Trans. Power Electron.* **2013**, *28*, 2151–2169. [CrossRef]
9. Bahn, O.; Marcy, M.; Vaillancourt, K.; Waaub, J.-P. Electrification of the Canadian road transportation sector: A 2050 outlook with TIMES-Canada. *Energy Policy* **2013**, *62*, 593–606. [CrossRef]
10. Pelletier, S.; Jabali, O.; Laporte, G. *Battery Electric Vehicles for Goods Distribution: A Survey of Vehicle Technology, Market Penetration, Incentives and Practices: CIRRELT-2014-43*; CIRRELT: Quebec, QC, Canada, 2014.
11. Lebeau, P.; Macharis, C.; van Mierlo, J. Exploring the choice of battery electric vehicles in city logistics: A conjoint-based choice analysis. *Transp. Res. Part E Logist. Transp. Rev.* **2016**, *91*, 245–258. [CrossRef]
12. Borden, E.J.; Boske, L.B. *Electric Vehicles and Public Charging Infrastructure: Impediments and Opportunities for Success in the United States*; Center for Transportation Research: Austin, TX, USA, 2013.
13. Gass, V.; Schmidt, J.; Schmid, E. Analysis of alternative policy instruments to promote electric vehicles in Austria. *Renew. Energy* **2014**, *61*, 96–101. [CrossRef]
14. Christensen, L.; Klauenberg, J.; Kveiborg, O.; Rudolph, C. Suitability of commercial transport for a shift to electric mobility with Denmark and Germany as use cases. *Res. Transp. Econ.* **2017**. [CrossRef]
15. van Duin, J.H.R.; Tavasszy, L.; Quak, H. Towards E(lectric)-urban freight: First promising steps in the electric vehicle revolution. *Eur. Transp.* **2013**, *54*, 1–19.
16. Taefi, T.T.; Kreutzfeldt, J.; Held, T.; Fink, A. Strategies to Increase the Profitability of Electric Vehicles in Urban Freight Transport. In *E-Mobility in Europe: Trends and Good Practice*; Leal Filho, W., Kotter, R., Eds.; Springer International Publishing: Cham, Switzerland, 2015; pp. 367–388.
17. Teoh, T.; Kunze, O.; Teo, C.-C. Methodology to Evaluate the Operational Suitability of Electromobility Systems for Urban Logistics Operations. *Transp. Res. Procedia* **2016**, *12*, 288–300. [CrossRef]
18. Davis, B.; Figliozzi, M. A methodology to evaluate the competitiveness of electric delivery trucks. *Transp. Res. Part E Logist. Transp. Rev.* **2013**, *49*, 8–23. [CrossRef]
19. Ehrler, V.; Hebes, P. Electromobility for City Logistics—The Solution to Urban Transport Collapse?: An Analysis Beyond Theory. *Procedia Soc. Behav. Sci.* **2012**, *48*, 786–795. [CrossRef]
20. Fuller, M. Wireless charging in California: Range, recharge, and vehicle electrification. *Transp. Res. Part C Emerg. Technol.* **2016**, *67*, 343–356. [CrossRef]
21. Deflorio, F.; Castello, L. Dynamic charging-while-driving systems for freight delivery services with electric vehicles: Traffic and energy modelling. *Transp. Res. Part C Emerg. Technol.* **2017**, *81*, 342–362. [CrossRef]

22. Teoh, T. *Suitability of Battery Electric Vehicles and Opportunity Charging for Urban Freight Transport: An Evaluation Framework*; Dr.-Ing: Munich, Germany, 2018.

23. Feng, W.; Figliozzi, M. An economic and technological analysis of the key factors affecting the competitiveness of electric commercial vehicles: A case study from the USA market. *Transp. Res. Part C Emerg. Technol.* **2013**, *26*, 135–145. [CrossRef]

24. Davis, B.; Figliozzi, M. Lifecycle evaluation of urban commercial electric vehicles and their potential emissions reductions Impacts. In *TRB 92nd Annual Meeting Compendium of Papers*; Transportation Research Board: Washington, DC, USA, 2013.

25. Gallo, J.-B.; Tomić, J. *Battery Electric Parcel Delivery Truck Testing and Demonstration: Public Interest Energy Research (PIER) Program*; Final project report CalHEAT: Pasadena, CA, USA, 2013.

26. Vonolfen, S.; Affenzeller, M.; Beham, A.; Wagner, S. Simulation-based evolution of municipal glass-waste collection strategies utilizing electric trucks. In Proceedings of the 3rd IEEE International Symposium on Logistics and Industrial Informatics, Budapest, Hungary, 25–27 August 2011.

27. Eisenhardt, K.M. Building Theories from Case Study Research. *Acad. Manag. Rev.* **1989**, *14*, 532–550. [CrossRef]

28. Den Boer, E.; Aarnink, S.; Kleiner, F.; Pagenkopf, J. *Zero Emissions Trucks: An Overview of State-of-the-Art Technologies and Their Potential*; CE Delft: Delft, The Netherlands, 2013.

29. eRoadArlanda. Electrified Roads Using Third Rail, Project Website. Available online: https://eroadarlanda.com/ (accessed on 27 May 2018).

30. Burke, A.F. Batteries and Ultracapacitors for Electric, Hybrid, and Fuel Cell Vehicles. *Proc. IEEE* **2007**, *95*, 806–820. [CrossRef]

31. National Environment Agency. Stiffer Fines for Idling Vehicle Engine Repeat Offences from 1 June 2016, Targeted Measures to Deter Repeat Offenders and Minimise Air Pollution. Advisories. 2016. Available online: http://www.nea.gov.sg/corporate-functions/newsroom/advisories/stiffer-fines-for-idling-vehicle-engine-repeat-offences-from-1-june-2016 (accessed on 27 May 2018).

32. NREL. Future Automotive Systems Technology Simulator, FASTSim. 2014. Available online: http://www.nrel.gov/transportation/fastsim.html (accessed on 13 April 2016).

33. Transportation Research Board and National Research Council. *Technologies and Approaches to Reducing the Fuel Consumption of Medium- and Heavy-Duty Vehicles*; National Academies Press: Washington, DC, USA, 2010.

34. Smith Electric Vehicles. *Smith Newton, Brochure*; Smith Electric Vehicles: Kansas City, MO, USA, 2011.

35. Smith Electric Vehicles. *Smith Edison, Brochure*; Smith Electric Vehicles: Kansas City, MO, USA, 2011.

36. Mercedes-Benz. *The Vito E-Cell, Brochure*; Mercedes-Benz: Stuttgart, Germany, 2011.

37. Renault. *Renault Kangoo Van ZE, Brochure*; Renault: Boulogne-Billancourt, France, 2013.

38. Boulder Electric Vehicle. The 500 Series, Available Configurations. 2013. Available online: http://www.boulderev.com/models.php (accessed on 6 May 2018).

39. Boulder Electric Vehicle. The 1000 Series, Available Configurations. 2013. Available online: http://www.boulderev.com/models.php (accessed on 6 May 2018).

40. Nissan. *Nissan e-NV200, Brochure*; Nissan: Yokohama, Japan, 2014.

41. Peugeot. *Peugeot New Partner, Prices, Equipment and Technical Specifications*; Peugeot: Paris, France, 2016.

42. Emoss BV. Electric Trucks. 2017. Available online: http://www.emoss.nl/en/electric-vehicles/full-electric-truck/ (accessed on 6 May 2018).

43. Emoss BV. Electric Delivery Vans. 2017. Available online: http://www.emoss.nl/en/electric-vehicles/electric-delivery-van/ (accessed on 6 May 2018).

44. Peters, J.F.; Baumann, M.; Zimmermann, B.; Braun, J.; Weil, M. The environmental impact of Li-Ion batteries and the role of key parameters—A review. *Renew. Sustain. Energy Rev.* **2017**, *67*, 491–506. [CrossRef]

45. Sears, J.; Roberts, D.; Glitman, K. A Comparison of Electric Vehicle Level 1 and Level 2 Charging Efficiency. In Proceedings of the 2014 IEEE Conference on Technologies for Sustainability (SusTech), Portland, OR, USA, 24–26 July 2014.

46. Idaho National Lab (INL). *PLUGLESS Level 2 EV Charging System (3.3 kW) by Evatran Group Inc., Results from Laboratory Testing as Installed on a 2012 Chevy Volt*; Idaho National Lab (INL): Idaho Falls, ID, USA, 2015.

47. Idaho National Lab (INL). Production EVSE Fact Sheet: DC Fast Charger: Hasetec. 2014. Available online: http://energy.gov/sites/prod/files/2014/02/f8/dcfc_hasetec.pdf (accessed on 6 May 2018).

48. Behrends, S.; Lindholm, M.; Woxenius, J. The Impact of Urban Freight Transport: A Definition of Sustainability from an Actor's Perspective. *Transp. Plan. Technol.* **2008**, *31*, 693–713. [CrossRef]
49. Verheijen, E.; Jabben, J. *Effect of Electric cars on Traffic Noise and Safety: RIVM Letter Report 680300009/2010*; RIVM: Bilthoven, The Netherlands, 2010.
50. Ellram, L.M. Total cost of ownership. *Int. J. Phys. Distrib. Logist. Manag.* **1995**, *25*, 4–23. [CrossRef]
51. Tomic, J.; Gallo, J.-B. Using Commercial Electric Vehicles for Vehicle-to-Grid. In Proceedings of the 26th Electric Vehicle Symposium 2012, Los Angeles, CA, USA, 6–9 May 2012; Curran Associates, Inc.: Red Hook, NY, USA, 2012.
52. Taefi, T.T.; Stütz, S.; Fink, A. Assessing the cost-optimal mileage of medium-duty electric vehicles with a numeric simulation approach. *Transp. Res. Part D Transp. Environ.* **2017**, *56*, 271–285. [CrossRef]
53. Energy Market Authority Singapore. Electricity Grid Emissions Factors and Upstream Fugitive Methane Emission Factor, Statistics. 2016. Available online: https://www.ema.gov.sg/cmsmedia/Publications_and_Statistics/Statistics/OTS12.pdf (accessed on 11 April 2016).
54. Mypower. Transmission Loss Factors. 31 March 2016. Available online: https://www.mypower.com.sg/About/Transmission_Loss_Factors.html (accessed on 11 April 2016).
55. Department for Environment, Food & Rural Affairs (DEFRA). 2012 Greenhouse Gas Conversion Factors for Company Reporting. 2013. Available online: https://www.gov.uk/government/publications/2012-greenhouse-gas-conversion-factors-for-company-reporting (accessed on 5 August 2017).
56. Ali, H.; Sanjaya, S.; Suryadi, B.; Weller, S.R. Analysing CO_2 emissions from Singapore's electricity generation sector: Strategies for 2020 and beyond. *Energy* **2017**, *124*, 553–564. [CrossRef]
57. Finenko, A.; Cheah, L. Temporal CO_2 emissions associated with electricity generation: Case study of Singapore. *Energy Policy* **2016**, *93*, 70–79. [CrossRef]
58. Bosshard, R.; Kolar, J.W. Inductive power transfer for electric vehicle charging: Technical challenges and tradeoffs. *IEEE Power Electron. Mag.* **2016**, *3*, 22–30. [CrossRef]
59. Karakitsios, I.; Palaiogiannis, F.; Markou, A.; Hatziargyriou, N. Optimizing the energy transfer, with a high system efficiency in dynamic inductive charging of EVs. *IEEE Trans. Veh. Technol.* **2018**, 1. [CrossRef]
60. Anseán, D.; González, M.; Viera, J.C.; García, V.M.; Blanco, C.; Valledor, M. Fast charging technique for high power lithium iron phosphate batteries: A cycle life analysis. *J. Power Sources* **2013**, *239*, 9–15. [CrossRef]
61. Nykvist, B.; Nilsson, M. Rapidly falling costs of battery packs for electric vehicles. *Nat. Clim. Chang.* **2015**, *5*, 329–332. [CrossRef]
62. Lin, J.; Zhou, W.; Wolfson, O. Electric Vehicle Routing Problem. *Transp. Res. Procedia* **2016**, *12*, 508–521. [CrossRef]
63. Juan, A.; Mendez, C.; Faulin, J.; de Armas, J.; Grasman, S. Electric Vehicles in Logistics and Transportation: A Survey on Emerging Environmental, Strategic, and Operational Challenges. *Energies* **2016**, *9*, 86. [CrossRef]
64. Snyder, J.; Chang, D.; Erstad, D.; Lin, E.; Rice, A.F.; Goh, C.T.; Tsao, A.-A. *Financial Viability of Non-Residential Electric Vehicle Charging Stations*; UCLA Luskin Center for Innovation: Los Angeles, CA, USA, 2012.
65. Nigro, N.; Welch, D.; Peace, J. *Strategic Planning to Implement Publicly Available EV Charging Stations: A Guide for Businesses and Policymakers*; Center for Climate and Energy Solutions: Arlington, VA, USA, 2015.

sustainability

MDPI

Article

Issues Concerning Declared Energy Consumption and Greenhouse Gas Emissions of FAME Biofuels

Ján Ližbetin *, Martina Hlatká and Ladislav Bartuška

Department of Transport and Logistics, Faculty of Technology, Institute of Technology and Business in České Budějovice, České Budějovice 37001, Czech Republic; hlatka@mail.vstecb.cz (M.H.); bartuska.vste@seznam.cz (L.B.)
* Correspondence: lizbetin@mail.vstecb.cz

Received: 26 June 2018; Accepted: 23 August 2018; Published: 25 August 2018

Abstract: The paper deals with the issue of greenhouse gas emissions that are produced by the road freight transport sector. These emissions affect the structure of the ozone layer and contribute to the greenhouse effect that causes global warming-issues that are closely associated with changing weather patterns and extreme weather events. Attention is drawn to the contradictions linked to FAME (Fatty Acid Methyl Esters) biofuels, namely the fact that although their use generates almost zero greenhouse gas emissions, their production requires high levels of energy consumption. The first part of the paper deals with the theoretical basis of the negative impacts of transport on the environment and the subsequent measurement of the extent of the harmful emissions generated by the road freight transport sector. In the methodical part of the paper, the calculation procedures and declared energy consumption and greenhouse gas emissions generated by transport services are analyzed according to the EN 16258 standard. The experimental part of the paper focuses on the application of the methodology to a specific shipment on a specified transport route, where the total energy consumption and production of greenhouse gas emissions is determined. These calculations are based on comprehensive studies carried out for a particular transport company that assigned the authors the task of determining to what extent the declared energy consumption and greenhouse gas emissions change when the type of fuel used is changed.

Keywords: energy consumption; greenhouse gas emissions; road freight transport; calculation; transport service

---

1. Introduction

Transport, as one of the fundamental parts of a logistics chain, has significant economic influence on the standards of living in the developed countries of the European Union. With the development of transport in the 1990s, came early warning signs of the negative impact of different means of transport on the environment. Initially, it concerned the impact on the environment in urban agglomerations, where emissions from (fossil fuel) engines and noise pollution started to reach permitted limits. The issue of reducing the negative impacts of transport subsequently started to be addressed by a wide range of governmental and non-governmental research organizations [1].

Over time, plans for reducing the negative environmental impact of transport were developed not only at the municipal and regional levels but also at the national and international levels. Scientists started to deal with the issue and developed studies that addressed a much wider range of negative aspects and their impact on the global environment. These works contain, inter alia, many proposals to address the situation, proposals which are considered more or less acceptable in terms of sustaining the growth of national economies and the standards of living of their inhabitants. This poses a fundamental dichotomy, i.e., how do we reduce the negative impacts of transport, as well as all other human activities on the environment, whilst trying to maintain current levels of economic growth and living standards [2,3]?

This article focuses on the specific issue of the declaration of energy intensity and greenhouse gas emissions of biofuels. Biofuels have much more positive results in terms of greenhouse gas emissions, but the efficiency of biofuel production is questionable. FAME (Fatty Acid Methyl Esters) are acids that are created during the transesterification of vegetable oils and animal fats that create biodiesel. FAME is the generic chemical term for biodiesel derived from renewable sources [4]. Therefore, the authors have identified a research question: "How does an increase of the bio-components in diesel fuel affect the energy intensity of biofuel production?" For this research work, the authors used the verified methodology EN 16258, which is characterized in Section 3.2. Using this methodology, the energy intensity of production was calculated for the individual shares of the bio-component [5,6]. In addition to the energy intensity of production, the authors also calculated the production of greenhouse gas emissions, but the calculation is illustrative only and aims to point to the environmental benefits of biofuels [7,8].

2. Literature Review

The issue of greenhouse gas emissions in association with freight transport and its sustainability have been extensively discussed in literature. For example, Quiros et al. [5] suggest that 70% of freight transport is conducted with the use of road vehicles that are responsible for producing 20% of greenhouse gas emissions in the area of transport. A study by Pan et al. [6] focuses on the possibilities for reducing energy consumption and greenhouse gas emissions through the consolidation of freight transport. In particular, they state: "It is well established that the consolidation of freight transport represents an effective way of improving the use of logistics resources." The authors explore the impact the pooling of supply chains at the strategic level might have on the environment. The suggested that pooling of supply networks represents a practical approach to CO_2 emissions reduction. The study presented by Woodcock et al. [7] produced similar results.

The majority of authors [8–13] suggest that the sustainability of freight transport lies in a more effective synchronization of individual kinds of transport; so-called synchromodality. As stated by Agbo and Zhang [8], synchromodality has the potential to increase the use of transport services. The advantages in terms of environmental sustainability lie in the reduced use of lorries and the subsequent reduction in greenhouse gas emissions, congestion, noise, etc.

As stated by several authors [14–17], the use of alternative fuels in road freight transport is another means by which to reduce energy consumption and greenhouse gas emissions. For example, according to Floden and Williamsson [17], the use of biofuels presents a practical way for developing a sustainable system of freight transport. In a study by de Jong et al. [16], measures are examined that could make future biofuel production more efficient.

In contrast to previously published results [8–11,16], the submitted article focuses on a different way of assessing the efficiency of biofuel use, namely from the point of view of the energy consumption required for its production. The methodology of the European Committee for Standardisation [18] was used as a basis for the assessment and was quoted from a study by Konecny and Petro [19].

The methodology is explained in Section 3.2. Having employed this method, the authors tried to point out the changes of the energy consumption of FAME biofuel production in regard to the increasing share of biofuels. The authors did not find this methodology and procedure in any professional literature.

3. Materials and Methods

3.1. Emissions Theory

According to Pohl [20], the negative impacts of transport on the environment can be divided into five basic categories, namely:

- emissions—air pollution associated with the imperfect combustion of fossil fuels;
- noise pollution—from combustion engines and the movement of vehicles along transport routes;
- vibrations—due to the movement of vehicles along transport routes;
- water pollution—from the leakage of working fluids, as well as the leakage of transported substances and fluids due to traffic accidents;
- traffic accidents—in terms of the people and animals killed.

Airborne pollutants are partially transported by air and therefore do not only influence the place where the emissions are generated. In order to fully analyze the negative impacts of transport, it is essential to fully understand the relationship "emission—transmission—deposition—immission".

1. **Emission**—this term describes the generation and release of harmful substances. Emissions are expressed in absolute terms, such as the weight of a specific airborne pollutant, or the pollutants generated by one vehicle in relation to (per) distance travelled.
2. **Transmission**—this term describes the spread of pollutants by air. Transmission depends on a variety of factors (type and amount of emissions, meteorological conditions, etc.).
3. **Deposition**—this term describes the deposition of pollutants at different sites on the Earth's surface due to transmission. Wet deposition includes precipitation in liquid form (e.g., rain, fog, etc.). Dry deposition includes pollutants that fall from the atmosphere due to transmission, mostly in the form of dust.
4. **Immission**—this term describes the concentration of pollutants in the air and their impact on humans and the environment, as well as on, for example, buildings. The scope of the impact directly depends on the concentration of the pollutant at the point and period of its activity. Immissions are expressed in absolute units of mass per volume (e.g., g/m^3).

When describing the impact of specific pollutants, it is necessary to analyze the complex path from emission to immission, since it is the only way to determine with precision the harmful effects, i.e., the relationship between cause and effect [21].

Gases that trap heat in the atmosphere are called greenhouse gases. The Kyoto Protocol highlights six fundamental gases that influence climate change the most: carbon dioxide (CO_2), methane (CH_4), nitrous oxide (N_2O), sulfur hexafluoride (SF_6), hydrofluorocarbons (HFCs), and perfluorocarbons (PFCs) [2,22].

The Kyoto Protocol [16–19] also states that these greenhouse gases must be converted into aggregate average emissions as expressed in CO_2e units (carbon dioxide equivalent). This conversion takes into account the different ability of the identified gases to cause the greenhouse effect, as well as their atmospheric lifetime. Although CO_2 is not the gas with the greatest ability to cause the greenhouse effect, it is the most significant anthropogenic greenhouse gas. It is for this reason that the other gases are converted into CO_2e (see Table 1).

Table 1. Global warming potential of greenhouse gases. GHG: greenhouse gas.

GHG	Chemical Formula	Atmospheric Lifetime (Years)	Global Warming Potential
Carbon dioxide	CO_2	50–200	1
Methane	CH_4	12 ($\pm$3)	21
Nitrous oxide	N_2O	120	310
Sulfur hexafluoride	SF_6	3200	23,900

Source: Authors based on Telang [3].

An example of the CO_2e calculation process follows. The calculation is based on the production of electricity by a diesel aggregate that consumes 100 L of diesel oil.

1. CO_2 emissions = Fuel consumption in unit of volume × CO_2 emission factor

2. CH_4 emissions = Fuel consumption in unit of volume $\times$ CH_4 emission factor
3. N_2O emissions = Fuel consumption in unit of volume $\times$ N_2O emission factor

Overall greenhouse gas (GHG) emissions in tCO_2e = (CO_2 emissions) + ($CH_4 \times 21$) + ($N_2O * 310$)

1. CO_2 emissions = 100×0.00265 [3,23]
2. CH_4 emissions = 100×0.00000036 [3,23]
3. N_2O emissions = 100×0.000000021 [3,23]

Overall GHG emissions = $0.265299393 + (0.000035819 \times 21) + (0.00000215 \times 310) = 0.2667$ tCO_2e

3.2. Methodology for the Calculation and Declaration of Energy Consumption and Greenhouse Gas Emissions for Transport Services According to the EN 16258 Standard

The EN 16258 standard sets out the methodology and requirements for the calculation and declaration of energy consumption and greenhouse gas emissions for transport services. The standard was developed for the purpose of unifying existing carbon footprint calculations and their comparison. In calculations of energy consumption and emissions in relation to vehicles, energy consumption and emissions relating to the production and distribution of fuels or electric energy are taken into account. This ensures that the standard assumes a "Well-to-Wheel" (WtW) approach in terms of calculations and declarations (see Figure 1).

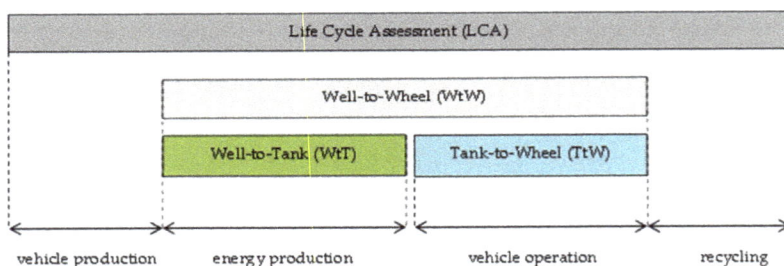

Figure 1. Vehicle life cycle assessment. Source: Authors based on the European Committee for Standardisation [18].

WtW therefore includes the aforementioned energy and emissions for the production of fuels or electric energy, Well-to-Tank (WtT), as well as the energy consumption and greenhouse gas emissions relating to the operation of vehicle Tank-to-Wheel (TtW) [18,19,24–28].

The Well-to-Wheel analysis therefore determines the consumption of fossil fuel energy and the production of CO_2e for driving conditions that correspond to the European homologation cycle. The standard further specifies individual processes and principles that are essential for the correct calculation of energy consumption and greenhouse gas emissions [29–32].

The operational processes of a vehicle must include the operation of all vehicle systems, including propulsion units (main engines), ancillary services, auxiliary equipment used for maintaining the temperature of the load area, and vehicle handling and transshipment systems.

The energy processes for the consumed fuel must include the extraction or primary energy production, refining, transformation, transport, and distribution of energy during all production stages. The principles for calculating the energy consumption and greenhouse gas emissions of transport services must take into consideration all vehicles used for providing transport services, including those that are subcontracted. Furthermore, it must include the overall fuel consumption of each energy carrier and all laden and unladen journeys [18].

3.2.1. Individual Steps of the Calculations for a Specified Transport Service

The calculation(s) for a specified transport service must include the following steps:

1. Identification of the various journeys (legs) that make up the specified transport service.
2. Calculation of energy consumption and greenhouse gas emissions for each leg of the specified transport service.
3. Sum of the results for each leg of the specified transport service [33,34].

The calculation of overall energy consumption and greenhouse gas emissions is performed as follows:

$$E_w \text{ (VOS)} = F \text{ (VOS)} \times e_w \tag{1}$$

$$G_w \text{ (VOS)} = F \text{ (VOS)} \times g_w \tag{2}$$

$$E_t \text{ (VOS)} = F \text{ (VOS)} \times e_t \tag{3}$$

$$G_t \text{ (VOS)} = F \text{ (VOS)} \times g_t \tag{4}$$

where is:

E_w (VOS)—Well-to-Wheel energy consumption vehicle operating system (VOS);
G_w (VOS)—Well-to-Wheel greenhouse gas emissions VOS;
E_t (VOS)—Tank-to-Wheel energy consumption VOS;
G_t (VOS)—Tank-to-Wheel greenhouse gas emissions VOS;
(VOS)—overall fuel consumption VOS;
e_w—Well-to-Wheel energy factor for fuel used;
g_w—Well-to-Wheel greenhouse gases factor for fuel used;
e_t—Tank-to-Wheel energy factor for fuel used;
g_t—Tank-to-Wheel greenhouse gases factor for fuel used.

The values for the energy and greenhouse gas factors are taken from the EN 16258 standard (2013).

3.2.2. Principles for Allocating the Share of Energy Consumption and Emissions per Unit of Cargo

The overall energy consumption and greenhouse gas emissions of the vehicle operation system (VOS) must be allocated to a unit of cargo. The EN 16258 standard applies tkm as the unit for determining transport performance, which is the product of the mass (weight) of the cargo carried and the kilometers travelled [35–37]. The mass is the weight of the cargo transported, including packaging, container(s), pallet, etc. The basic formulas for allocating the cargo are as follows:

$$S \text{ (leg)} = (T \text{ (leg)}) / (T \text{ (VOS)}) \tag{5}$$

$$E_w \text{ (leg)} = E_w \text{ (VOS)} \times S \text{ (leg)} \tag{6}$$

$$G_w \text{ (leg)} = G_w \text{ (VOS)} \times S \text{ (leg)} \tag{7}$$

$$E_t \text{ (leg)} = E_t \text{ (VOS)} \times S \text{ (leg)} \tag{8}$$

$$G_t \text{ (leg)} = G_t \text{ (VOS)} \times S \text{ (leg)} \tag{9}$$

where is:

S (leg)—factor for calculating the share of energy consumption and emissions of the vehicle operation system (VOS) to be allocated to a specified transport service;
T (leg)—transport performance for leg of the specified transport service;
T (VOS)—transport performance VOS.

4. Results

The results presented in this article are taken from a study conducted for a private transport company [38] that entrusted the authors with the task of finding out how a change in the kind of fuel used by its lorries for international road freight transport might influence their energy consumption and greenhouse gas emissions. The calculations were made for a specified transport route, namely Aschaffenburg to Domoradice, as determined by the transport company. The results of the calculations with regard to energy consumption and greenhouse gas emissions according to EN 16258 for the route are summarized in Table 2.

Table 2. Calculated results for diesel.

Diesel with 6% Biocomponent						
Distance (km)	Consumed Fuel (L)	Cargo Weight (t)	E_w (MJ/L)	E_t (MJ/L)	G_w (kgCO$_2$e/L)	G_t (kgCO$_2$e/L)
582	169	17,251	7469.8	6033.30	534.04	424.19

Source: Authors.

Table 3 shows the energy consumption and greenhouse gas emissions figures on the basis of the use of 100% FAME biodiesel. The route is the same but with slight nuances in the distance travelled due to the lorry landing at a different hall.

Table 3. Calculated results for FAME (Fatty Acid Methyl Esters) biofuel.

FAME Biodiesel						
Distance (km)	Consumed Fuel (L)	Cargo Weight (t)	E_w (MJ/L)	E_t (MJ/L)	G_w (kgCO$_2$e/L)	G_t (kgCO$_2$e/L)
596	184	16,830	12,604.00	6035.20	353.28	0.00

Source: Authors.

Figure 2 shows a clearer comparison of the results. The results clearly show that the share of energy consumed for FAME production is much higher than that for diesel. This fact is logical when it is taken into consideration that this type of fuel is made from industrial crops. In this case, it does not concern extraction but the cultivation of crops and the process of their transformation into a substance that can be used, for example, as a fuel for compression-ignition engines.

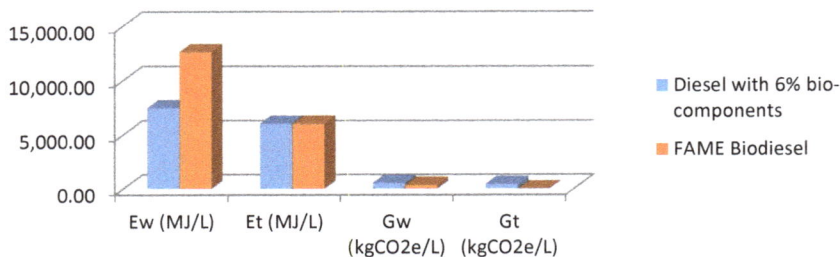

Figure 2. Comparison of diesel with 6% biocomponent and FAME biodiesel. Source: Authors.

What is even more important is the fact that there are no greenhouse gas emissions during combustion in compression-ignition engines. Greenhouse gas emissions are generated only in the production and distribution of FAME, which is also partly influenced by the natural production of CO_2 during the cultivation of the plants. However, it should be stressed that this article only deals with greenhouse gas emissions, i.e., carbon dioxide (CO_2), methane (CH_4), nitrous oxide (N_2O), sulfur hexafluoride (SF_6), hydrofluorocarbons (HFCs), and perfluorocarbons (PFCs) [39–42].

5. Discussion

Having conducted their study of energy consumption and greenhouse gas emissions for various kinds of fuels, the authors were motivated by previous existing literature [13–16,43–50] to apply the methodology to the calculation of the energy consumption related to FAME biofuel production and the EN 16258 standard.

The results (see Figure 2) reveal a relatively large increase in energy consumption for FAME biodiesel. The applied method enables a quick and easy analysis of the energy consumption and greenhouse gas emissions related to the production of mixed fuels.

The calculations that follow are based on the following assumption: that the same model of vehicle goes from point A to point B and always consumes 100 L of diesel, but that the volume of the biocomponent in the diesel differs—from pure diesel (0% biocomponent) up to 100% FAME biodiesel.

The calculations were made for all available volume shares of diesel/biodiesel [51,52]. In order to calculate the energy consumption on the basis of fuel production and distribution, so-called WtT (Well-to-Tank), the difference in the numbers for total energy consumption (WtT) and the energy consumed for the transport service (TtW) needed to be explored. The results for WtT are presented in MJ units in Table 4.

Table 4. Energy consumption analysis for FAME biofuels.

Biocomponent (in %)	0%	1%	2%	5%	8%	10%	15%	20%	50%	85%	100%
TtW e_t [MJ]	3590	3590	3580	3570	3570	3560	3540	3530	3440	3330	3280
WtW e_w [MJ]	4270	4300	4320	4400	4480	4530	4660	4790	5560	6460	6850
TtW g_t [kgCO₂e]	267	264	262	254	246	240	227	214	134	40	0
WtW g_w [kgCO₂e]	324	323	321	317	313	311	304	298	258	212	192
WtT [MJ]	680	710	740	830	910	970	1120	1260	2120	3130	3570

Source: Authors based on [17,22,38].

The following Figure 3 compares an increase in energy consumption for WtT production of mixed fuel diesel/biodiesel in relation to a percentage volume share of biocomponents with a decreasing production of greenhouse gas emissions from the entire life cycle of WtW fuels.

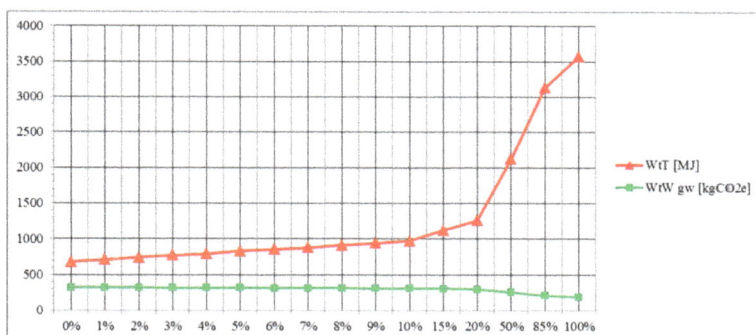

Figure 3. Comparing results of analysis on FAME biofuels energy consumption. Source: Authors.

It is clear from Figure 3 that under current technologies the production of biofuels, namely FAME biofuels, is unacceptably energy-intensive. This results in very poor competitiveness in the fuel market. Fossil fuels are currently considered environmentally unsuitable and, as a result, are presently listed as non-renewable energy sources. Biofuels have the character of a renewable source of fuel, but their huge drawback is their overly "expensive production" (which is shown in Figure 3). It will therefore be necessary to look for a more energy-efficient biofuel production process in order to be competitive with other types of fuels.

6. Conclusions

If we compare the lowest greenhouse gas emissions of 100% FAME biodiesel and the lowest energy consumption necessary for the production of 100% diesel (fuel with 0% biocomponent), it is possible to draw an interesting conclusion. To reduce greenhouse gas emissions by 69% compared to common diesel, we must produce biofuel, the production and distribution of which would consume more than four times more energy than diesel.

Biofuels, therefore, have considerably more favorable parameters than conventional fossil fuels in terms of greenhouse gas emissions, but their production is inefficient in terms of energy intensity. The high energy intensity of the FAME biofuel production is negatively reflected in its fuel price. The price is still higher than the price of conventional fuels, which results in a lack of interest in using more environmentally acceptable types of fuel. There is still a large group of customers who do not take into account the production of greenhouse gas emissions but only fuel costs.

This opens up a new area of research, namely into the search for possibilities to lower this level of energy consumption to an acceptable level so that 100% FAME biodiesel becomes competitive on the fuel market, thereby contributing to the reduction of the negative impacts of transport on the environment and ensuring sustainable transport development.

In conclusion, the authors want to highlight the need for a new approach in the assessment of energy consumption and greenhouse gas emissions in transport due to the high energy consumption linked to the production of FAME biofuels. Within this context, the presented results should open up a broader discussion into the sustainability of biofuels as a whole.

Author Contributions: M.H. conceived and designed the calculations; L.B. and J.L. analyzed the data; J.L. and L.B. wrote the paper.

Funding: This research received no external funding.

Acknowledgments: The authors would like to thank to the GW Logistics company for practical measurements and computations.

Conflicts of Interest: The authors declare that there are no conflicts of interest.

References

1. Duffy, A.; Crawford, R. The effects of physical activity on greenhouse gas emissions for common transport modes in European countries. *Transp. Res. Part D* **2013**, *19*, 13–19. [CrossRef]
2. Schmied, M.; Knörr, W. *Calculating GHG Emissions for Freight Forwarding and Logistics Services in Accordance with EN 16258*; European Association for Forwarding, Transport, Logistics and Customs Services (CLECAT): Bruxelles, Belgium, 2012.
3. Telang, S. How to Calculate Carbon Dioxide Equivalent Emissions from Different GHG Sources? Available online: http://greencleanguide.com/2012/06/05/how-to-calculate-carbon-dioxide-equivalent-emissions-from-different-ghg-sources/ (accessed on 8 November 2015).
4. FAME Biodiesel—Everything You Need to Know about Fame Diesel. Available online: https://www.crownoil.co.uk/everything-need-know-fame-biodiesel/ (accessed on 17 July 2018).
5. Quiros, D.C.; Smith, J.; Thiruvengadam, A.; Huai, T.; Hu, S.H. Greenhouse gas emissions from heavy-duty natural gas, hybrid, and conventional diesel on-road trucks during freight transport. *Atmos. Environ.* **2017**, *168*, 36–45. [CrossRef]
6. Pan, S.; Ballot, E.; Fontane, F. The reduction of greenhouse gas emissions from freight transport by pooling supply chains. *Int. J. Prod. Econ.* **2013**, *143*, 86–94. [CrossRef]

7. Woodcock, J.; Edwards, P.; Tonne, C.; Armstrong, B.G.; Ashiru, O.; Banister, D.; Beevers, S.; Chalabi, Z.; Chowdhury, Z.; Cohen, A.; et al. Health and Climate Change 2 Public health benefits of strategies to reduce greenhouse-gas emissions: Urban land transport. *Lancet* **2009**, *374*, 1930–1943. [CrossRef]

8. Agbo, A.A.; Zhang, Y.W. Sustainable freight transport optimisation through synchromodal networks. *Cogent Eng.* **2017**, *4*. [CrossRef]

9. Wang, M.; Thoben, K.D. Sustainable Urban Freight Transport: Analysis of Factors Affecting the Employment of Electric Commercial Vehicles. In *Dynamics in Logistics, Proceedings of the 5th International Conference LDIC, Bremen, Germany, 22–26 February 2016*; Book Series: Lecture Notes in Logistics; Springer: Berlin, Germany, 2017; pp. 255–265. [CrossRef]

10. Schliwa, G.; Armitage, R.; Aziz, S.; Evans, J.; Rhoades, J. Sustainable city logistics—Making cargo cycles viable for urban freight transport. *Res. Transp. Bus. Manag.* **2015**, *15*, 50–57. [CrossRef]

11. Bouhana, A.; Zidi, A.; Fekih, A.; Chabchoub, H.; Abed, M. An ontology-based CBR approach for personalized itinerary search systems for sustainable urban freight transport. *Expert Syst. Appl.* **2015**, *42*, 3724–3741. [CrossRef]

12. Pena, D.; Tchernykh, A.; Radchenko, G.; Nesmachnow, S.; Ley-Flores, J.; Nazariega, R. Multiobjective Optimization of Greenhouse Gas Emissions Enhancing the Quality of Service for Urban Public Transport Timetabling. In Proceedings of the 4th International conference on Engineering and Telecommunication (EN&T), Moscow, Russia, 29–30 November 2017; pp. 114–118.

13. Oberscheider, M.; Zazgornik, J.; Henriksen, C.B.; Gronalt, M.; Hirsch, P. Minimizing driving times and greenhouse gas emissions in timber transport with a near-exact solution approach. *Scand. J. For. Res.* **2013**, *28*, 493–506. [CrossRef]

14. Schade, B.; Wiesenthal, T.; Gay, S.H.; Leduc, G. Potential of Biofuels to Reduce Greenhouse Gas Emissions of the European Transport Sector. In *Transport Moving to Climate Intelligence: New Changes for Controlling Climate Impacts of Transport after the Economic Crisis*; Springer: Berlin, Germany, 2011; pp. 243–269.

15. Reijnders, L.; Huijbregts, M.A.J. Climate Effects and Non-greenhouse Gas Emissions Associated with Transport Biofuel Life Cycles. In *Biofuels for Road Transport: A Seed to Wheel Perspective*; Springer: Berlin, Germany, 2009; pp. 101–127.

16. De Jong, S.; Hoefnagels, R.; Wetterlund, E.; Pettersson, K.; Faaij, A.; Junginger, M. Cost optimization of biofuel production—The impact of scale, integration, transport and supply chain configurations. *Appl. Energy* **2017**, *195*, 1055–1070. [CrossRef]

17. Floden, J.; Williamsson, J. Business models for sustainable biofuel transport: The potential for intermodal transport. *J. Clean. Prod.* **2016**, *113*, 426–437. [CrossRef]

18. European Committee for Standardisation. *EN 16258: Methodology for Calculation and Declaration of Energy Consumption and GHG Emissions*; European Committee for Standardisation (CEN): Brussels, Belgium, 2012.

19. Konecny, V.; Petro, F. Calculation of selected emissions from transport services in road public transport. In Proceedings of the 18th International Scientific Conference, LOGI 2017, České Budějovice, Czech Republic, 19 October 2017; Volume 134.

20. Pohl, R. *Úvod do Dopravní a Manipulační Techniky: Městské, Silnční, Vodní, a Vzdušné Dopravní Prostředky*; Asociace Dopravních Inženýrů Spolku Česká Technika Při ČVUT: Praha, Czech Republic, 1997; ISBN 80-238-2067-2.

21. Becker, U.; Gerike, R.; Winter, M. *Základy Dopravní Ekologie*; Ústav pro Ekopolitiku: Praha, Czech Republic, 2008; ISBN 978-80-87099-05-6.

22. Overview of Greenhouse Gases. Available online: https://www.epa.gov/ghgemissions/overview-greenhouse-gases (accessed on 17 July 2018).

23. EFDB. The Intergovernmental Panel on Climate Change. 2015. Available online: http://www.ipcc-nggip.iges.or.jp/EFDB/main.php (accessed on 14 November 2015).

24. Lizbetin, J.; Stopka, O.; Nemec, F. Methodological Assessment of Environmental Indicators in Combined Transport in Comparison with Direct Road Freight Transport. In Proceedings of the International Conference Transport Means 2016, Kaunas, Lithuania, 5–7 October 2016; pp. 151–155.

25. Nadolski, R.; Ludwinek, K.; Staszak, J.; Jaskiewicz, M. Utilization of BLDC motor in electrical vehicles. *Przeglad Elektrotechniczny* **2012**, *88*, 180–186.

26. Tol, R.S.J. The Economic Impacts of Climate Change. *Rev. Environ. Econ. Policy* **2017**, *12*, 4–25. [CrossRef]

27. Nathanail, E.; Gogas, M.; Adamos, G. Smart interconnections of interurban and urban freight transport towards achieving sustainable city logistics. *Transp. Res. Procedia* **2016**, *14*, 983–992. [CrossRef]
28. Kijewska, K. The Importance of the City Logistics Manager for the Sustainable Development of Urban Freight Transport. In Proceedings of the Carpathian Logistics Congress (CLC 2013), Cracow, Poland, 9–11 December 2013; pp. 84–89.
29. Ruesch, M.; Hegi, P.; Haefeli, U.; Matti, D.; Schultz, B.; Rutsche, P. Sustainable goods supply and transport in conurbations: Freight strategies and guidelines. *Procedia Soc. Behav. Sci.* **2012**, *39*, 116–133. [CrossRef]
30. Mattila, T.; Antikainen, R. Backcasting sustainable freight transport systems for Europe in 2050. *Energy Policy* **2011**, *39*, 1241–1248. [CrossRef]
31. Branza, G.; Nistor, C.; Popa, L.; Surugiu, G. The impact of technological changes on sustainable intermodal freight transport in Europe. In Proceedings of the 6th International Conference on the Management of Technological Changes, Alexandroupolis, Greece, 3–5 September 2009; Volume 1, pp. 29–32.
32. Black, W.R. Sustainable solutions for freight transport. In *Globalized Freight Transport: Intermodality, E-Commerce, Logistics and Sustainability*; Edward Elgar Publishing: Cheltenham, UK, 2007; pp. 189–216. ISBN 978-1-84542-502-9.
33. Konecny, V.; Kostolna, M. Deklarovanie spotreby energie a emisií skleníkových plynov z dopravných služieb. *Železničná Doprava a Logistika* **2014**, *2*, 41–49.
34. Iglinski, H.; Babiak, M. Analysis of the potential of autonomous vehicles in reducing the emissions of greenhouse gases in road transport. *Procedia Eng.* **2017**, *192*, 353–358. [CrossRef]
35. Meunier, D.; Quinet, E. Valuing greenhouse gases emissions and uncertainty in transport cost benefit analysis. *Transp. Res. Procedia* **2015**, *8*, 80–88. [CrossRef]
36. Petro, F.; Konecny, V. Calculation of emissions from transport services and their use for the internalisation of external costs in road transport. *Procedia Eng.* **2017**, *192*, 677–682. [CrossRef]
37. Roso, V.; Brnjac, N.; Abramovic, B. Inland Intermodal Terminals Location Criteria Evaluation: The Case of Croatia. *Transp. J.* **2015**, *54*, 496–515. [CrossRef]
38. Fabera, P. Výpočet a Deklarace Emisí Skleníkových Plynů Nákladní Silniční Dopravy ve Společnosti GW Logistics A.s. Ph.D. Thesis, The Institute of Technology and Business in České Budějovice, Budějovice, Czech Republic, 2016.
39. Cristea, A.; Hummels, D.; Puzzello, L.; Avetisyan, M. Trade and the greenhouse gas emissions from international freight transport. *J. Environ. Econ. Manag.* **2013**, *65*, 153–173. [CrossRef]
40. Stojanovic, D.; Velickovic, M. The Impact of freight Transport on greenhouse gases emissions in Serbian Cities—The Case of Novi Sad. *Metal. Int.* **2012**, *17*, 196–201. [CrossRef]
41. Bartuska, L.; Gross, P.; Nemec, F. Measurement of the Efficiency of the Combustion Engine Using a Mixture of Hydrogen-oxygen Gas. In Proceedings of the International Conference Transport Means 2016, Kaunas, Lithuania, 5–7 October 2016; pp. 481–486.
42. Vermeulen, A.T.; van Loon, M.; Builtjes, P.J.H.; Erisman, J.W. Inverse transport modeling of non-CO_2 greenhouse gas emissions of Europe. In *Air Pollution Modeling and Its Application XIV, Proceedings of the 24th NATO/CCMS International Technical Meeting on Air Pollution Modelling and Its Application, Boulder, CO, USA, 15–19 May 2000*; Springer: Berlin, Germany, 2001; pp. 631–640. ISBN 0-306-46534-5.
43. Zamora-Cristales, R.; Sessions, J.; Marrs, G. Economic implications of grinding, transporting, and pretreating fresh versus aged forest residues for biofuel production. *Can. J. For. Res.* **2017**, *47*, 269–276. [CrossRef]
44. Licht, F.O. Biofuels in transport in 2016. *Int. Sugar J.* **2016**, *118*, 200–203. [CrossRef]
45. Hao, H.; Geng, Y.; Li, W.Q.; Bin Guo, B. Energy consumption and GHG emissions from China's freight transport sector: Scenarios through 2050. *Energy Policy* **2015**, *85*, 94–101. [CrossRef]
46. Fedorko, G.; Molnar, V.; Strohmandl, J.; Vasil, M. Development of Simulation Model for Light-Controlled Road Junction in the Program Technomatix Plant Simulation. In Proceedings of the International Conference Transport Means 2015, Kaunas, Lithuania, 22–23 October 2015; Volume 169, pp. 466–499.
47. Garcia-Alvarez, A.; Perez-Martinez, P.J.; Gonzalez-Franco, I. Energy Consumption and Carbon Dioxide Emissions in Rail and Road Freight Transport in Spain: A Case Study of Car Carriers and Bulk Petrochemicals. *J. Intell. Transp. Syst.* **2013**, *17*, 233–244. [CrossRef]
48. Odhams, A.M.C.; Roebuck, R.L.; Lee, Y.J.; Hunt, S.W.; Cebon, D. Factors influencing the energy consumption of road freight transport. *Proc. Inst. Mech. Eng. Part C J. Mech. Eng. Sci.* **2010**, *224*, 1995–2010. [CrossRef]

49. Simikic, M.; Tomic, M.; Savin, L.; Micic, R.; Ivanisevic, I.; Ivanisevic, M. Influence of biodiesel on the performances of farm tractors: Experimental testing in stationary and non-stationary conditions. *Renew. Energy* **2018**, *121*, 677–687. [CrossRef]

50. Cho, H.U.; Park, J.M. Biodiesel production by various oleaginous microorganisms from organic wastes. *Bioresour. Technol.* **2018**, *256*, 502–508. [CrossRef] [PubMed]

51. Kolb, I.; Wacker, M. Calculation of energy consumption and pollutant emissions on freight transport routes. *Sci. Total Environ.* **1995**, *169*, 283–288. [CrossRef]

52. Vaishnav, P. Greenhouse Gas Emissions from International Transport. *Issues Sci. Technol.* **2014**, *30*, 25–28. [CrossRef]

![sustainability logo] *sustainability*

MDPI

Article

Sustainable Timber Transport—Economic Aspects of Aerodynamic Reconfiguration

Erik Johannes [1], Petter Ekman [1], Maria Huge-Brodin [2] and Matts Karlsson [1,*]

[1] Division of Applied Thermodynamics and Fluid Mechanics, Department of Management and Engineering, Linköping University, Linköping 581 83, Sweden; erijo972@student.liu.se (E.J.); petter.ekman@liu.se (P.E.)
[2] Division of Logistics and Quality Development, Department of Management and Engineering, Linköping University, Linköping 581 83, Sweden; maria.huge-brodin@liu.se
*   Correspondence: matts.karlsson@liu.se; Tel.: +46-13-281199

Received: 27 April 2018; Accepted: 6 June 2018; Published: 12 June 2018

Abstract: There is a need to reduce fuel consumption, and thereby reduce CO_2-emissions in all parts of the transport sector. It is also well known that aerodynamic resistance affects the fuel consumption in a major way. By improving the aerodynamics of the vehicles, the fuel consumption will also decrease. A special type of transportation is that of timber, which is performed by specialized trucks with few alternative uses. This paper follows up on earlier papers concerning Swedish timber trucks where aerodynamic improvements for timber trucks were tested. By mapping the entire fleet of timber trucks in Sweden and investigating reduced fuel consumption of 2–10%, financial calculations were performed on how these improvements would affect the transport costs. Certain parameters are investigated, such as investment cost, extra changeover time and weight of installments. By combining these results with the mapping of the fleet, it can be seen under which circumstances these improvements would be sustainable. The results show that it is possible through aerodynamics to lower the transportation costs and make an investment plausible, with changeover time being the most important parameter. They also show that certain criteria for a reduced transportation cost already exist within the vehicle fleet today.

Keywords: timber trucks; fuel consumption; aerodynamic design; financial consequences

1. Introduction

According to the latest reports, greenhouse gas emissions continue to rise in the world [1]. In Sweden through the 1990s greenhouse gas emissions have increased by 20% [2] and demand for reduced emissions and increasing fuel prices are ongoing challenges for the whole transport sector. A large majority of Sweden's cargo is transported on the road [3] and a special case of that is the timber transports.

The total fleet of timber trucks consists of over 1600 registered timber trucks that transported 60 million tonnes of round wood last year. These trucks represent 2% of all heavy trucks in Sweden [4]. As these trucks are for a special purpose (timber transport only) the empty running is at least 50%, resulting in a fill rate under 50%. This is not only when measured by weight, which ordinary trucks also can achieve, but also measured by volume, which makes timber transports unique [5]. The half part of the transport distance that is performed by empty trucks makes the fleet unique and any changes in design must include not only all problems with an irregular cargo situation (piles of timber) but also the empty vehicles [5]. The average drag coefficient can sometimes also be higher with an empty truck than a loaded one which creates unique situations. Here it is also important to note that drag coefficient is not the same as energy efficiency [5]. A lot of effort has over many decades been made by truck manufacturers to reduce the emissions from the engine itself over many decades. However little progress has been made when it comes to other efforts of reducing the fuel consumption of the timber trucks, and reports show that fuel consumption has remained around 0.58 L/km for the last years, or 0.030 L/tonneskm, seen in Figure 1 [6]. Hence, alternative measures need to be taken. From a technology perspective, the fleet needs to become greener through means other than more efficient engines.

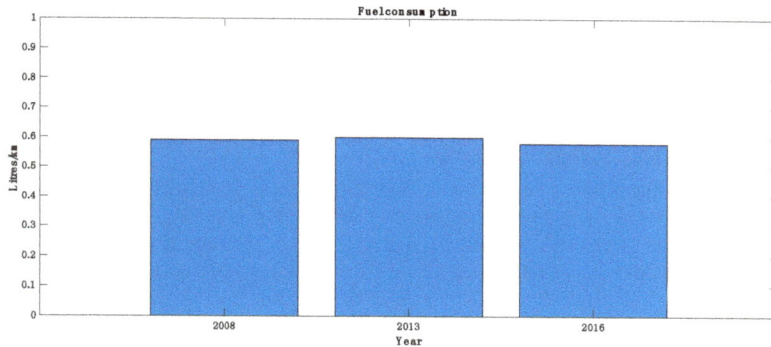

Figure 1. Presents the fuel consumption of timber trucks over the last 10 years. As can be seen, no real development has taken place.

Through earlier research it has been shown that by improving the aerodynamics of the timber trucks the drag coefficient can be reduced, and with that the fuel consumption [5,7]. This reduces the CO_2-emissions from the trucks, and lower fuel costs also benefit the haulers [5]. Fuel efficiency is also a first priority for carriers since it is a major part (30%) of the total operating cost. Earlier research has addressed aerodynamics improvements for trucks and implications from reduced drag on fuel consumption, however holistic analyses of the economic consequences is still missing. Reduced fuel consumption will also lead to reduced cost for the road carriers, but aerodynamics investments will affect the trucks more than just reduced fuel consumption. For example, aerodynamics installment will affect the cost of the trucks and the weight of the trucks as well. The implications this will have on the total transport cost and economic calculations beyond reduced fuel cost is yet to be investigated, but they will have a crucial part in the success of aerodynamic installments and improvements.

This study aims to further earlier research focused on technical, aspects and investigate and conceptualize the economic implications aerodynamic improvements will have on transport costs. The study addresses more parameters of aerodynamic improvements than just the fuel consumption and aims to understand how these parameters will affect transport costs. It addresses the opportunity to introduce aerodynamic improvements in the timber truck fleet as a means to reduce fuel consumption and thereby reduce emissions. Through mapping the entire fleet of vehicles, numerical simulations and economic calculations, the fleet is assessed and the possibility for feasible investment, here described in terms of aerodynamic improvements, is investigated. This paper discusses challenges and opportunities related to making the fleet more sustainable through reconfiguration.

Research Background

This study takes off from prior research in aerodynamic re-design of timber trucks based on a combination of flow simulations and wind tunnel experiments. In addition, this analysis takes inspiration from prior findings from green logistics research.

The flow simulations were done using Computational Fluid Dynamics (CFD). The simulations were performed in the commercially available software ANSYS Fluent 18.0 (ANSYS Inc., Canonsburg, PA, USA). The model has shown good agreement with experimental data for similar studies [8–11].

In CFD simulations the domain consisted of a rectangular tunnel making it a "virtual wind tunnel" where the model was placed 3.5 and 5 truck lengths from the inlet and outlet, respectively, to reduce possible effects from the boundary conditions. The frontal area of the truck covers less than 1% of the domain cross section area to minimize the blockage effect.

The free-stream was modeled with a velocity inlet with a prescribed uniform velocity profile, with turbulence intensity 0.1%, a zero pressure-outlet and free-slip condition for the top and side surfaces of the domain. The grid consisted of triangle surface mesh connected to a Cartesian grid. Refinement of the mesh was done in regions where large gradients were expected. Between 6 and 16 prisms layers were added on all the no-slip surfaces to accurately capture the near wall flow. The two-equation k-ε realizable turbulence model with Enhanced Wall Treatment (EWT) was used for the RANS (Reynolds Averaged Navier-Stokes) simulations.

The ground and the surfaces of the truck (model) were modeled with no-slip condition, and to replicate road conditions moving ground and rotating wheel conditions (rotational velocity boundary conditions) were used. The truck was simulated at a Reynolds number of 4.9 million, based on the truck height, which corresponds to the truck driving at 80 km/h and all simulations were performed at a yaw angle of 5°.

The tractive resistance of a ground vehicle can be divided into four separate parts [12], Equation (1).

$$F_T = F_{ROLL} + F_{ACC} + F_{CLIMB} + F_{DRAG} \tag{1}$$

where F_{ROLL} is the rolling-, F_{ACC} the acceleration-, F_{CLIMB} the hill climb- and F_{DRAG} the aerodynamic drag. For a vehicle driving at constant speed on a relatively flat road, the rolling resistance and aerodynamic drag are the dominating forces. While the rolling resistance varies linearly with the speed of the vehicle, the aerodynamic drag varies with the square of this speed, Equation (2), which increases the importance of it at higher speeds.

$$F_{DRAG} = C_D \cdot 0.5 \cdot \rho_{air} \cdot U^2_\infty \cdot A \tag{2}$$

where C_D is the drag coefficient, ρ_{air} the density of air, U_∞ the free-stream velocity and A the reference area. The results are presented in Table 1. Two representative cases are presented in Table 1. The Baseline truck (BLT) is modelled after a standard, contemporary, road-going configuration, whereas the Bulkhead Shield and the Side Skirts represents possible additions for the Baseline configuration. All C_D reductions are computed using CFD applying the methodology described above; the estimated fuel consumption reduction is estimated as one-third of the C_D reduction according to industry practice.

Table 1. Presents the results from different configurations.

Configuration	C_D	C_D Reduction	Estimated Fuel Consumption Reduction
Baseline	0.70	-	-
Bulkhead Shield	0.63	−10.4%	3.5%
Side Skirts	0.52	−24.8%	8.3%

Total pressure equal to zero represents the large pressure drop, hence energy loss in the flow. The larger region, the larger energy loss for the flow and thereby more drag created. Figures 2–4 presents examples of the aerodynamic simulations where reduction in purple area (=total pressure equal to zero) is clearly seen.

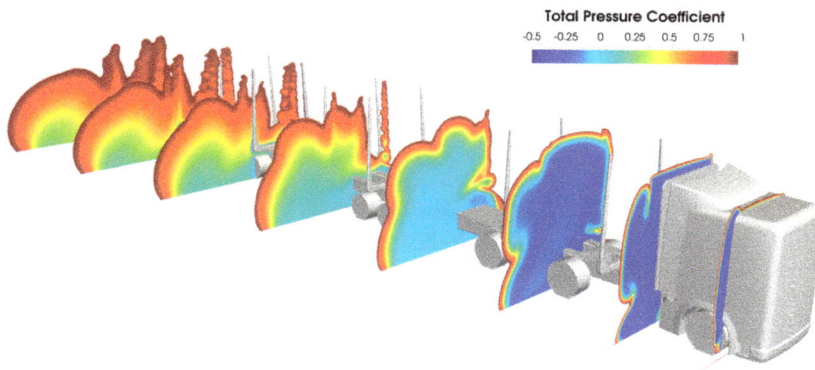

Figure 2. Total pressure equal to zero for the Baseline unloaded timber truck at 5° yaw (C_d = 0.7). Note a quite significant wake behind the cab.

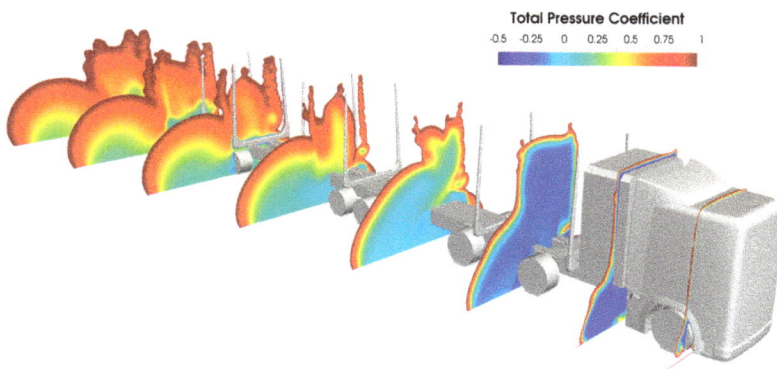

Figure 3. Total pressure equal to zero for the unloaded timber truck fitted with a bulkhead shield at 5° yaw (C_d = 0.63). Note a reduced wake both in the immediate vicinity of the cab and along the trailer.

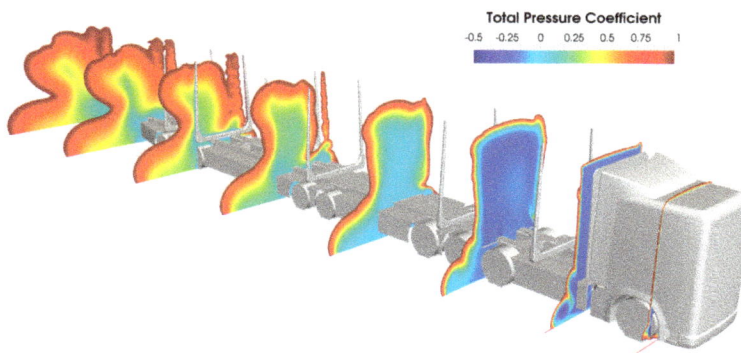

Figure 4. Total pressure equal to zero for the unloaded timber truck fitted with side skirts at 5° yaw (C_d = 0.52). Note a smaller wake that is reduced along the whole of the vehicle.

Green logistics research includes, among other areas, reducing freight transport externalities [13]. While this area comprises measures for reducing noise, as well as unwelcome environmental effects, this study relates mainly to the latter.

In general, road carriers, as well as other logistics service providers (LSPs), have been quite slow and reactive in adopting various green initiatives [14]. Important drivers for adopting green initiatives among LSPs include top management engagement [15], customer demands, legislation and the engagement among employees [14]. Barriers to adopting green initiatives include lack of customer demands, lack of knowledge and insecure investments [16].

Technology in itself is not often a driver or barrier for road carriers wanting to go greener. Overall technological development has in many ways already reduced CO_2-emissions from freight transport, while demands on fast and agile logistics contribute to accelerate them [17]. It is evident that technology can reduce CO_2-emissions from freight [18], however the insecure pay-off times—partly due to short-term contracts with customers—in many cases hinder the implementation of technological solutions for greener freight transport [18]. While green logistics research among other areas address corporate environmental strategies [13], the strategic investment in greening technology is sparsely addressed in prior research.

2. Materials and Methods

In order to collect the data needed for this project, a number of different methods and tools have been used. The initial phase of the project consisted of gathering and the categorizing of the data needed to map the current fleet of timber trucks, including vehicle data and travel distances.

This combined data was used as input for the financial calculations performed to analyze the effects of these aerodynamic improvements. Environmental effects in term of reduced CO_2-emissions will be considered as proportional to reductions in fuel consumptions and will be discussed along with the numerical analyses. In order to understand the strategic and logistics consequences of investments in aerodynamics technology, a set of semi-structured interviews were conducted with different timber carriers to obtain ideas and concerns about these possible aerodynamic improvements.

2.1. Mapping of the Fleet

The study includes all registered timber trucks in Sweden, hence no representative selection was needed. The size of the fleet, 1662 trucks, makes is a suitable fleet to investigate and work with registration data for every single truck.

The initial phase of the project was to categorize the entire fleet of timber trucks, including such data as engine size and where they were registered. This was done to be able to use the results from the economic calculation and try to match them with the results from the mapping and see where in the country the changes of the fuel consumption would have the largest effect.

The mapping consisted of the gathering of data from different databases (publicly available) and also the compiling of these data into a special purpose database. The vehicle data was received from Transportstyrelsen, a state department in Sweden responsible for the all traffic. It is at Transportstyrelsen a Swedish citizen register his/her vehicle and therefore the department is a valuable resource with information about all the vehicles in Sweden. All the data had been collected by Transportstyrelsen.

The data received (per 19 June 2017) consisted of more than 30 different parameters for each and every one of the 1662 timber trucks registered. This data represents a relevant subset of all the data that is handled by Transportstyrelsen. The data was disconnected from the individual vehicles, hence there is no possibility to trace specific data sets back to a specific vehicle. The parameters covered a variety of types of information about the timber trucks, for example from the year it was manufactured to the number of axis and length between them. All the parameters and the information of the trucks used in this report can be found in Table 2.

Table 2. Shows the data used from Transportstyrelsen in this project.

Vehicle Data		Technical Data	
Manufacturer	Engine effect (kw)	Height (mm)	ECO-car
Vehicle type	Fuel	Number of axis	CO_2-emissions
Year	Cylinder volume (cc)	Distance between axis	Particle value for fule
Date of registration	Total weight (kg)	Rim dimensions	Emissions
Coach work code	Tax weight (kg)	Tire dimensions	Tank volume
Energy consumption	Trailer weight (kg)	Coupling device	EEG
	Length (mm)	Taxrate	
	Year of production	Emission-classification	
	EURO-classification	Gearbox	

In the subsequent part the fleet was categorized according to distance travelled. Data was received from Skogsbrukets Datacentral (SDC) and Skogforsk of the distances transported from every one of the 21 counties in Sweden [19]. SDC is an economic association that connects the forestry industry in Sweden, with over 500 forest industry groups connected. SDC regularly publish reports about the forestry industry, timber transports included. Skogforsk is the Swedish research institute for the forestry industry. With the help of that data the length and weight of the transports were analysed. SDC collects data about where transport has taken place but not who or what vehicle did that transport. Data about average distance and total weigh transported was received for three categories, wood used as primarily biofuels, timber and pulpwood for each county. The data from each category were then summarized for each county and average distance and total weight was concluded.

The emissions were calculated using data received from SDC and an emission factor [20]. By using the length of the average transport in the county and multiplying it with the total weight transported in that county the transport work (tonnes-km) was calculated. By then multiplying the transport work with the emission factors, the emissions are calculated. It is important to note that this method was chosen since no more specific data about every single transport in a county was received. The method is then an estimate of the real-life situation, but it is a rough number and works as an indication.

2.2. Financial Calculations and Simulations

To be able to assess the economic result of the improved aerodynamics a Skogforsk program, TransAm, was used. The program is developed by Skogforsk and is an Excel-application for investment calculation of timber trucks [21]. The trucks can be modelled with a trailer and an on-board crane and the cost for each can be set by the user. The costs that can be set for the different versions of trucks are investment cost, salvage value, costs for service and reparation of trucks and wheels but also driver wages and taxes and interest on capital.

The transport cost is of importance when using TransAm and its price function is an important part of the financial calculation [22]. TransAm calculates the transport cost (Swedish kronor per tonnes (SEK/tonnes)) for three user set distances and the cost is calculated as a linear regression of the sum of the fixed and variable costs for the transport distances. The transport cost is calculated from a couple of variables: maximum speed, changeover time, break time between transports, fuel consumption and transport distance. These parameters together with the cost for the vehicle can then be changed for the sensitivity analyses and a new transport cost in SEK/tonnes is calculated compared to a user set base-line scenario [23]. In this report the investment cost, fuel consumption, gross weight and changeover time were investigated besides the change in fuel consumption by the aerodynamic improvements.

The base-line truck (BLT) for this project was a 64 tonnes gross weight timber truck with the standard machinery and 44.8 tonnes net load weight. The truck was fully equipped, which means that it has an on-board crane. The fuel consumption of the BLT was set to 0.58 L/km with a fuel price of 10 SEK/L and the changeover time was set to five minutes and the time for loading of was 40 min with an average speed of 80 km/h for the whole distance. It was also assumed that the BLT-truck drives fully loaded one way and empty on the way back.

To be able to categorize which trucks the improvement of aerodynamics would be useful for, different parameters were analysed to scope out the different requirements that the nature of the trucks and carriers had to have. The different cases are presented in Table 3. The parameters used in this study includes the (extra) weight and (extra) changeover time and the investment cost. The numbers used here are representative values ranging from a very light-weight and simple aero-shield to a fully re-designed truck-trailer outfit, including any combination thereof. The tests in Table 3 represents a comprehensive parameter sweep in order to capture the general behavior of responses due to single parameter variations and are based on a multitude of possible aerodynamically sound concepts generated in a previous pilot study [5].

Table 3. The different parameters investigated in each test.

Test	Parameters	Values Tested
Test 1	Fuel reduction	2%, 4%, 6%, 8% and 10%
Test 2	Weight of instalment	100 kg, 500 kg and 1000 kg
Test 3	Extra changeover time	5 min, 10 min and 15 min extra
Test 4	Investment cost	10,000 SEK, 50,000 SEK and 100,000 SEK

Combination of the parameters in Table 3 were also investigated in to additional cases presented in the result section and in Table 4. These combinations are set up in order to stress-test possible aerodynamic concepts with real-world financial aspects.

Table 4. The different parameters investigated in each case.

Case	Parameters
Case 1	Highest investment cost, highest weight, no extra changeover time
Case 2	Lowest investment cost, lowest weight, no extra changeover time, 5 min extra changeover time and 15 min extra changeover time
Case 3	Highest investment cost, highest weight, 15 min extra changeover time and 10% fuel reduction

These different parameters and cases were then, as mentioned, used to map what requirements a timber truck was to have to be profitable with the investment.

2.3. Interviews

A set of 4 interviews with road carriers were performed. The interviews were semi-structured and based on a thematic interview protocol with the themes around the business model for how road carrier contracts are designed. The interviews were conducted by phone. The results from the interviews are used in the analysis and discussion to enrich the analysis of the calculations by adding a company perspective.

3. Results

In this section the results of the project are presented. The results are presented according to the different phases in the project. First, general results about the mapping of the vehicle fleet are presented followed by a more extensive presentation of the economical results from TransAm.

3.1. The Mapping

Today the fleet of timber trucks in Sweden consists of 1662 vehicles spread out across all the counties in the country. All the counties are represented and the most number of registered trucks can be found in Västernorrlands' county, as seen in Figure 5.

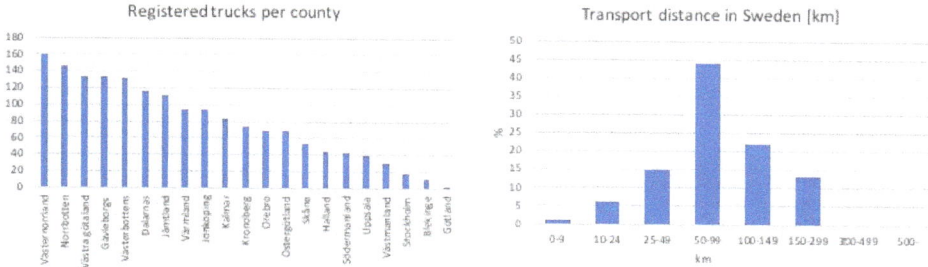

Figure 5. (**left**) Presents the number of registered timber trucks per county; (**right**) Presents the transport distance in Sweden.

The distance of the timber transports is short, where the majority the transports are no longer than 100 km. A reason for this is that a big part of the timber is only transported within the county, due to the strategic placing of mills close to the resources [24]. The average distance for timber transport was 81.9 km [24] and almost 50% of all transports were between 40–99 km, but more than 75% are also longer then 50 km. Not so surprisingly there is a correlation between which county has transported the most weight and the counties with the most registered trucks, presented in Figure 5. There is no overcapacity, and the counties with the most trucks are also among the top counties with the most weight. The county with the most shipped weight according to transport work was Västernorrland. The smallest amount on shipped weight took place in Gotland, the county with also the smallest amount of registered trucks, presented in Figure 6.

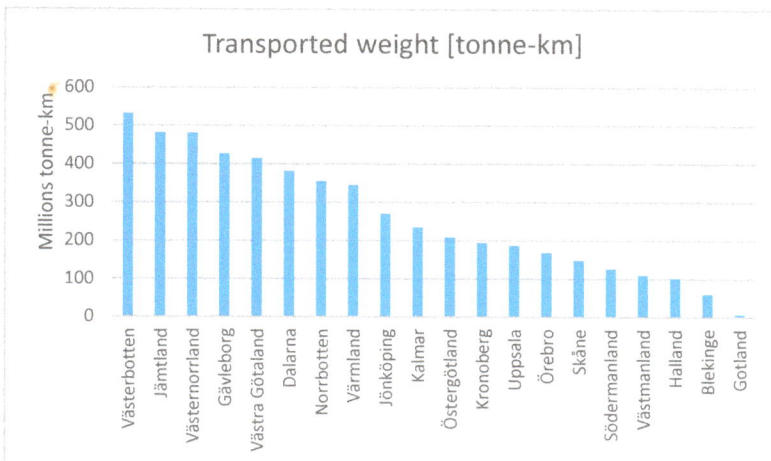

Figure 6. The distribution of counties in order of which county transported the longest average distance and the total weight transported in that county.

For the registered timber trucks, a number of trends can be seen. An unsurprising trend is that trucks and the freight trains are getting heavier the newer the truck is. This is presented in Figure 7. In the figure it can be seen that the biggest percentage of the freight trains weigh in average about 70–90 tonnes. It can also be seen that the development the latest years show that the freight trains are getting heavier. The trucks that have been left out of the figure were older models and lighter trucks, resulting in 1614 samples. The weight of the freight trains in this figure is the max weight combined with the trailer weight for each truck.

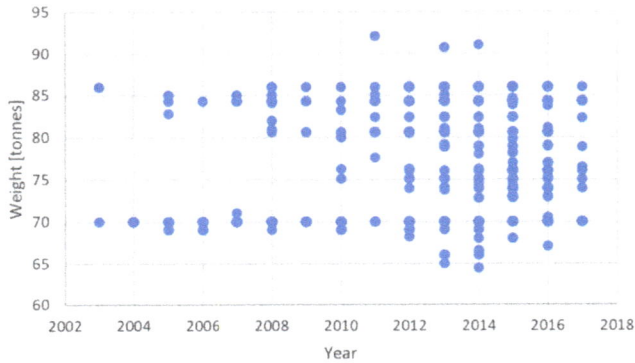

Figure 7. The maximum weight of freight trains. The figure represents the development since 2000. The figure includes 1614 trucks.

In correspondence to the bigger trucks the engines have also gotten larger during the years, which is presented in Figure 8.

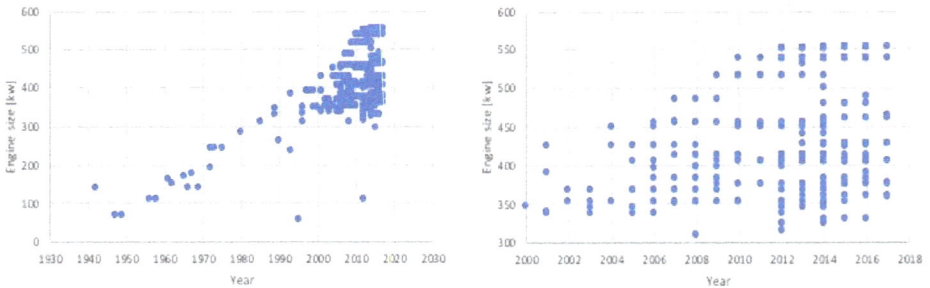

Figure 8. Presents the development of engine size. (**left**) Presents the engine size of the whole fleet as of today and what year the trucks are from and (**right**) presents the development of the trucks from 2000–2020. At (**left**) 1662 trucks are included and at (**right**) 1614 are included.

The engines of the fleets are also up to date which can be seen in Figure 9 where the EURO-classes are presented.

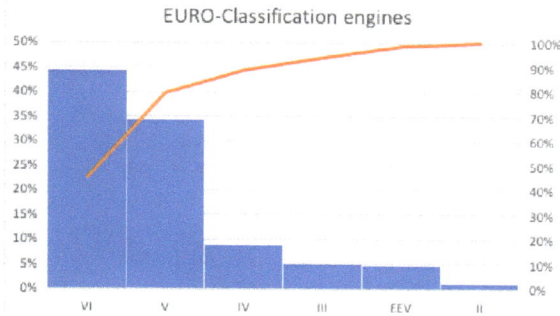

Figure 9. The EURO-classifications of all timber trucks. The stacks present the share of the fleet and the orange line is the accumulated shares.

80% of the trucks have the latest models of engines, of which the majority (almost 45% of all trucks) are equipped with EURO-VI-engines. Even though the engines have developed, and the fleet now mostly consists of newer engines, the fuel consumption of the trucks has not been improved, as presented earlier.

3.2. Financial Calculations and Simulations

In this section the economic results according to TransAm are presented. The results of the aerodynamic changes are presented in how they would affect the total cost of the transport in Swedish kronor per tonnes (SEK/tonnes).

The reduction in fuel consumption that was investigated was from 2–10% with 2% intervals, and the results are presented in Figure 10.

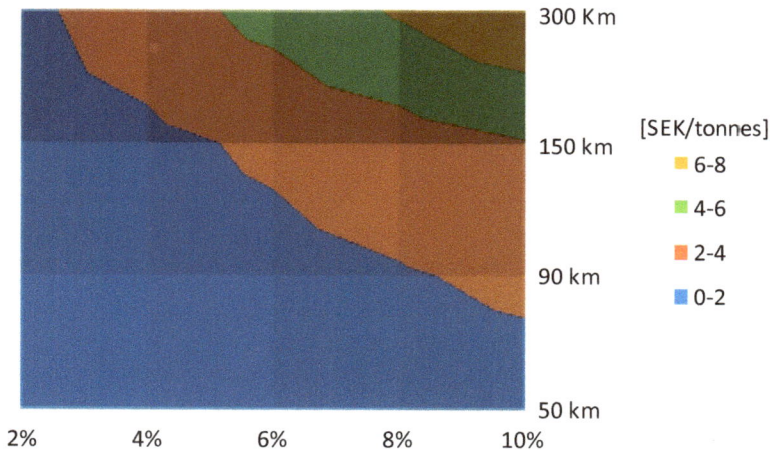

Figure 10. Presents the cost reduction depending on reduction of fuel consumption. The values represent how big the decrease in transport cost is in Swedish kronor per tonnes (SEK/tonnes). The percent represents reduction in fuel consumption and the km the length of the transport A negative value represents an increase in transport cost and a positive value a decrease. For example, it can be seen that the length has to be longer then 150 km with a fuel reduction of 10% to reach a reduced cost of 4 SEK/tonnes or more.

The biggest reduction in cost happens when the fuel consumption is the most improved, 10%. It can also be seen that even an improvement of 2% generates a decrease in the total cost of the transport. The decrease of the total cost is a linear function and it increases with the distance of the transport, which can be seen by the biggest reductions taking place in transports with a distance 300 km.

The result presented in Figure 10 is only depending on the reduction in fuel consumption for the timber trucks, but the aerodynamic kits will also affect other parameters of the trucks. In Figure 11 the results regarding the weight of the aerodynamic kit is presented. The weight will affect how much the timber trucks can haul and also therefore how profitable they will be able to be.

For weights less than or equal to 100 kg the investment will be profitable at all distances and with all tested reductions in fuel consumption, therefore the figure for that is not presented.

When the weight of the installment increases, to 1000 kg in Figure 11 (right), the results show that a fuel reduction of around 6% will lead to a decrease in transport cost for all distances. A lower reduction than 5% will lead to an increase in transport cost. Naturally, the results also show that the longer the distance, the larger the decrease in transport cost. The decrease in transport cost is also lower when the weight of the instalment is higher, which can be seen by comparing the two.

In Figure 11, an unprofitable economical result is presented. The transport cost will increase for the carriers, but the overall positive effect the fuel reduction has on the environment is not assessed in this analysis. The greenhouse gas emissions will decrease with the fuel reduction, leading to a decrease in external costs and even though the transport costs increase this decrease in external cost can justify the investment.

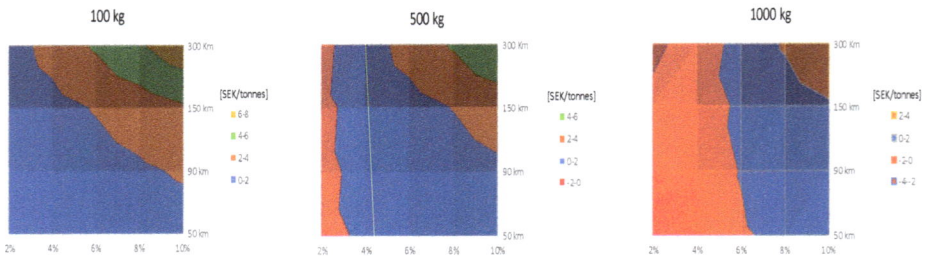

Figure 11. Presents the cost reduction as a function of weight. (**left**) Presents the cost reduction if the weight of the installments is 100 kg; (**middle**) Presents the cost reduction if the weight of the installments is 500 kg; (**right** Presents the cost reduction if the weight of the installments is 1000 kg. Depending on the weight of the installment the truck will be able to load less. The values represent how big the decrease in transport cost is in SEK/tonnes. A negative value represents an increase in transport cost.

The unanimous response from the interviews was that the changeover time would be the most important factor according to profitability. Therefore, an increase in changeover time was analyzed through TransAm and the results are presented in Figure 12.

As can be seen, the changeover time has a big impact on the transport cost. An increase in changeover time can be explained by, for example, extra time to set up the aerodynamic kits. An increased changeover time of just 5 min will lead to increased transport cost, even with a fuel reduction of 8% for distances of 50 km. An even longer time means that the fuel reduction needs to be bigger or the transport distance longer for the investment to be profitable. The results show that an increase in changeover time of 15 min will lead to the investment being unprofitable for distances below 90 km, even with a fuel reduction of 10%. For distances of 300 km the fuel reduction needs to be at least 4% for the investment to be profitable.

5 minutes   15 minutes

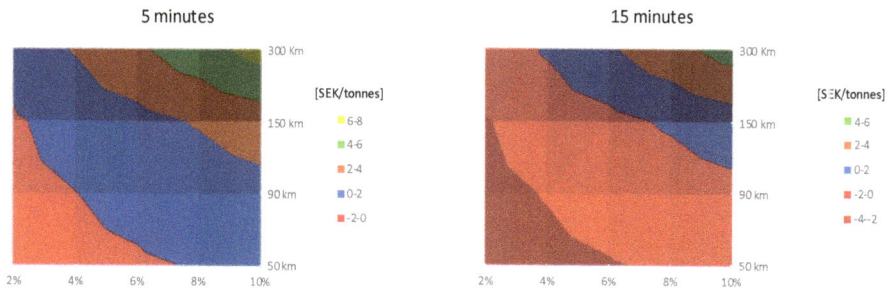

Figure 12. Presents the cost reduction as a result of increased changeover time. (**left**) The cost reduction as a result of five minutes extra changeover time; (**right**) The cost reduction as a result of 15 min extra changeover time. The values represent how big the decrease in transport cost is in SEK/tonnes. A negative value represents an increase in transport cost and a positive value a decrease.

The investment cost of the aerodynamics kits will also affect the transport cost. How much of the impact it has is presented in Figure 13 (left). Only the highest investment cost investigated, 100,000 SEK, did only result in increased transport cost at the minimal distance and minimal reduction in fuel consumption. Therefore, the lower investment costs are not presented since they in all cases lead to a decreased transport cost.

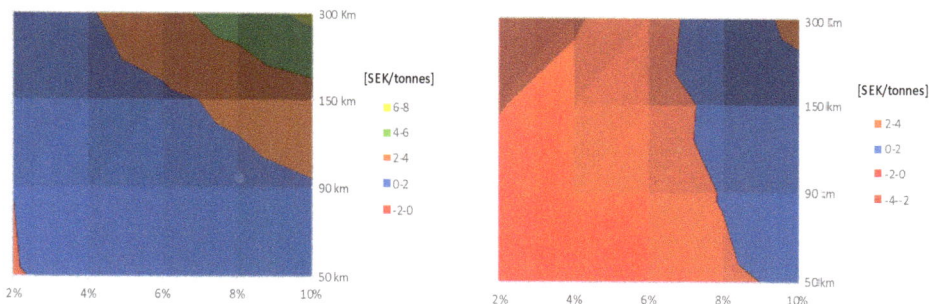

Figure 13. Presents the cost reduction of two different cases. (**left**) Presents the cost reductions with the highest investment cost of 100,000 SEK; (**right**) Shows the cost reduction with the highest investment cost and weight. No extra changeover time is included. The values represent how big the decrease in transport cost is in SEK/tonnes. A negative value represents an increase in transport cost and a positive value a decrease in transport cost.

To be able to see how these parameters would impact the result in combination with each other additional cases were also investigated. The first case (Case 1 in Table 4) investigated was the highest investment cost, the highest weight, but no extra changeover time; the results are presented in Figure 13 (right).

The results show the investment cost does not impact the transport cost in a major way. Even the biggest investigated investment cost of 100,000 SEK will still lead to a cost reduction at a decreased fuel consumption of 2% and at the shortest distance, presented in Figure 13 (left). Though in Figure 13 (middle) the impact of weight of installment is shown. The weight impacts the result and it can be seen for the investment to be profitable for all distances a fuel reduction of around 9% is necessary.

To be able to determine the importance of changeover time, a case was analyzed with or without extra changeover time to see how the results would be affected (Case 2 in Table 4). The extra changeover time was analyzed with the lowest investment cost and weight of the installment, the results can be seen in Figure 14.

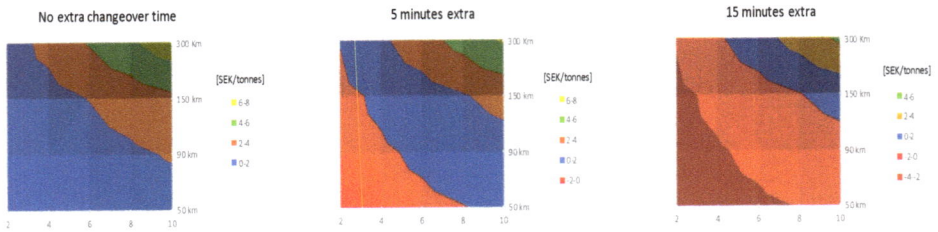

Figure 14. Presents result from Case 2, how the changeover time affects the result. (**left**) Presents the result with no extra changeover time, lowest investment cost and weight of installment; (**middle**) Presents the result with 5 min extra changeover time, lowest investment cost and weight of installment; (**right**) Presents the result with 15 min extra changeover time, lowest investment cost and weight of installment. The values represent how big the decrease in transport cost is in SEK/tonnes. A negative value represents an increase in transport cost and a positive value a decrease.

The results show that the changeover time impacts the result. As can be seen in Figure 14 (middle), for the lowest distance the reduction in fuel consumption needs to be around 8% for the investment not to lead to an increase in cost. If it is lower than 8%, the aerodynamic improvements will not be sufficient enough to make up for the lost time that is needed while mounting the aerodynamic kits. For the longest distance, 300 km, the aerodynamics improvements will lead to a decrease in transport cost from the beginning, but it is approximately 2 SEK/tonnes lower than without the extra changeover time.

With the same settings, how a 15min longer changeover time would affect the cost reduction was also investigated. With an extra 15 min changeover time for assembly the investment will not be profitable for distances shorter than 90 km. For distances of 150 km the fuel reduction has to be greater than 7% and for distances of 300 km and longer the fuel reduction must to a minimum be 4%, otherwise will it lead to an increase in cost. The results highlight the importance of the changeover time.

The worst possible case was also investigated. The meaning of worst possible case is the highest investment cost, highest weight and the longest extra changeover time. The results show that even with a 10% fuel reduction the investment will lead to increased cost for all investigated distances. The investment will then not be strictly economical profitable.

To be able to understand what sort of effect this would have on the entire fleet and not to a single carrier, a hypothetical case was analyzed. The county with the most transported weigh was Västerbotten, as seen in Figure 5. If a fuel consumption of 0.030 L/tonneskm is assumed and all cases of tested fuel reductions is tested this will lead to a decrease in total fuel consumption presented in Table 5.

Table 5. Presents the total results for the County of Västerbotten with assumed fuel reduction. The fuel price was assumed 10 SEK/L and the emission factor 2.82 kg CO_2-eq/L.

Fuel Reduction	2%	4%	6%	8%	10%
Decreased fuel (thousand litres)	319	638	957	1257	1557
Decreased emissions (tonnes CO_2-eq)	900	1800	2700	3500	4400

As can be seen, over a million litres of fuel can be saved only in the County of Västerbotten with aerodynamic improvements which will lead to decreased emissions as well.

4. Discussion & Analyses

As presented in the results section, it is possible, through investments in aerodynamics, for road carriers to lower their transportations costs by reducing their fuel consumption. The investment will in many cases lower their operating transportation cost and it then becomes a question whether the pay-off time of the investment becomes short enough for the road carrier's willingness to invest in them. This is something this study has not taken in consideration, but that needs to be considered in a real-life investment situation.

The incentives for making a green investment depend on many factors, and the economic factor can be a barrier towards investing. There are ways to lower this barrier and share the risk of the investment. Today a timber truck usually consists of a truck purchased from one company, a trailer purchased from another company, and the trailer usually has stakes from a third company. Many actors are involved in a complete truck and this provides an opportunity for split the risk between the actors. In the extended supply chain—there are five primary parameters included in a contract between a road carrier and its customer where the risk can be shared by the seller and the buyer according to according to Eng-Larsson (2017), which are presented in Table 6 [25].

The contracts between the timber carriers and their customers are today are mostly based on how much timber is transported and how long the transport is. The pricing of the transport is based on the volume that is delivered and a tariff is used for prices according to distance. From a risk viewpoint this puts more risk on the road carriers for the parameters price structure, volume commitment and performance-based payment which makes it more unlikely they would make green investments. To reduce the risks for the road carriers, the pricing should avoid being based on achievement and instead set a number of transports and volume that is supposed to be transported.

Table 6. How the risk is split between buyers and sellers in a contract [20].

Contract Component	More Risk is Allocated to the Seller when Using	More Risk is Allocated to the Buyer when Using
Price structure	Price table	Price specified across all dimensions
Price indexing	No indices or surchagre	Indices that capture all costs that may vary over the contract horizon
Performance-based payment	Payment contingen on performance	No performance-based payment
Volume commitent	Frame agreements with loosely specifiec volumes	Fixed volume and frequency of shipment
Contract period	Short period	Long period

The contracts are longer than the individual rides and usually last for a couple of years which is good from a risk viewpoint: this lowers the risk for the road carrier. Compared to the general transport market situation [18], the timber transport market hence stands out as stable, with longstanding contracts that support investments. For price indexing, lesser risk is also put on the road carrier. The price that usually varies is the fuel price, and for that an index is used which changes every month according to the fuel price, hence the road carrier is reimbursed at a reasonable level. The prices the road carriers use will in this way reflect the actual price that the road carriers pay for the fuel which reduces their risk.

The road carriers running a timber truck business are mostly very small companies, typically one to a few trucks and a few drivers. Such companies in general fear investments to a higher degree than larger firms who can better spread the risk of an investment. However, this study clarifies that the road carriers don't hesitate to invest in the latest engine technology, which was surprising for the researchers. This investment can be interpreted as a positive attitude towards new technology, which would be an important driver for implementing aerodynamic technology, and a novel driver compared to previous studies [9–11]. However, as mentioned, the contracts between the road carrier and the forest owner needs to be more designed towards decreasing the risk for the road carrier in line with what has been mentioned above. This is in line with prior research results regarding general LSPs [13,18].

The results show that changeover time is the parameter that impacts the result the most. One reason for this is the already tight time schedule the carriers are working on. The investment cost and weight will affect the result, but not with such an amplitude as the changeover time. Even with the lowest investment cost and weight the smallest added changeover time lead to barely profitable results for distances lower then 90 km at a 10% fuel reduction. With 15 min extra changeover time, the investment would lead to an increase in transport cost for all transports shorter then 90 km and a small decrease in cost for distances close to 150 km. This shows the importance of the changeover time and that producing an aerodynamic kit that does not have to be manually put on is important for the profitability of the investment. The importance of the changeover time was also something that the road carriers mentioned in the interviews, where all interviewees stated the changeover time to be the most important and critical parameter. The timber trucks already today operate on a tight time schedule and even more changeover time would make the business less profitable.

By combining the results of the mapping and the economical result from TransAm it is also possible to decide where in Sweden the investment would be most profitable. Only some investments investigated are profitable at short distances and these types of investments would be the only one possible for, for example, Gotland, with the shortest average transport distance. By combining the results, investments can also be ruled out, for example, an investment that needs distances longer then 90 km can almost be ruled out for all counties since the longest average transport distance is 100 km. This creates an investment window where it is possible to see where and if an investment would be profitable for a certain county and transport distance.

As discussed earlier, some results that are economically unprofitable from a company perspective will still be profitable results from a societal perspective because of the improved environmental effects and the decrease in external cost the fuel reduction will lead to. The value of environmental parameters is far beyond comparing them with only company level economical results and they cannot be seen in the presented financial calculations. The timber fleet today, as presented, is already influenced by early adopters who take responsibility for a green fleet which can be seen by the majority of Euro VI-engines, far greater than the share of Euro VI-engines in the complete heavy truck fleet, which was only 27% (in relation to 45% for timber trucks in this study) [26]. The decreased external cost and positive impacts of the environment from the unprofitable economic results and investments will be of more importance for the progressive fleet and the environmental results will overshadow the unprofitable economic results. But, as presented, aerodynamic investments can also lead to decreased transport cost, meaning that the carriers will both make a profitable economic investment but also a profitable environmental investment.

Both on a company and a societal level, a sustainability assessment can be valid. While economic sustainability is important for the road carriers, they can also benefit from standing out as environmentally prominent [14,16]. In addition, by reducing the cost from lower fuel consumption, the economic sustainability for the companies will improve and therefore also the sustainability for the employees and drivers. By also taking in consideration that the installments can't be too heavy and can't be manually mounted every time leading to increased changeover time, which the results presented show, the work situation for the drivers will be less demanding and more sustainable. It is equally important to reflect on how the aerodynamic installments will affect the sustainable living of employees and companies, and not just the environment per se.

This article has shown that it is possible through reconfiguration and aerodynamic improvements to reduce the fuel consumption of timber trucks and make the fleet greener through technology. Fuel consumption can be reduced to such a level that is will be profitable for the road carriers to go through with the investments. It is possible to make aerodynamic kits within the parameters tested in this study, and as the mapping shows, it is also possible to find areas and transports in the countries which have the characteristics that are needed for a particular investment to reach profitability.

The author would like to stress that the same method can be used for larger data sets for even bigger vehicle fleets. The method of generating large data sets for vehicle fleets can be used for all kinds of operations.

5. Conclusions

This paper set out to investigate the opportunities when implementing aerodynamic equipment with timber trucks in Sweden.

The analysis has demonstrated a range of opportunities. In many cases there are possibilities to both lower the fuel consumption and reach reduced operating transport costs. The transport distances today are also of such a distance that are required for the improvements to be economically beneficial. In all, these examples show viable options for making timber truck operations more sustainable.

The analysis also revealed a range of challenges, mainly associated with the road carrier. The risk of the investment in aerodynamic equipment is an important barrier. However, this barrier can be lowered by the sharing of risks among the many actors involved in the operations, but also by prolonging the contract periods between the timber road carriers and their customers.

This study adds to prior research, as it takes a holistic perspective on the trade-offs between environmental and financial aspects regarding reconfiguration of trucks. While environmentally related benefits like reduced fuel consumption reduce the environmental impact and potentially lower the climate effects, reconfiguration also supports the business of road carriers. Where lower fuel consumption saves costs, it simultaneously also adds to strengthening the environmental profiles of the companies, thus securing long term profitability. Together with this, the studied reconfigurations also demonstrate the potential of improving the working conditions for the truck drivers in the short term, as well as in the longer time perspective.

The calculations made with TransAm provide a set of variables to consider in any road carrier's business model in general, and in particular when environmental investments are considered. We believe that a similar investment calculation would be a viable way to improve long-term decision making among road carriers. For instance, similar calculations would be made based on long-term effects of investments in eco-driving by both educational efforts and a sustainable follow-up system [27].

With respect to the research design, in this study we were able to base our calculations on the total population of timber trucks in Sweden. That implies that the results reflect not a representation but the full population for the investigation. We believe that similar analyses can be performed on other types of vehicles as well. However, timber trucks are dedicated vehicles, whereas other types of trucks can serve multiple purposes and carry the combination of many different types of goods. Hence, a similar approach to a wider range of vehicles proposes many new challenges for analysis.

Timber trucks are very common in Sweden, where the forest industry is a major industry. Therefore, the results of this research, even those of incremental magnitude, have a considerable impact on the overall fuel consumption and related greenhouse gas emissions. Comparable countries, with the forest industry as a major industry, are, for example, Finland and Canada. It would be interesting to expand this research and to compare the different timber truck fleets' performance and potential in different countries.

This paper focuses on technology measures and their consequences in terms of sustainability. Other avenues for future research includes, in general, to take a look into logistics and transport planning and its consequences. In the case of timber trucks, their alternative use is very restricted, so the opportunities to identify complementary cargo for the empty running is limited. Nevertheless, the results from this study applied to general trucks, and in combination with an analysis of logistics opportunities and consequences, would be a viable way to combine technology with logistics for greening freight transport.

Sustainability **2018**, *10*, 1965

Managerial implications from this research relate mainly to the road carriers with timber trucks, where the results offer advice for implementation of aerodynamic equipment and the financial consequences thereof. Some advice to the customers of the road carriers would be to offer contracts of longer time frames as this would lower the barriers for investments, which in turn would benefit both the road carriers and their customers, financially and environmentally, in the longer time perspective.

In terms of policy implications, this research demonstrates the potential effects of investments in aerodynamic equipment on timber trucks. Any policy measures that would ease the burden of an investment in such equipment would increase the speed of making timber trucks more sustainable.

One exciting implication of the present study is that if the timber truck manufacturer(s) would take these aerodynamic re-designs into account already when designing the next generation vehicles, the fuel reduction (and hence CO_2 reduction) presented here would be essentially free of charge. Thus, a comprehensive combined understanding of the vehicle fleet, operating conditions, technological possibilities and financial barriers will lead the way to greener transport.

Author Contributions: E.J. collected the data on timber trucks, categorized the data, performed most of the analysis which entitled the selection of analysis set-up and led the writing process. P.E. provided the research background on aerodynamic design and computations and illustrations of related drag of timber trucks. M.H.-B. analysed the data results from a logistics perspective, contributed to the writing process and revised the manuscript. M.K. headed the research group, initiated the study, supported the analysis, and revised the manuscript.

Acknowledgments: The research presented in this paper is part of the ETTaero2-project, funded by Energimyndigheten (the Swedish Energy Agency). The computations were performed on resources provided by the Swedish National Infrastructure for Computing (SNIC) at NSC.

Conflicts of Interest: The authors declare no conflict of interest. None of the funding sponsors, industry or research organisations had any role in the research design, selection of data, computations or interpretations of the results.

References

1. Intergovernmental Panel on Climate Change (IPCC). *Climate Change 2014: Mitigation of Climate Change. Contribution of Working Group III to the Fifth Assessment Report of the Intergovernmental Panel on Climate Change*; Edenhofer, O., Pichs-Madruga, R., Sokona, Y., Farahani, E., Kadner, S., Seyboth, K., Adler, A., Baum, I., Brunner, S., Eickemeier, P., et al., Eds.; Cambridge University Press: Cambridge, UK; New York, NY, USA, 2014.
2. Izzo, M.; Myhr, A. *Lastbilars Klimateffektivitet och Utsläpp—Rapport 2015:12; (Trucks Environmental Efficiency and Emissions—Report 2015)*; Trafikanalys: Stockholm, Sweden, 2015. (In Swedish)
3. Trafikanalys. *Varuflödesundersökning; (Commodity Flow)*; Trafikanalys: Stockholm, Sweden, 2016. (In Swedish)
4. Trafikanalys. *Lastbilars Klimateffektivitet och Utsläpp; (Trucks Climate Efficiency and Emissions)*; Trafikanalys: Stockholm, Sweden, 2015.
5. Karlsson, M.; Gårdhagen, R.; Ekman, P.; Söderblom, D.; Löfroth, C. *Aerodynamics of Timber Trucks—A Wind Tunnel Investigation*; SAE Technical Paper; No. 2015-01-1562; SAE International: Warrendale, PA, USA, 2015. [CrossRef]
6. Brunberg, T.; Johansson, F.; Löfroth, C. *Dieselförbrukning hos Virkesfordon under 2016; (Diesel Consumption in Forest Trucks 2016)*; Skogforsk: Uppsala, Sweden, 2016. (In Swedish)
7. Cordis. *CONVENIENT—Report Summary*; CENTRO RICERCHE FIAT SCPA: Torino, Italy, 2016.
8. Ekman, P.; Gårdhagen, R.; Virdung, T.; Karlsson, M. *Aerodynamic Drag Reduction—From Conceptual Design on a Simplified Generic Model to Full-Scale Road Tests*; SAE Technical Paper; No. 2015-01-1543; SAE International: Warrendale, PA, USA, 2015. [CrossRef]
9. Krastev, V.; Bella, G. *On the Steady and Unsteady Turbulence Modeling in Ground Vehicle Aerodynamic Design and Optimization*; SAE Technical Paper; No. 2011-24-0163; SAE International: Warrendale, PA, USA, 2011. [CrossRef]
10. Anbarci, K.; Acikgoz, B.; Aslan, R.A.; Arslan, O.; Icke, R.O. Development of an Aerodynamic Analysis Methodology for Tractor-Trailer Class Heavy Commercial Vehicles. *SAE Int. J. Commer. Veh.* **2013**, *6*, 441–452. [CrossRef]

11. Taherkhani, A.R.; de Boer, G.N.; Gaskell, P.H.; Gilkeson, C.A.; Hewson, R.W.; Keech, A.; Thompson, H.M.; Toropov, V.V. Aerodynamic Drag Reduction of Emergency Response Vehicles. *Adv. Automob. Eng.* **2015**, *4*. [CrossRef]

12. Hucho, W.-H. *Aerodynamics of Road Vehicles*; Society of Automotive Engineers, Inc.: Warrendale, PA, USA, 1998; ISBN 978-0-7680-0029-0.

13. McKinnon, A. Environmental Sustainability: A New Priority for Logistics Managers. In *Green Logistics: Improving the Environmental Sustainability of Logistics*, 3rd ed.; McKinnon, A., Browne, M., Piecyk, M., Whiteing, A., Eds.; Kogan Page: London, UK, 2015.

14. Isaksson, K. Logistics Service Providers Going Green—A Framework for Developing Green Service Offerings. Linköping Studies in Science and Technology Dissertations No. 1600. Ph.D. Thesis, Linköping University, Linköping, Sweden, June 2014.

15. Isaksson, K.; Evangelista, P.; Huge-Brodin, M.; Liimatainen, H.; Sweeney, E. The adoption of green initiatives in logistics service providers—A strategic perspective. *Int. J. Bus. Syst. Res.* **2017**, *11*, 349–364. [CrossRef]

16. Evangelista, P.; Huge-Brodin, M.; Isaksson, K.; Sweeney, E. Purchasing Green Transport and Logistics Services: Implications from the Environmental Sustainability Attitude of 3PLs. In *Outsourcing Management for Supply Chain Operations and Logistics Services*; Folinas, D., Ed.; IGI Global: Hershesy, PA, USA, 2013.

17. Aronsson, H.; Huge-Brodin, M. The environmental impact of changing logistics structures. *Int. J. Logist. Manag.* **2006**, *17*, 394–415. [CrossRef]

18. Huge-Brodin, M. The role of Logistics Service Providers in the Development of Sustainability-Related Innovation. In *Supply Chain Innovation for Competing in Highly Dynamic Markets: Challenges and Solutions*; Evangelista, P., McKinnon, A., Sweeney, E., Esposito, E., Eds.; IGI Global: Hershesy, PA, USA, 2011; pp. 215–223.

19. Asmoarp, V.; Skogforsk. Personal communication, 2017.

20. Energimyndigheten. *Drivmedel och Biodrivmedel 2015; (Fuels and Biofuels 2015)*; Energimyndigheten: Eskilstuna, Sweden, 2016. (In Swedish)

21. Rådström, C. Virkesflöde och val av Hjulsystem på Virkesfordon Inom Region Iggesund, Holmen Skog. (Wood Flow and Choice of Wheel Systems on Timber Trucks at Regions Iggesund, Holmen Skog). Master's Thesis, Sveriges Lantbruksuniversitet, Umeå, Sweden, 2014. (In Swedish)

22. Johansson, F.; von Hofsten, H. *HCT-Kalkyl—An Interactive Cost Calculation Model for Comparing Trucks of Different Sizes*; Skogforsk: Uppsala, Sweden, 2017.

23. Röhfors, G. *Däckutrsustnings Påverkan På Miljö och Driftsekonomi vid Rundvirkestransport; (The Tire Equipment's Effect on Environment and Operating Cost When Log Haulin)*; Sveriges Lantbrukuniversitet: Uppsala, Sweden, 2015. (In Swedish)

24. Trafikanalys. *Skogens Transporter—En Trafikslagsövergripande Kartläggning; (The Transports of the Forest—An Intermodal Survey)*; Trafikanalys: Stockholm, Sweden, 2015. (In Swedish)

25. Eng-Larsson, F. Risk-sharing Green Transport Investments. In *Greening Logistics*; Björklund, M., Huge-Brodin, M., Eds.; Studentlitteratur: Lund, Sweden, 2017; pp. 37–49.

26. Trafikanalys. *Prognoser för Fordonsflottans Utveckling i Sverige; (Forecast for the Vehicle Fleet in Sweden)*; Trafikanalys: Stockholm, Sweden, 2017. (In Swedish)

27. Huge-Brodin, M.; Martinsen Sallnäs, U.; Karlsson, M. Sustainable Logistics Service Providers—A strategic perspective on green logistics service provision. In Proceedings of the 20th Annual Logistics Research Network Annual Conference (LRN 2015), Derby, UK, 9–11 September 2015.

MDPI

St. Alban-Anlage 66

4052 Basel

Switzerland

Tel. +41 61 683 77 34

Fax +41 61 302 89 18

www.mdpi.com

Sustainability Editorial Office

E-mail: sustainability@mdpi.com

www.mdpi.com/journal/sustainability